2nd International Conference on Advances in Energy Technology, Environmental Engineering and Materials Science (AETEEMS 2018)

IOP Conference Series: Materials Science and Engineering
Volume 948

El Jadida, Morocco
13 - 14 December 2018

ISBN: 978-1-7138-3197-6
ISSN: 1757-8981

Printed from e-media with permission by:

Curran Associates, Inc.
57 Morehouse Lane
Red Hook, NY 12571

Some format issues inherent in the e-media version may also appear in this print version.

Institute of Physics
Dirac House, Temple Back
Bristol BS1 6BE UK

Phone: 44 1 17 929 7481
Fax: 44 1 17 920 0979

techtracking@iop.org

Additional copies of this publication are available from:

Curran Associates, Inc.
57 Morehouse Lane
Red Hook, NY 12571 USA
Phone: 845-758-0400
Fax: 845-758-2633
Email: curran@proceedings.com
Web: www.proceedings.com

TABLE OF CONTENTS

COMPTATIONNAL METHODS

ELECTRONICS

ENERGY AND ENERGY EFFICIENCY

MECHANICS

IOP Conf. Series: Materials Science and Engineering **948** (2020) 011001 doi:10.1088/1757-899X/948/1/011001

Advances in Energy Technology, Environmental Engineering and Materials Science

A.Hajjaji[1],*, M. El jouad[1], Y. Kouari[2] and Y Boughaleb[1]

[1] Laboratory of Engineering Sciences for Energy, National School of Applied Sciences, Chouaib Doukkali University, El Jadida, Morocco,
[2] Laboratory of Physics of Condensed Matter and Renewable Energy Faculty of Science and Technology, Mohammedia, Morocco
*Corresponding author: hajjaji.a@ucd.ac.ma

This special edition produced as part of the 2nd International Conference on Advances in Energy Technology, Environmental Engineering and Materials Science (AETEEMS 2018) will be held in El Jadida, Morocco, on 13-14 December 2018. It was devoted to experimental and simulation studies in materials sciences for energy harvesting and the environment in addition to some work in electronics and mechanics. It provides a platform for scientists, researcher's students and academic researchers to share their experiences and research's findings in materials science, new energy technologies and environmental engineering.

The International Conference will be held every two years as a platform for researchers to share their perspectives and experiences in materials science, new technologies in the field of energy and environmental engineering, and as related fields. The strategic nature of the issues addressed at this conference will provide scientific and socio-economic benefits for the industrial fabric and the development of intangible goods. This cognitive capital participates in the most recent indicators used at the international level to measure the development of states.

The conference AETEEMS is organized jointly between the Laboratory of Physics of Condensed Matter and Renewable Energy - FST of Mohammedia and the Laboratory of Engineering Sciences for Energy – ENSA of El Jadida).

The present edition includes works on different themes such as Comptationnal methods [1-5], electronics [6-7], Energy and energy efficiency [8-14], Environment [15-17], Materials sciences [18-26] and mechanics [27-32]

References

1. Elidrysy et al. 2020 IOP Conf. Ser.: Mater. Sci. Eng. 948 012001
2. L. Amallah et al. 2020 IOP Conf. Ser.: Mater. Sci. Eng. 948 012002
3. K. Machrouhi et al. 2020 IOP Conf. Ser.: Mater. Sci. Eng. 948 012003
4. A. Bendaraa et al. 2020 IOP Conf. Ser.: Mater. Sci. Eng. 948 012004
5. M. El Mliles et al. 2020 IOP Conf. Ser.: Mater. Sci. Eng. 948 012005
6. K. El Batal et al. 2020 IOP Conf. Ser.: Mater. Sci. Eng. 948 012006
7. A. Galadi et al. 2020 IOP Conf. Ser.: Mater. Sci. Eng. 948 012007
8. O, Mesbahi et al. 2020 IOP Conf. Ser.: Mater. Sci. Eng. 948 012008
9. H El Achouby et al. 2020 IOP Conf. Ser.: Mater. Sci. Eng. 948 012009
10. O. Mrajji et al. 2020 IOP Conf. Ser.: Mater. Sci. Eng. 948 012010
11. F. Eddiai et al. 2020 IOP Conf. Ser.: Mater. Sci. Eng. 948 012011
12. S. Touili et al. 2020 IOP Conf. Ser.: Mater. Sci. Eng. 948 012012
13. A. Erradi et al. 2020 IOP Conf. Ser.: Mater. Sci. Eng. 948 012013
14. Y. Elhmamsy et al. 2020 IOP Conf. Ser.: Mater. Sci. Eng. 948 012014
15. S. Rochd et al. 2020 IOP Conf. Ser.: Mater. Sci. Eng. 948 012015
16. A. Naamane et al. 2020 IOP Conf. Ser.: Mater. Sci. Eng. 948 012016
17. Y, Raji et al. 2020 IOP Conf. Ser.: Mater. Sci. Eng. 948 012017
18. M.Tahiri et al. 2020 IOP Conf. Ser.: Mater. Sci. Eng. 948 012018
19. T Garmim et al. 2020 IOP Conf. Ser.: Mater. Sci. Eng. 948 012019
20. F Z Bouragba et al. 2020 IOP Conf. Ser.: Mater. Sci. Eng. 948 012020
21. R. Anoua et al. 2020 IOP Conf. Ser.: Mater. Sci. Eng. 948 012021
22. Y. Chaouqi et al. 2020 IOP Conf. Ser.: Mater. Sci. Eng. 948 012022
23. O. Mbarka et al. 2020 IOP Conf. Ser.: Mater. Sci. Eng. 948 012023
24. K. Oumghar et al. 2020 IOP Conf. Ser.: Mater. Sci. Eng. 948 012024
25. A, Ziti et al. 2020 IOP Conf. Ser.: Mater. Sci. Eng. 948 012025
26. K. Sbiaai et al. 2020 IOP Conf. Ser.: Mater. Sci. Eng. 948 012026
27. K Essaadaoui et al. 2020 IOP Conf. Ser.: Mater. Sci. Eng. 948 012027
28. A. Hader et al. 2020 IOP Conf. Ser.: Mater. Sci. Eng. 948 012028
29. I. Achik et al. 2020 IOP Conf. Ser.: Mater. Sci. Eng. 948 012029
30. Z. Malki et al. 2020 IOP Conf. Ser.: Mater. Sci. Eng. 948 012030
31. M. Rouway et al. 2020 IOP Conf. Ser.: Mater. Sci. Eng. 948 012031
32. M, Tanasehte et al. 2020 IOP Conf. Ser.: Mater. Sci. Eng. 948 012032

IOP Conf. Series: Materials Science and Engineering **948** (2020) 011002 doi:10.1088/1757-899X/948/1/011002

Peer review declaration

All papers published in this volume of IOP Conference Series: Materials Science and Engineering have been peer reviewed through processes administered by the Editors. Reviews were conducted by expert referees to the professional and scientific standards expected of a proceedings journal published by IOP Publishing.

Type of peer review: Single-blind

Conference submission management system: www.sciencesconf.org

Number of submissions received: 100 papers

Number of submissions sent for review: 60 papers

Number of submissions accepted: 32 Papers

Acceptance Rate (Number of Submissions Accepted / Number of Submissions Received X 100): 32%

Average number of reviews per paper: 2 reviewers

Total number of reviewers involved: 20 reviewers

Any additional info on review process: We have sent all the papers to more than 60 reviewers but in most of the time we have not had a response to review the articles. This caused a considerable delay in the procedure.

Contact person for queries:

Abdelwahed Hajjaji (hajjaji.a@ucd.ac.ma)
Mohamed El Jouad (eljouad.m@ucd.ac.ma)

IOP Conf. Series: Materials Science and Engineering **948** (2020) 012001 doi:10.1088/1757-899X/948/1/012001

3D anisotropic Ising model with Monte Carlo simulation

A. Elidrysy[1,*], S. Harir[1,2] , A. Zouhair[1] and Y.Boughaleb[1]

1 Laboratory of Bio-Geosciences and Materials Engineering, Higher Normal School,

Hassan II University, Casablanca, Morocco.

2 Department of Physics and Chemistry, Regional Center for the professions of education and

training, Casablanca – Settat, Morocco.

* Corresponding author: anouar.elidrysy-etu@etu.univh2c.ma

Abstract: Magnetization, susceptibility and Curie temperature of 3D anisotropic Ising model for ½ spins system have been investigated by making use of the Monte Carlo simulation. The considered model is parametrized by exchange interaction parameter *J*, anisotropy parameter γ. The obtained results allowed us to show that for strong coupling case in z axis ($\gamma > 1$), the critical temperature is widely to the conventional value of the 3D materials ($T_c = 6{,}32J$). Thus, an optimal integration of interplanar atoms in z-axis can generate materials that retain the ferromagnetic property even at high temperature.

Keywords: Monte Carlo simulation, Ising model, Anisotropy, phase transition, critical temperature

1. Introduction

The Ising model [1], is a classical model of statistical physics, it provides a simplified microscopic description of ferromagnetic. It has been introduced for the description of magnetism in ferromagnetic materials [2]. But the resolution of this model is not always obvious as a two-dimensional example under nonzero external field conditions and three-dimensional have not yet received an exact analytical solution [3].

There are several approximate theoretical methods that have tried to solve the Ising model such as the average field method [4,6], the series development [7-8], the renormalization group method [9-10] and the finite cluster approximation [10,14]; it has been improved qualitatively and quantitatively the results obtained in the framework of the field method, this is due to the nature of the approximations used. but none of these methods have could give acceptable results. Numerical simulation techniques will therefore be indispensable tools to better understand systems such as the Ising model.

In the present paper we have analyzed the Ising model by a numerical method. The latter facilitates the calculation of the partition function when the model considered is placed on a finite size lattice [15], although this method makes it possible to calculate the values of physical quantities such as; the magnetization, the energy, the specific heat and the susceptibility for a given temperature, note that the accuracy of the results obtained depends on the size considered.

The numerical method with which we worked is the Monte Carlo method [16], this method is used to simulate complex physical phenomena, in several scientific. MC simulation based on the metropolis method is carried out assuming periodic boundary condition for three-dimensional square lattices of $L\times L\times L$ with linear lattice size. The difficulty encountered in Monte Carlo simulations with local updates is the critical slowing down near a phase transition. Considerable progress in the development of eγ cient non-local Monte Carlo algorithms which overcome critical slowing down to a large extent [17-18].

2. Model and formalism

We consider spin $\frac{1}{2}$ Ising model on a cubic lattice with anisotropic nearest neighbor interactions in Z axis as displayed in Fig.1.

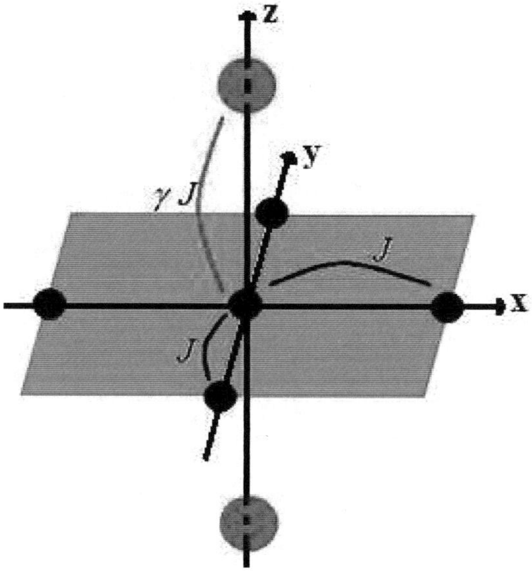

Fig. 1: Different exchange interaction in 3D anisotropic Ising model.

The Hamiltonian of the system is given by the following expression:

$$H = -J\sum_{<i,j>}\sigma_i\sigma_j - \gamma.J\sum_{<k,l>}\sigma_k\sigma_l - h\sum_i\sigma_i$$

where J is the exchange interaction parameter between two nearest-neighbor spins, $\sum_{<i,j>}$ is summation stand for nearest neighbors spins in OXY plan, $\sum_{<k,l>}$ is summation stand for nearest neighbors spins in Z axis, γ is the anisotropy parameter and h is the external longitudinal magnetic field. For $\gamma = 1$ case, the considered is equivalent to usual Ising model.

To study the anisotropy effect on some local magnetic properties of the considered system, we use Monte Carlo simulation. This simulation calculation based on Metropolis algorithm, free boundary conditions are applied in the X, Y and Z-directions. Configurations are generated by selecting the sites in sequence through the lattice and making single-spin-flip attempts, which are accepted or rejected according to some probability based on Boltzmann statistics. Starting from different initial conditions, all the Monte Carlo simulations are averaging over different configurations for each starting initial condition. The simulations are carried out for N spins, where N is the number of all spin in the lattice.

The aimintation is defined by :

$$<M>=\frac{1}{N}<\sum_i^N\sigma_i>$$

whereas the susceptibility χ is given by :

$$\chi = \frac{\partial M}{\partial H} = \frac{(\Delta M)^2}{NK_BT} = \frac{<M^2> - <M>^2}{NK_BT}$$

with $\beta = \frac{1}{K_B.T}$, T is the absolute temperature and K_B is the Boltzmann constant taken to be

$K_B = 1$ for simplicity.

3. Results and discussion

3.1. Usual Ising model case

The fig.2 and 3 show the variation of aimantation M as a function of temperature T in weak and strong coupling regime, respectively.

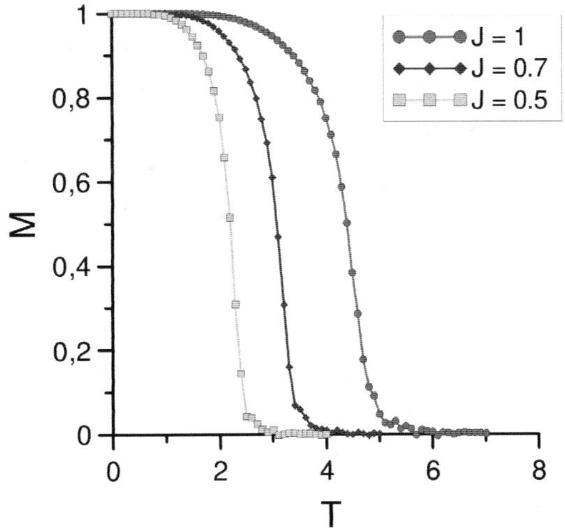

Fig. 2: Magnetization M as a function of temperature T in weak coupling regime at $h=0$

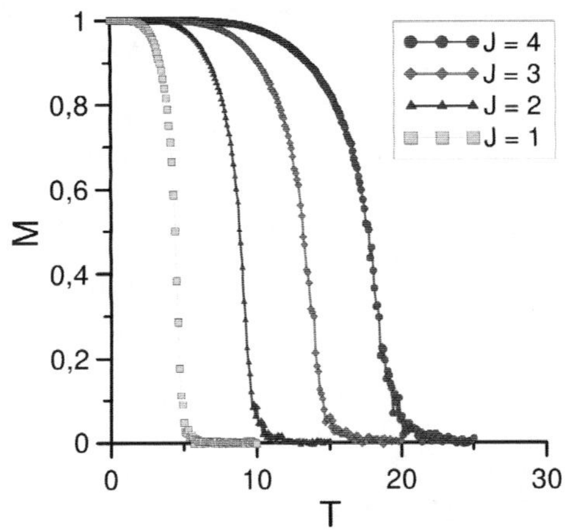

Fig. 3: Magnetization M as a function of temperature T in strong coupling regime at $h=0$

The two figures show that the magnetization decreases as a function of *T* before being canceled at a critical temperature *Tc* which represents in this point a phase transition para-magnetic ferromagnetic. The figures show also that this critical temperature increases with *J*. This to precise the J dependency of *Tc*, we note in table 1 the value of *Tc* for different values of *J*.

J	*Tc*	*Tc/J*
0.5	3.099	6.18
0.7	4.389	6.27
1	6.690	6.69
2	12.90	6.45
3	18.60	6.2
4	24.20	6.05

Table 1: Critical temperature for 3D Ising model for different *J*

The obtained results show that the quotient $\frac{T_C}{J}$ is particularly constant, which responds to a number of published results [19-20] which shows that the ratio $\frac{T_C}{J}$ is constant and depends only on the geometry of a spin system. For a one dimensional spin chain, in which no phase transition occurs , Ising initially proposed in his doctoral thesis that there is also no phase transition in higher dimensions which turned out to be a misinterpretation. The Ising model on a two-dimensional without magnetic field was analytically $T_c = 2.269185J$ [21] whereas, at 3D system the critical value is $5,8J$. Thus, we plot in Fig. 4 the variations of *M* as a function of *T/J*. The curves shows that show the *M* is independent for J where it vanishes at $\frac{T_C}{J} = 6,30$

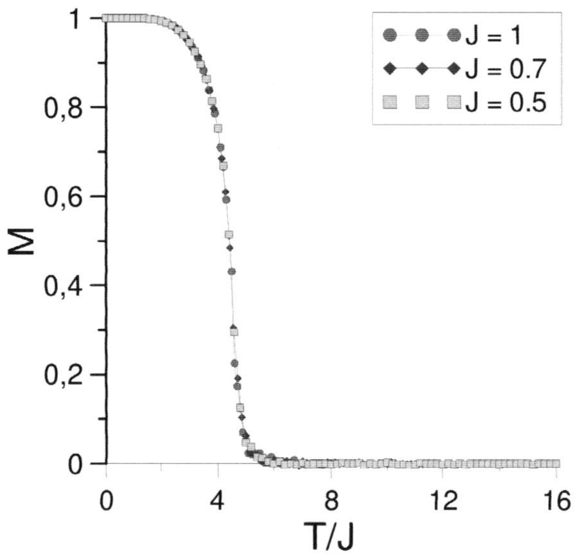

Fig. 4 : Magnetization *M* as a function of *T/J* at *h=0*

To study the effect of external field *h* on same local properties of 3D system ½ spins, we plotted in Fig. .a and Fig. 5. b the variations of aimantation *M* as a function of *h* for T<Tc case and T>Tc, respectively. In paramagnetic phase, the curves of Fig. 4.b shows that the aimantation *M* increases smoothly as a function h, especially in the weak coupling regime (*J=0.5*), whit an inflexion point near (0;0).

IOP Conf. Series: Materials Science and Engineering **948** (2020) 012001 doi:10.1088/1757-899X/948/1/012001

Moreover, the curves show that the spins of our consideration of the magnetic field depends on the fact that the effects of interactions between the spins the nearest neighbors favors spin alignment even at low magnetic field. Thus, the magnetization response is larger in the high coupling limit, where the magnetization field is almost independent to *J*. This means that the magnetization response of a ferromagnetic system is independent of the collective effects of spins.

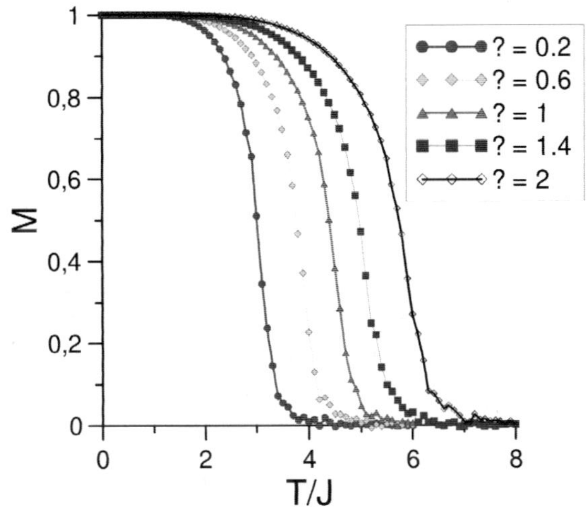

Fig. 5.a: Magnetization *M* as a function of external field *h* in ferromagnetic phase

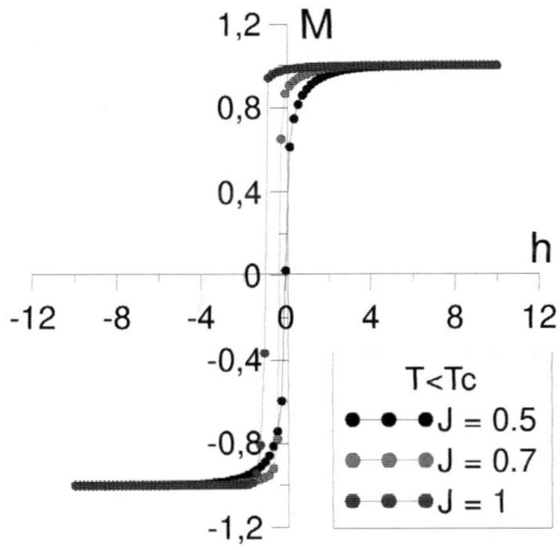

Fig. 5.b: Magnetization *M* as a function of external field *h* in parramagnetic phase

3.2. Anisotropic 3D Ising model case

To show the anisotropy effect on some local properties of 3D spins system, we study the behaviors of aimantation *M* and the susceptibility χ as a function of *T/J* and anisotropy parameter γ.

First, in Fig. 6 and Fig. 7, we plot the variations of *M* and χ for different values of γ at $h=0$, respectively.

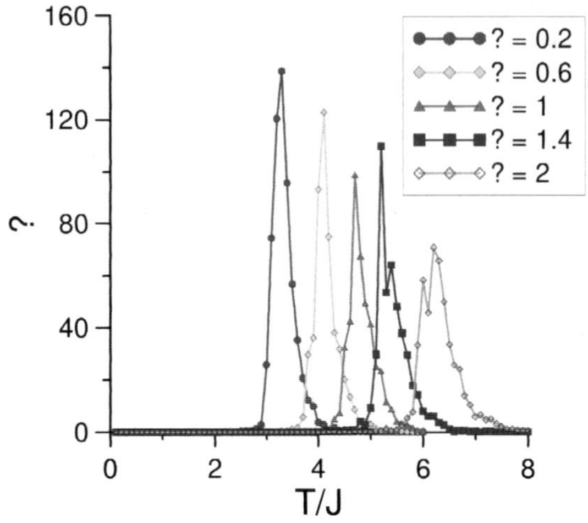

Fig. 6 : Magnetization M as a function of temperature T/J for different γ at h=0

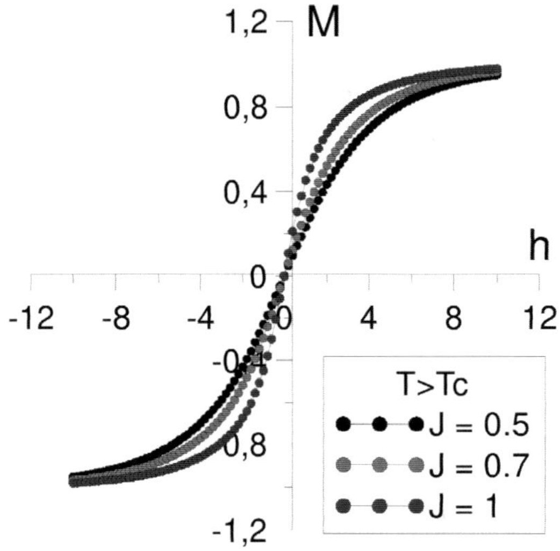

Fig.7 : susceptibility χ as a function of temperature *T/J* for different $\gamma$and h=0

For weak coupling case in z-axis (γ = *0.2*), the corresponding curves show that phase transition occurs at $T_c = 4,1J$. Noting that this critical temperature is particularly very close to one of bidimensional $\frac{1}{2}$ spins system. Thus, in this limit of weak coupling, one can consider that the collective effects between the spin of the oz axis are particularly non-existent. In this cubic lattice can be modelled in non-correlated square lattices.

Whereas, for strong coupling case in z axis (γ = 2 case), the corresponding curves show that the critical temperature becomes more important where it exceeds widely the conventional value of the three- dimensional materials ($T_c = 6,32J$), thus, an optimal integration of interplanar atoms in z-axis can generate materials that retain the ferromagnetic property even at high temperature.

Finaly, the Fig.8 shows *M* as a function of *h/J* for different values of γ in ferromagnetic phase. The obtained curves show that the aimantation *M* varies between -1 and +1 according to an a scending step at a critical value of spins inversion h_0, where $|h_0|$ decreases with γ.

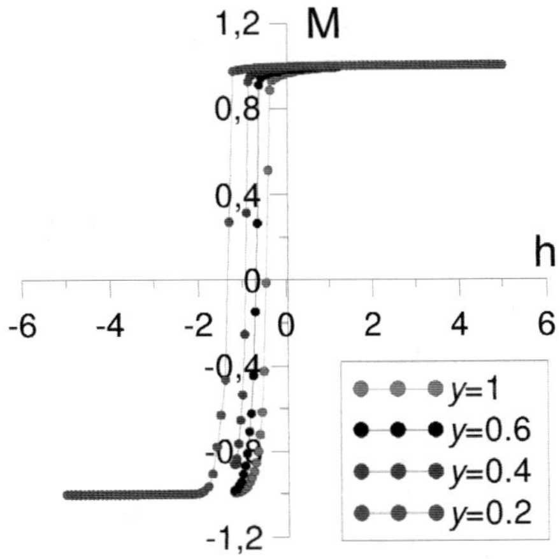

Fig.8: Magnetization M as a function of external field for different γ.

4. Conclusion

In this paper, we have studied z -anisotropy effect on some local magnetic properties of ½ spins system described by 3D anisotropic Ising Model. In particular, we have considered *N* correlated ½ spins in 3D cubic lattice with anisotropic nearest neighbor interactions in Z axis. Thus, the ingredients of adapted Ising model are exchange interaction parameter *J*, anisotropy parameter γ and external magnetic field *h*. Monte Carlo simulation allowed us to study the behavior of some local properties such as aimantation *M* and the susceptibility χ as a function of J , γ , *h* and temperature *T*.

The obtained results shows that 3D anisotropic Ising model exhibits a phase transition at critical temperature T_C which is clearly correlated anisotropy parameter γ. For weak anistropy ($\gamma \rightarrow 0$), the phase transition occurs at critical temperature which is particularly very close to one of bidimensional $\frac{1}{2}$ spins system. Whereas, for strong coupling case in z axis ($\gamma > 0$) the critical temperature widely the conventional value of the 3D materials ($T_c = 6,32J$), thus, an optimal integration of interplanar atoms in z-axis can generate materials that retain the ferromagnetic property even at high temperature.

REFERENCES

[1] E.Ising, Z. Phys. 31 (1925) 253-258.
[2] S. Bouhou, I. Essaoudi, A. Ainane, A. Oubelkacem, R. Ahuja, F. Dujardin, Journal of Super conductivity and Novel Magnetism 28 (2015) 885.
[3] M. Diaconu, R. Puscasu, A. Stancu, Journal of Optoelectronics and Advanced Materials Vol. 5, No. 4 (2003) 971 – 976.
[4] T. Kaneyoshi, Physica A . 353(2005) 297-308.
[5] T. Kaneyoshi and T. Aoyama, J. Magn. Magn. Mater. 96 (1991) 67-76.
[6] M. Saber, Phys. Status Solidi B.178 (1993) 99 –103.
[7] H. P. Griffiths, D. W. Wood, J. Phys. C. 7 (1974) 54 -58.
[8] H. P. Griffiths, D. W. Wood, J. Phys. C. 7 (1974) 4021- 4036.
[9] J. Oitmaa and R. W. Gibberd, J. Phys. C. 6 (1973) 2077 – 2088.
[10] R. Kremar, A. Gendiar, K. Ueda, T. Nishino, J. Phys. A: Math. Theor. 41 (2008) 125001
[11] N. Boccara, Phys. Lett. A 94 (1983) 185.

[12] A. Benyoussef and N. Boccara, J. Phys. 44 (1983) 1143.
[13] Azhari, M., Benayad, N., Mouhib, M.: Journal of physics conference series. Vol. 758. (2016).
[14] M. Azhari, N. Benayad and M. Mouhib, Superlattices Microstruct. 79, 96 (2015).
[15] M. E. J. Newman et G. T. Barkema, Monte Carlo Methods in Statical Physics, Oxford University press, 2002.
[16] N. Metropolis and S. Ulam, J. Am. Stat. Assoc. 44, 335 (1949).
[17] A.D. Sokal, Nucl. Phys. B (Proc. Suppl.) 20, 55 (1990)
[18] L. B. Drissi and S. Zriouel, J. Stat. Mech. Theory Exp. 5, 053206 (2016).
[19] M. E. FISHER, http://iopscience.iop.org/0034-4885/30/2/306
[20] I. Onsager. Rev.65 (3-4) (1944) 117-149, doi :10.1103/PhysRev.65.117.
[21] T. Preis, P. Virnau, W. Paul, and J. Schneider, Journal of Computational Physics, Vol. 228, No. 12. (01 July 2009) 4468-4477.

Langevin Dynamic Simulation of self-propelled particles in two-dimensional systems

L. Amallah[a, *], A. Hader[a,c], R. Bakir [a], H. Sbiaai[a], I. Tarras[a] and Y. Boughaleb[a,b]

[a] *LBGIM, Ecole Normale Supérieure, University Hassan II. Casablanca, Morocco*
[b]*LPMC , University Chouaib Doukkali. El Jadida, Morocco*
[c] *Centre régional des métiers d'éducation et de formation Casablanca- Settat/ établissement Settat, Morocco*

* Corresponding author: layla.amallah-etu@etu.univh2c.ma
ORCID ID: 0000-0002-4536-9253

Abstract: Collective motion of self-propelled particles is one of basic phenomenon observed in large spectra of biological system behavior due to the correlated process evolution in space and time. In this manuscript, we study numerically the kinetic properties and the correlation process in complex systems evolves out equilibrium phase by employing the Langevin dynamics. In this model, we have adopted one zone of orientation where the system evolves spontaneously in presence of quenched stochastic noise. The results show that the system evolves to its equilibrium phase by reaching one orientation. Hence, this evolution is characterized by a correlation process increasing in time but with decelerate profile. However the obtained profile of the correlation function per time unity shows that the collective motion in complex system, can be characterized by a characteristic time when the system change the acceleration of the correlation process. Our result shows that this characteristic time decreases exponentially with the quenched noise. In the additional crossover time at when the system reaches its equilibrium phase, scales with quenched noise as power law. This result is more consisting with the one of Vicsek model

Keywords: Collective motion; Flocks; Self-propelled particles; Noise; Orientation zone; Langevin dynamics; Correlation process.

1. Introduction

Collective phenomena are ubiquitous in nature on many different scales, at any time and space such as: mammal herds [1], fish schools [2], bird flocks [3], insect swarms [4], bacterial colonies [5] and even motion of molecular [6], these biological systems have many similarities with the behaviors observed in the crowds of pedestrians [7]. The captivating characteristics of these organisms are their ability to align and move together in the same direction, in order to avoid collision with nearest neighbours or predators [8].The omnipresence of these phenomena has attracted attention of biologists and physicists, for understanding the collective movements of these groups of animals [9]. Enormous studies [10-13] have been consecrated to study and analyze the dynamic properties of complex living systems [14], by using computational or simulation approaches, the aim of this study to explain how these groups of animals move together, align their motion with their neighbours, find food or escape and disperse predators [15]. Therefore, the notion of self-propelled particles was used to pattern the collective motion of these biological phenomena [16]. Moreover, a major aspect in the study of these collective behaviors is that an individual's behavior is influenced by the behavior of other members of the group [17]. This distinguishing between each individual and the group of individuals leads to important phenomena; like self-organization and emergence of the collective motion [9]. These complexes self-propelled systems have wide applications in our lives for example: the control of human interactions in crowded situations [17], or an artificial swarm of autonomous vehicles [18].

The several collective models have been developed for self-propelled particles, which depend on interactions between particles. Reynolds it was the first model proposed to describe the collective behavior of groups of animals; especially flocks, herds and schools [19]. A decade later, Vicsek et al introduced a minimalist model which describes the collective motion of self-propelled particles in tow-dimensional space, under the effect of noise [20]. In the Vicsek model or flock model the individuals (particles) move with constant velocity and at each time step the individuals align with the average orientation of their neighbours. The results showed that the dynamics of these systems in the presence of noise; is governed by a second-order phase transition from disordered to an ordered state [20]. Couzin et al model, also called three-zones model [21], In this model all individuals have a specific area of motion direction, each zone has a specific rule: the first structural rule is that each individual must have a minimum distance in order to avoid collision with other neighbours, the second rule requires that individuals attract and orient themselves to avoid isolation, when all individuals were greater than the minimum separation, the third structural rule each individual moves with his neighbours in the same direction in order to maintain group cohesion [21]. These behavioral rules are used to reproduce the collective movement observed in nature of fish schools and bird flocks [2-3].
In this study, we propose a new pattern to study the collective behavior of these living systems, by using the Langevin dynamics; it is a computational approach to model the dynamics of complex systems. In our work we investigated the dynamic properties and correlation process of self-driven particle in two-dimensional systems; through the Langevin dynamic simulations, as a stochastic differential equation. Also, this dynamic has been used to simulate Brownian motion [22-23]. Our investigation involves noise η, density of system ρ and orientation zone of radius R as control parameter, and velocity as order parameter to discuss the different kinetic properties and correlation process of the system studied. In order to confirm the validity of our pattern, the results show that the system velocity increases with time and reaches to finite value at the equilibrium phase, which describe the real behavior of these living systems in nature.

This paper is organized as follows. After, this introduction in the first part to the second section, in which we discuss our model based in the use of Langevin Dynamics. In the third section, we present and we discuss the results of our numerical simulations. In the last section, we summarize the main conclusion.

2. The Model

To better understand the collective behavior of complex systems, we investigate the dynamic proprieties and correlation process of interacting self-propelled particle in two-dimensional systems, by employing the Langevin dynamic simulations.
In our pattern, we consider a group of indiscernible particles distribute randomly in a two-dimensional space, with periodic boundary conditions. In which the density of system is given by $\rho = {}^{N}/_{L^2}$. Let m_i, x_i and v_i be respectively the mass, position and velocity of each particle *i*. The collective motion of self-driven particles is described by the stochastic equation below, which is the Langevin dynamics:

$$m_i \frac{dv_i}{dt} = -m_i \eta v_i + F_i(t) \qquad (1)$$

Where η represents the quenched noise is related to the random force $F_i(t)$ by the fluctuation-dissipation theorem according to:

$$\prec F_i(t) \succ = 0 \text{ and } \prec F_i(t)\, F_i(t') \succ = 2m_i \eta K_B T \delta(t-t') \tag{2}$$

Where K_B is Boltzmann's constant.
Initially each particle (individual) move with some random perturbation η added in the motion of particles , the particle i seek to maintain a minimum distance from its local neighbors, in order to avoid collisions with the other individuals. However, the number of particles in orientation zone is given by N_R with ($j = 1 \ldots.. N_R$ *avec* $j \neq i$). At each time step t, the particles adopt the average direction $\theta_i(t)$, is the mean of the N_R individual's orientation in the orientation zone of radius R:

$$\theta_i(t) = \frac{1}{N_R} \sum_{j=1}^{N_R} \theta_j(t) \tag{3}$$

With noise η affect the individual orientation $\theta_i(t+1)$ at time $t+1$ as: $\theta_i(t+1) = \theta_i(t) + \delta\theta$ and $\delta\theta$ is chosen randomly from the interval $[^{-\eta}/_2, ^{\eta}/_2]$. Thus, to study the time evolution of the collective behavior of self-propelled particles in tow-dimensional system, in order to determine the kinetic proprieties and correlation process of the studied system, under effect of orientation zone for radius R. We calculate the order parameter which is the normalized average velocity v_a, giving by:

$$v_a = \frac{1}{N_R} \left| \sum_{i=1}^{N_R} v_i \right| \tag{4}$$

The time evolution depending on the variation of control parameters namely: the density of system ρ, the quenched noise η and the orientation zone for radius R. we remark that the particle density ρ and the quenched noise η; play an important role and their variations can affect significantly the collective behavior of the system.

3. Results and Discussions

In order to investigate the collective motion of N self-propelled particles in tow-dimensional system, we consider square lattice of size $L \times L$, on which we distribute randomly the above particles. Each particle i move with the same initial velocity v_0, with a random orientation θ_i. The system evolves dynamically according to the above model. In our investigation, we adopted a cyclic boundary condition. At each time t, we calculate the time evolution of the both velocity and direction $\theta_i(t)$ by using the Langevin equation 1. However, the direction $\theta_i(t)$ of the particle i is calculated by the mean of the orientation of all its neighboring ones within orientation zone of radius R. in the additional, the system evolves in presence of different of quenched noise η. The physical parameter, which describes the self-propelled dynamic of the used system, is the mean velocity system defined as $v_a = \frac{1}{N_R} \sum_{i=1}^{N_R} v_i$.
In figure 1, we represent the time evolution of the mean velocity system for system density $\rho = 0.5$ and radius value $R = 5$. The obtained profile shows that the system evolves with two different regimes separated by characteristic time t_c. In the first regime where $t < t_c$, the particles evolves with correlated and self-criticality process until reaching its equilibrium phase in the second regime when $t > t_c$. Hence the obtained profile of the mean velocity system is well fitted by $v_a(t) = v_l(1-e^{-t/t_c})$, where v_l is the phase equilibrium velocity. This result is more consisting with the Vicsek model [24]. Figure 2, represents an image system calculated for noise value $\eta = 0.2$ and at time $t = 0.14\, t_f$ where t_f is the time when the system reaches its phase equilibrium. This image proves that system evolves with clusters of different sizes separated by characteristic length ξ, which exhibits a maximal value at crossover time, which scales with the one of the equilibrium phase [25-26]. One of the characteristic parameter of the kinetic system, is the scale time t_c. So, in order to determinate the effect of the quenched noise η on dynamic system, we plot in figure 3, the characteristic time t_c versus quenched noise. The results on log-log plot, shows that t_c scales with

IOP Conf. Series: Materials Science and Engineering **948** (2020) 012002 doi:10.1088/1757-899X/948/1/012002

quenched noise as power law: $t_c \infty \eta^{\beta}$, where the exponent value is $\beta = -0.9$. Hence, the quenched noise has an effect to accelerate the phase equilibrium phase system. These results are more consisting with the ones obtained by the Vicsek model [24].

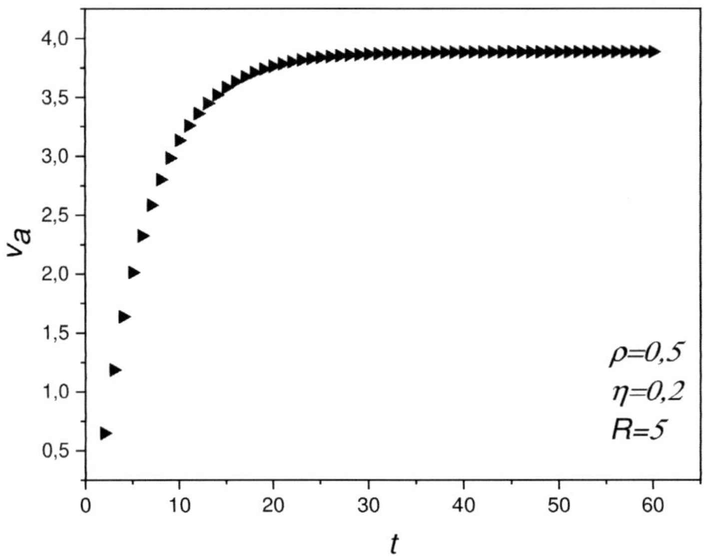

Figure 1: Time evolution of the mean velocity system for fixed values of density and noise.

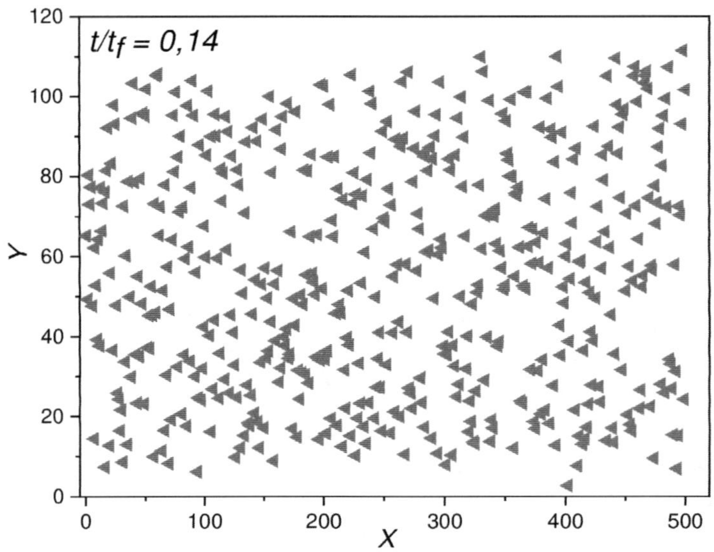

Figure 2: Image system calculated at time $t = 0.14\ t_f$ for fixed values of density and noise.

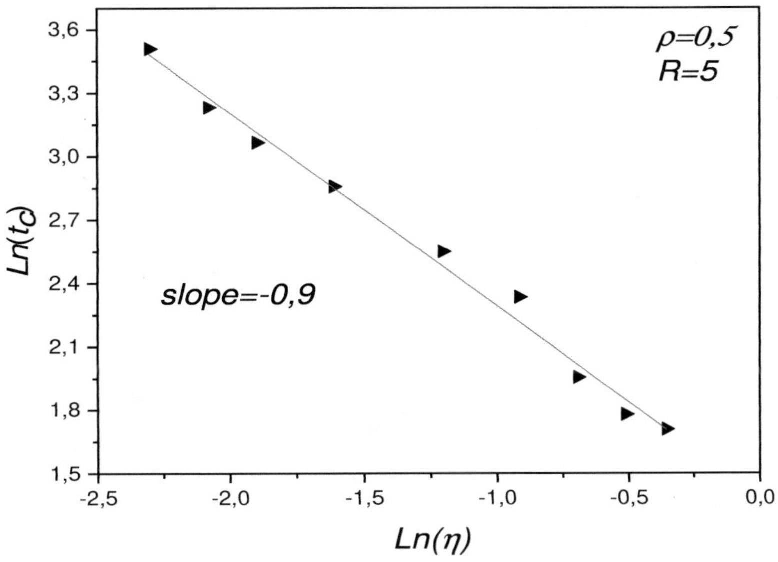

Figure 3: Log-Log plot of the characteristic time t_c versus quenched noise η.

In reality the systems increases its mean velocity in the out equilibrium phase but with decreasing its mean acceleration $a = {dv}/{dt}$. In the figure 4, we represent the time evolution of the mean acceleration system for different values of quenched noise. So, for higher quenched noise value, the system reaches very wickedly its equilibrium phase. In the out equilibrium, phase the mean acceleration system decreases exponentially with the quenched noise η.

IOP Conf. Series: Materials Science and Engineering **948** (2020) 012002 doi:10.1088/1757-899X/948/1/012002

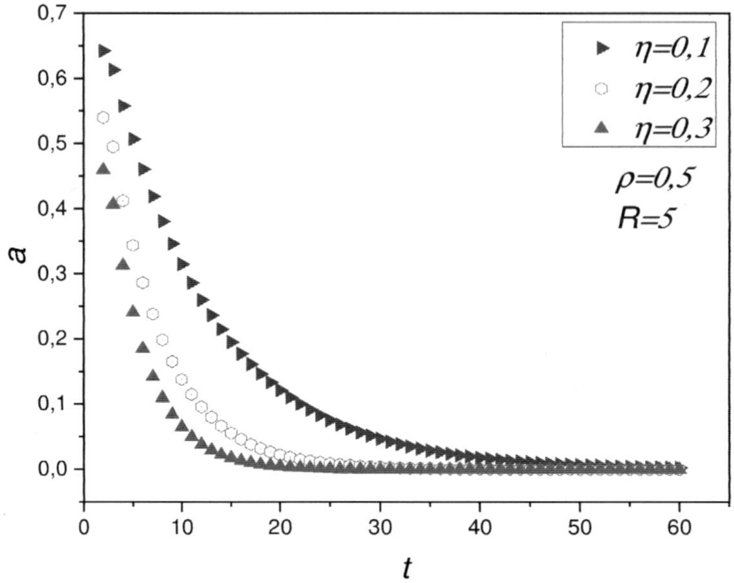

Figure 4: the time evolution of the acceleration system for different values of quenched noise.

Correlation motion is the basic phenomenon in all biological system [27-30] as an alternative for selection. The idea is to represent available objects by motion in the system, have users identify a target by mimicking its specific motion. The resulting interaction has compelling properties, as users are guided by oriented motion. To quantify the correlation process in the collective motion, we investigated the correlation function of the mean velocity system v_a defined as: $G\langle v_a(t)v_a(t+\tau)\rangle_t$. Corresponding results calculated for system density $\rho = 0.5$ and for two different values of quenched noise are presented in the figure 5. The results show that particles evolve to its equilibrium phase with increasing correlation process due by reaching homogenate orientation. However, in the earlier time, the correlation process is more accelerated, but after crosser over time $t_s$corresponding to inflexion point, the acceleration of the correlation process decreases until reaching zero value in the phase equilibrium. To highly this result, we plotted in figure 6 the derivate of the correlation function $dG/_{dt}$ versus time. In reality, the obtained profile of $dG/_{dt}$ is well fitted by Gaussian function, which is characterized by point inflexion at cross over time according to results of figure 5. Hence, at the phase equilibrium system $dG/_{dt} = 0$.

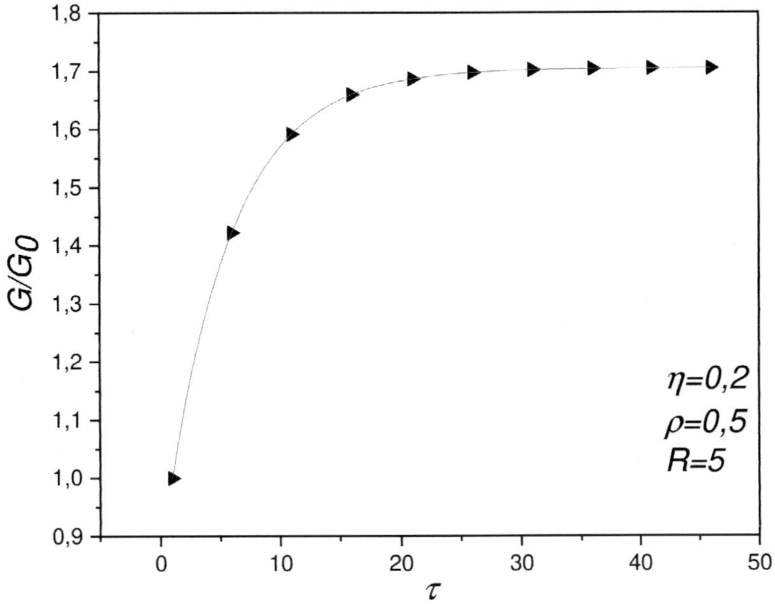

Figure 5: The time correlation function system for fixed noise value.

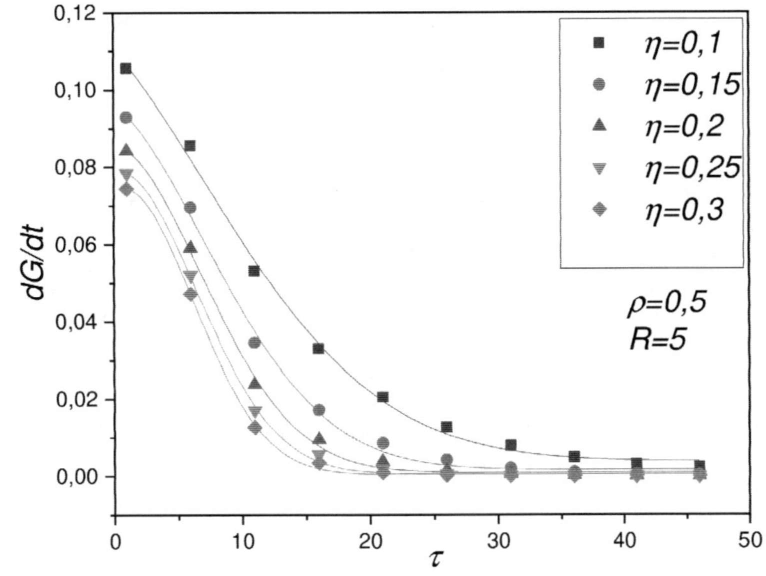

Figure 6: The correlation functions per time unity versus time for different quenched values.

To check the effect of the quenched noise on the cross over time t_e, we plot in figure 7, $t_e$versus noise η. As, we observe, the characteristic time of the correlation function G decreases exponentially with quenched noise. So, for higher quenched noise value, the correlated motion time of system decreases and the system reach very quickly its phase equilibrium. This result is more consisting with the one of [25].

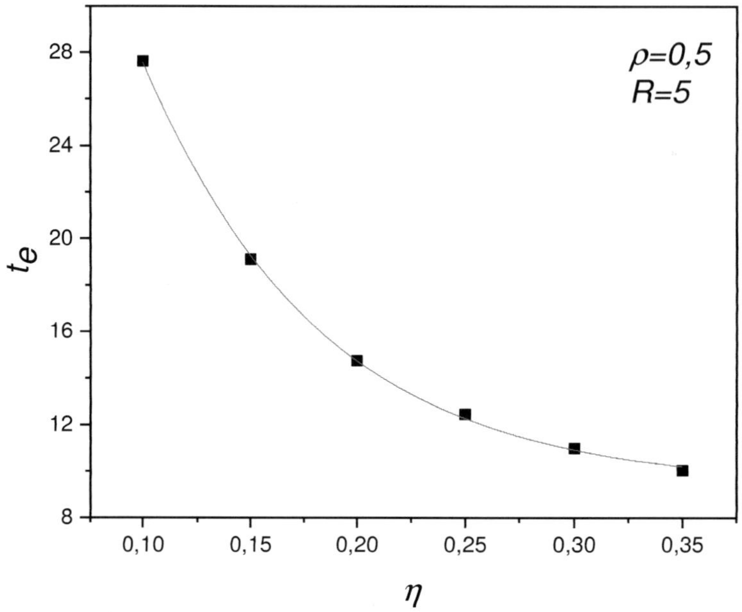

Figure7: The characteristic time t_e versus quenched noise η.

4. Conclusion

In summary, we simulate the dynamical properties and the correlation process of self-propelled particles in two-dimensional systems, by employing the Langevin model. However, we have considered *N* particles evolving from out equilibrium phase to its equilibrium one by reaching a homogenate orientation according to a specific correlation process. In the additional, we have adopted a single orientation zone in which the particles assume a mean orientation direction and we have modeling the interaction process by quenched noise. Our results show the mean velocity system increases exponentially with time until reaching a static velocity in the equilibrium phase after characteristic time, which scales as a power law with quenched noise. Moreover, the collection motion of the considered system is characterized by correlation process, which changes the acceleration at crossover time decreasing exponentially with quenched noise. Moreover, obtained results with Langevin model are more consisting with the ones obtained by Vicsek model.

IOP Conf. Series: Materials Science and Engineering **948** (2020) 012002 doi:10.1088/1757-899X/948/1/012002

References

[1] J. K. Parrish, W. M. Hamner, Animal Groups in Three Dimensions. Cambridge University Press (1997).
[2] Katz Y et al. Inferring the structure and dynamics of interactions in schooling fish. Proc. Natl Acad. Sci. USA . (2011).
[3] Attanasi A et al. Information Transfer and Behavioural Inertia in Starling Flocks.Nat. Phys. (2014).
[4] A. C. Mailleux, J.-L. Deneubourg, C. Detrain, Animal behaviour 59 (2000).
[5] M.P. Brenner, L.S. Levitov, E.O. Budrene. Physical mechanisms for chemotactic pattern formation by bacteria. Biophysical Journal 74 (1998).
[6] M. Badoual, F. Julicher, J. Prost. Bidirectional cooperative motion of molecular motors. Proceedings of the National Academy of Sciences of the United States of America 99 (2002) 6696.
[7] D. Helbing, I. Farkas, T. Vicsek. Simulating dynamical features of escape panic. Nature (2000).
[8] C. K. Hemelrijk, H. Kunz, Behavioral Ecology 16 (2005).
[9] Vicsek T and Zafeiris A . Phys. Rep. 517 71 (2012).
[10] I. Tarras, N. Moussa, M. Mazroui, Y. Boughaleb, A. Hajjaji. Modern Physics Letters B (2013).
[11] T. Iliass, D. Cambui, International Journal of Modern Physics B 30 (2016) 1650002.
[12] I. Tarras, N. Moussa, M. Mazroui, Y. Boughaleb, International Journal of Computer Applications 46 (2012).
[13] R. Bakir, I. Tarras, L. Amallah, H. Sbiaai, A. Hader, M. Mazroui, Y. Boughaleb, Chin. J. Phys. (2017).
[14] E. Ben-Jacob, I. Cohen, H. Levine. Cooperative self-organization of microorganisms. Advances in Physics 49 (2000).
[15] I.L. Bajec and F.H. Heppner. Organized flight in birds. Animal Behaviour (2009).
[16] Couzin I. D, Collective cognition in animal groups, Trends in Cognitive Sciences (2009).
[17] T. Vicsek and A. Zafeiris. Collective motion. Physics Reports (2012).
[18] Martinez S, Cortes J and Bullo F. IEEE Contr. Syst. Mag. (2007).
[19] C. W. Reynolds. Flocks, herds and schools: A distributed behavioural model. Computer Graphics (1987).
[20] Vicsek. T. A. Czirók, E. Ben-Jacob, I. Cohen, and O. Shochet, Novel type of phase transition in a system of self-driven particles, Phy.Rev. Let (1995).
[21] Iain D Couzin, Jens Krause, Richard James, Graeme D Ruxton, and Nigel R Franks. Collective memory and spatial sorting in animal groups. Journal of theoretical biology (2002).
[22] M. Mazroui, A. Asaklil, Y. Boughaleb. Conductivity of two-dimensional systems of interacting Brownian particles within the effective potential description. Surface Science (1998).
[23] M. Mazroui, Y. Boughaleb. Surface Diffusion in Systems of Interacting Brownian Particles. International Journal of Modern Physics B (2001).
[24] A. Czirok, H. E. Stanley and T. Vicsek, J. Phys. A: Gen. (1997).
[25] L. Amallah, A. Hader, R. Bakir, I. Achik, H. Sbiaai, I. Tarras, and Y. Boughaleb, Sensor Letters. (2018).
[26] L. Amallah, A. Hader, M. Tanasehte, Y. Hariti, H. Sbiaai, Y. Boughaleb. Kinetic and scaling behavior of collective motion in biomaterials. Materials Today: Proceedings (2020).
[27] I. D. Couzin, J. Krause, N. R. Franks and A. S. Levin, Nature (2005).
[28] H. Chate, F. Ginelli, G. Gregoire, F. Peruani and F. Raynaud, Eur. Phys. J. B (2008).
[29] J. Krause and G. D. Ruxton, Living in Groups. Oxford University Press (2002).
[30] I. D. Couzin and J. Krause, Adv. Stud. Behav. (2003).

NUMERICAL STUDY OF WATER CIRCULATION IN SHALLOW BASIN USING LATTICE BOLTZMANN METHOD

K. MACHROUHI[1], M. M. CHARAFI[1,*], A. HASNAOUI[1]

[1]*LS3M polydisciplinary faculty of Khouribga, Sultan Mly Slimane University 25000, Morocco.*

[*]*Corresponding author:* Charafi.moulaymustapha@usms.ac.ma

Abstract: In the present work, a code based on the Lattice Boltzmann method (LBM) has been developed to study the hydrodynamic field in a shallow basin with complicated geometry. It is a non-regular border basin with dikes built inside. The LBM method was therefore used to solve the Navier Stokes equations with boundary conditions which express that the components of the horizontal two-dimensional velocity field are zero at closed boundaries (sealing and non-slip conditions), while the flow is imposed at the entrance of the basin, and a zero speed gradient is imposed at the free exit of the basin (condition of continuity of the flow). The validation of the developed code was verified by comparing the results obtained with those of Charafi et al. [5], where the finite difference method with a MacCormack scheme was used. A good agreement of the results was noted. A series of simulations was made for different values of the Reynolds number, and showed good performance of the code. The main advantage of the LBM method seen in this work lies in the simplicity of its implementation. However, the model suffers from some limitations such as its validity for low Reynolds numbers (Re = 400)

Keywords: Shallow water flow, Numerical methods, Lattice Boltzmann Method(LBM), Finite difference method (FD), dikes.

1 Introduction

The Lattice Boltzmann method is increasingly used to describe the behavior of flowing fluids, whether they are incompressible or compressible. By a Chapman-Enskog expansion, it leads to the Euler equation at order 1 and to the Navier-Stokes equation at order 2. Its main advantage is that it allows to describe fairly easily free surface flows, flows in complex media (porous media), flows with heat transfers (conduction, convection and phase changes), as well as flows of multi-component mixtures. The LBM method therefore differs from other simulation methods by its simplicity of use [1] - [3]. In the present work, it was a question of making investigations on the capacity of the Lattice Boltzmann method to simulate the problems of shallow water in a complex geometry. Hence the idea of developing a numerical code to simulate a flow in a shallow basin with an irregular border and equipped with two dikes. The said basin was the subject of a study carried out by Charafi et al [5], where the MacCormack scheme was used to solve the Saint-Venant equations discretized using the finite difference method. We will therefore use the results of this study to validate our code.
After a description of the Navier-Stokes equations in section 2, section 3 will be devoted to the formulation of the problem using the LBM method, and to the presentation of the boundary conditions. In section 4, the validation of the developed code will be confirmed by comparing the results with those of Charafi et al [5]

IOP Conf. Series: Materials Science and Engineering **948** (2020) 012003 doi:10.1088/1757-899X/948/1/012003

2. Navier–Stokes Equation

In fluid mechanics, flows are governed by the Navier - Stokes (NS) equations (equations of continuity and momentum conservation). For a two-dimensional incompressible flow, the conservative form of the N-S equation can be written in Cartesian coordinates, without bodily force, as follows [4] :
The x – momentum equation:

$$\frac{\partial \rho u}{\partial t} + \frac{\partial(\rho uu)}{\partial x} + \frac{\partial(\rho uv)}{\partial y} = -\frac{\partial p}{\partial x} + \frac{\partial}{\partial x}\left(\mu\frac{\partial u}{\partial x}\right) + \frac{\partial}{\partial y}\left(\mu\frac{\partial u}{\partial y}\right) \quad (1)$$

The y – momentum equation :

$$\frac{\partial \rho v}{\partial t} + \frac{\partial(\rho uv)}{\partial x} + \frac{\partial(\rho vv)}{\partial x} = -\frac{\partial p}{\partial y} + \frac{\partial}{\partial x}\left(\mu\frac{\partial v}{\partial x}\right) + \frac{\partial}{\partial x}\left(\mu\frac{\partial v}{\partial y}\right) \quad (2)$$

The continuity equation :

$$\frac{\partial(\rho u)}{\partial x} + \frac{\partial(\rho v)}{\partial x} = 0 \quad (3)$$

3. Formulation by the LBM method

The LBM method comes from the kinetic theory of gases of Boltzmann. The basic idea of kinetic theory is that fluids are made up of a large number of particles which move in random motion [7]. The exchange of momentum and energy is obtained by elastic collisions between the particles and the propagation of these particles. This process is modeled by the Boltzmann transport equation [4]:

$$\frac{\partial f_k(x,t)}{\partial t} + c_k\frac{\partial f_k(x,t)}{\partial x} = \Omega \quad (4)$$

Where the distribution function (x,c,t) of a simple particle represents the probability of finding at time t a particle at position x and having a speed c. The collision operator Ω characterizes the rate of evolution of the distribution function f due to collisions between particles. The second term on the left represents the modification of the function of distribution due to the movement of the particles (in the case where they are not subjected to any force). [8]

$$f_k(x+c_k\Delta t, t+\Delta t) - f_k(x,t) = \frac{-\Delta t}{\tau_f}\left(f_k(x,t) - f_k^{eq}(x,t)\right) \quad (5)$$

where f_k is the distribution function of particles; Δx is the lattice size; Δt is the time step; τ_f is the single relaxation time which controls the rate of approach to the local equilibrium; c_k is the velocity vector of a particle in the (k) link, which is defined by [4]:

$$c_0\begin{pmatrix}0\\0\end{pmatrix}; c_1\begin{pmatrix}1\\0\end{pmatrix}; c_2\begin{pmatrix}0\\1\end{pmatrix}; c_3\begin{pmatrix}-1\\0\end{pmatrix}; c_4\begin{pmatrix}0\\-1\end{pmatrix}; c_5\begin{pmatrix}1\\1\end{pmatrix}; c_6\begin{pmatrix}-1\\1\end{pmatrix}; c_7\begin{pmatrix}-1\\-1\end{pmatrix}; c_8\begin{pmatrix}1\\-1\end{pmatrix} \quad (6)$$

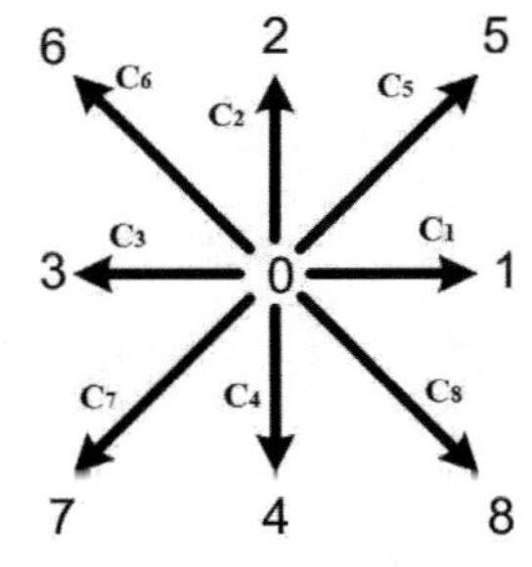

Figure 1 Model D2Q9

f_k^{eq} is the local equilibrium distribution function defined as:

$$f_k^{eq} = \rho\omega_k\left(1 + 3(\overrightarrow{C_k}.\vec{V}) + \frac{9(\overrightarrow{C_k}.\vec{V})^2}{2} - \frac{3\vec{V}^2}{2}\right)(7)$$

The fluid viscosity is related to the relaxation frequency by and the kinematic viscosity ν is:

$$\nu = \frac{\Delta x^2}{3\Delta t}\left(\tau_f - 0.5\right)(8)$$

The weight coefficients are [4]

$$\omega_0 = \frac{4}{9}, \omega_{1-4} = \frac{1}{9} \text{and } \omega_{5-8} = \frac{1}{36}$$

The density ρ is the momentum density :

$$\rho = \sum_{k=0}^{8} f_k, \rho V = \sum_{k=0}^{8} f_k C_k(9)$$

On the boundaries (see Figure 2), the fluid speed being zero, the dynamic boundary conditions are defined as follows:

$$\rho u = f_1 + f_5 + f_8 - f_3 - f_6 - f_7 = 0\,(10)$$

$$\rho v = f_5 + f_2 + f_6 - f_7 - f_4 - f_8 = 0(11)$$

Solid boundary conditions: We may apply no-slip, slip or semi-slip boundary conditions. For no-slip conditions, the standard bounce-back scheme can be used

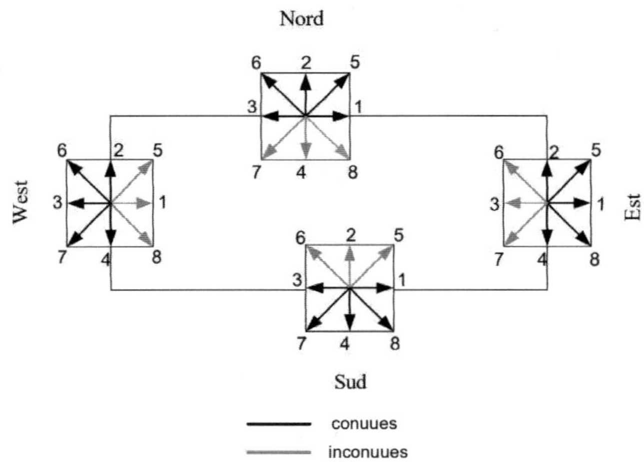

Figure 2 Configuration of boundary conditions

Inflow boundary conditions: If the velocity are known, the unknown distribution function f_k at the boundary can be decided with the method described by Zou and He[6].For example in **Figure 2**, given the velocity at the inflow boundary, after streaming, the unknown f_2, f_5 and f_6 can be decided as:

$$f_2 = f_4 + \frac{2}{3}\rho v \quad (12)$$

$$f_5 = f_7 - \frac{1}{2}(f_1 - f_3) + \frac{1}{6}\rho v + \frac{1}{2}\rho u \quad (13)$$

$$f_6 = f_8 + \frac{1}{2}(f_1 - f_3) + \frac{1}{6}\rho v - \frac{1}{2}\rho u \quad (14)$$

Outflow boundary conditions: In some problems, the outlet velocity is unknown. The normal practice is to use extrapolation for the unknown distribution functions. For instance, if the east boundary conditions (right-handside)of **Figure 2**represents the outlet condition, thenf_3,f_6, and $f_7$need to be calculated at the east boundary, i =n. A second-order polynomial can be used as

$$f_3 = 2.f_{3,n-1} - f_{3,n-2} \quad (15)$$

$$f_6 = 2.f_{6,n-1} - f_{6,n-2} \quad (16)$$

$$f_7 = 2.f_{7,n-1} - f_{7,n-2} \quad (17)$$

4. Validation

To validate the developed code, several simulations were carried out on the case of a shallow basin with irregular borders. The performance of the LBM method to solve the Naviers stokes equations in the case of a shallow flow has therefore been examined, by comparing the results obtained with those found by Charafi et al [5].

4.1 Brief description of the studied case

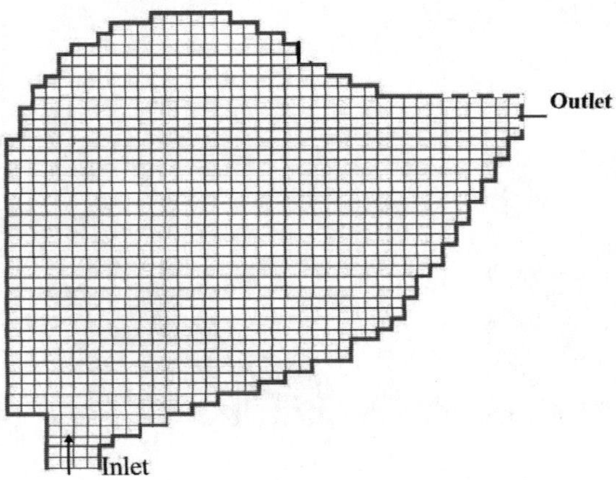

Figure3 geometry of the basin studied

Figure 3 shows the geometry of the basin with an inlet, an outlet and an irregular border. A set of meshes was used in our code in order to describe the same field by defining the nodes at the borders.

4.2 Results and discussions

IOP Conf. Series: Materials Science and Engineering **948** (2020) 012003 doi:10.1088/1757-899X/948/1/012003

As illustrated in FIG. 4, an almost perfect agreement was found by comparing the results of this work and those of Charafi et al [5], and this for different Reynolds numbers. In these simulations, we can notice the influence of the borders which therefore generate a circulation area which begins to mobilize downstream. Figure 4 therefore presents a flow model up to 1000 s of simulation which illustrates the stability of the LBM method and its efficiency in the field of flows and more particularly in the case of complex geometries for different values of the number of Reynolds: Re = 90, 150, and 240. The structure of the flow determined numerically using the two methods LBM and FDM therefore shows figures which are almost identical, meaning that the two methods give the same results.

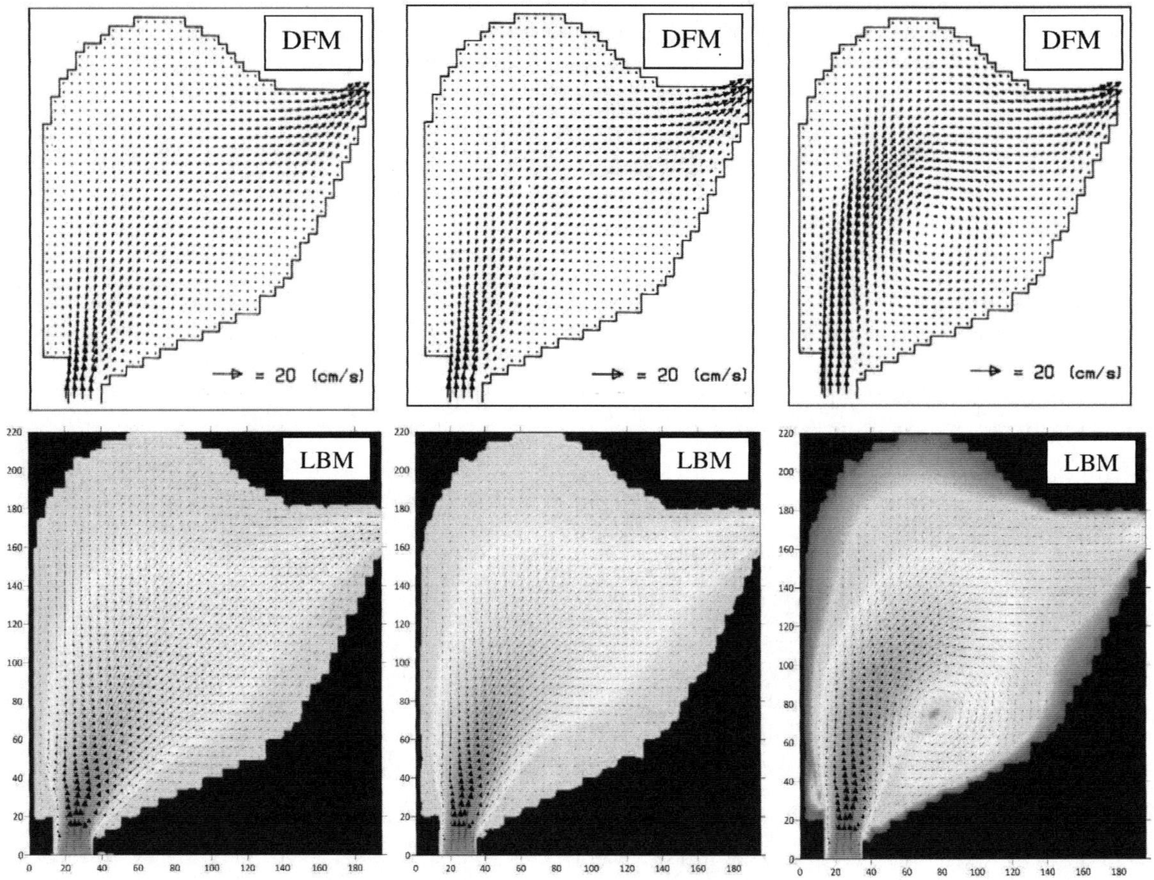

Figure 4. Hydrodynamic field obtained using the two methods for Re = 90, 150, and Re = 240.

4.3 Effect of dikes construction

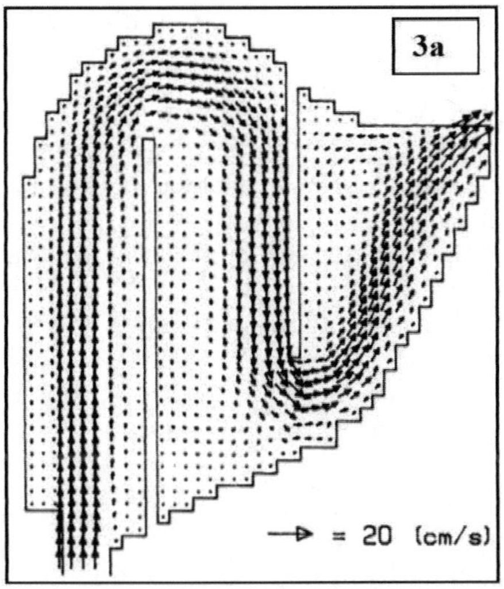

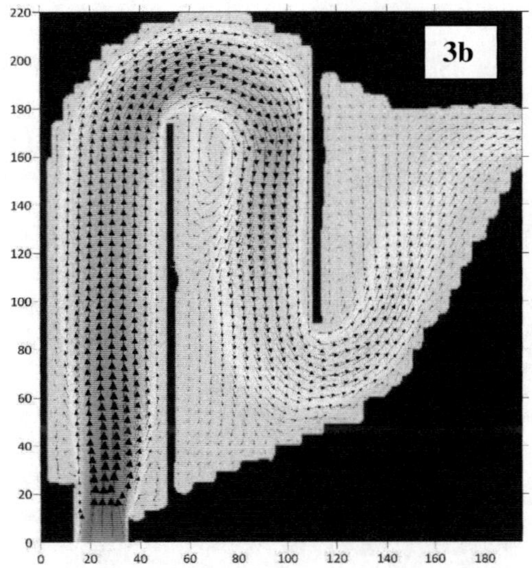

Figure 5. Effect of dikes in model of the basin:(a) circulation pattern with a dike using McCormack-FDM scheme:(b) circulation with a dike using LBM.

After modifying the structure of the basin by adding dikes, the speed field illustrated still shows the perfect agreement of the results obtained using the LBM method with those obtained using the McCormack-FDM scheme method (Charafi et al. [5]). The purpose of adding the dikes was to assess their impact on increasing the length of stay in the flow. The shape of the hydrodynamic field indeed confirms this increase since the length of the advective path of the flow increases.

5. Conclusion

In this work, a code based on the LBM method was used to study the hydrodynamic field in a shallow basin, with irregular border and dikes built inside it. The validation of the code was verified by comparing the results obtained with those of Charafi et al. [5], where the finite difference method - MacCormack scheme was used. Several simulations for different Reynolds numbers were carried out and a good agreement of the results was noted, as well in the case of a basin without dikes as a basin with dikes. The simulation with the addition of dikes made it possible to obtain a shape of the hydrodynamic field which shows an increase in the length of the advective path of the flow, and consequently the increase in its residence time.

References

[1] K. Zhao et al., “Spatial variations of concentrations of copper and its speciation in the soil-rice system in Wenling of southeastern China,” Environ. Sci. Pollut. Res., vol. 21, no. 11, pp. 7165–7176, 2014.

[2] Z. Guo, B. Shi, and N. Wang, “Lattice BGK Model for Incompressible Navier-Stokes Equation,” J. Comput. Phys., vol. 165, no. 1, pp. 288–306, 2000.

[3] R. Begum and M. A. Basit, “Lattice Boltzmann method and its applications to fluid flow problems,” Eur. J. Sci. Res., vol. 22, no. 2, pp. 216–231, 2008.

[4] A. A. Mohammed, Lattice Boltzmann Method: Fundamentals and Engineering Applications with Computer Codes, Springer, New York, 2011,.

[5] M. M. Charafi, A. Sadok, A. Kamal, and A. Menai, “Numerical modeling of water circulation and pollutant transport in a shallow basin,” Int. J. Mod. Phys. C, vol. 11, no. 4, pp. 655–664,

2000.
[6] Q. Zou and X. He, "On pressure and velocity boundary conditions for the lattice Boltzmann BGK model," Phys. fluids, vol. 9, no. 6, pp. 1591–1598, 1997.
[7] P. J. DELLAR, *An interpretation and derivation of the lattice Boltzmann method using strang splitting*, Computers and Mathematics with Application, vol. 65, P. 129-141, 2013
[8] M. Yoshino , Y. Hotta , T. Hirozane and M. Endo, "A numerical method for incompressible non- Newtonian fluid flows based on the lattice Boltzmann method,"
Department of Mechanical and Department of Electrical Systems Engineering, Shinshu University, 4- 17-1 Wakasato, Nagano 380-8553, Japan,2007.

IOP Conf. Series: Materials Science and Engineering **948** (2020) 012004 doi:10.1088/1757-899X/948/1/012004

NUMERICAL STUDY OF NATURAL CONVECTION IN SQUARE CAVITY USING LATTICE BOLTZMANN METHOD: OBSTACLES EFFECT

A. BENDARAA[1,*], M. NOUNI[1], M. OUHROUM[1], M. M. CHARAFI[1], A. HASNAOUI[1]

[1]LS3M polydisciplinary faculty of Khouribga, Sultan Mly Slimane University 25000, Morocco.

[*]Corresponding author: bendaraa.anass@usms.ac.ma

Abstract: In this work, the natural convection is studied in a closed square cavity differentially heated, using the Lattice Boltzmann numerical Method. Completely filled with a newtonian fluid, the cavity is managed by two rigid walls in right and left sides, and is provided with two rectangular adiabatic obstacles interposed on the mid-horizontal walls. The height of the two obstacles is varied from 10% to 40% of the side (L). The calculation code was performed for a laminar flow and an incompressible fluid with a Prandtl number Pr = 0.71 (the case of air). The computation code thus developed was used to analyse the effect of the length of obstacles in the mid-square cavity. The obtained results were validated by comparing with those of a previously published works. It has been deduced that the addition of the obstacles influences the thermal and hydrodynamic profiles of the flow inside the cavity and its thermal performances.

Keywords: *Natural convection, Numerical methods, Lattice Boltzmann Method, Finite difference method (FD),* obstacles.

1. Introduction

The heat transfer phenomenon by natural convection in confined spaces have been and continue to be the subject of numerous research activities in the fundamental field as well as in the experimental fields. Many works and applications studied will be able to satisfy the requirements of many industrial applications, in particular in building, solar collectors, cooling of electronic components, nuclear engineering, heating and ventilation, Etc. Natural convection is fundamental interest to mathematicians, physicists, chemists and biologists, and many others scientists. As a result, it has become essential for any engineer, whatever the field in which he will be called, to have good knowledge of the fundamental laws of thermal science.

To solve this type of problems, we must resort to numerical methods [1] if we cannot solve them analytically. Among these most used methods, we can cite, finite volumes (FVM), finite elements (FEM), finite differences (FDM) methods. In this work, we have used the Lattice Boltzmann Method to solve the problem of convection [2], [3]. We will present a contribution in the use of this new method which is based on the thermal mesoscopic approach of the Boltzmann equation on a Lattice, applied to the problem of heat transfer by natural convection within a closed square cavity and differentially heated. The cavity is equipped with two adiabatic obstacles. It is considered that the two vertical walls of the cavity are isothermal and the horizontal ones are adiabatic. The obstacles effect has become the subject of significant investigation by researchers in the field giving example of [4] who have studied the effect of fins on convective flow.

The Lattice Boltzmann method, commonly called LBM [5], is a recent new alternative for the numerical simulation of physical phenomena which emerged in the last decade. The method which is based on the resolution of the Boltzmann equation with the hydrodynamic approach, it describes the movement of a fluid at a mesoscopic level (speed, pressure, density) by the distribution of these different particles [6]. This method is well suited to simulation because it represents a very interesting new approach relative to classical methods, such as finite volumes, finite differences and finite elements which are based on the discretization of the Navier- Stokes equations. The Lattice Boltzmann method has proven advantages by its efficiency in the solution of fluid mechanics problems lies mainly in the simplicity of the modelling and the simulation of complex flow phenomena, it is very suitable in the implementation which allows an easy and efficient parallel calculation. The study presented in this work aims to analyse the phenomenon of heat transfer by natural convection in laminar mode, in a square cavity with two horizontal adiabatic walls and two vertical isothermal and differentially heated walls [7]. The cavity completely filled with Newtonian fluid is provided with two adiabatic obstacles of rectangular shape, inserted on the horizontal walls. The study is carried out for Rayleigh numbers from 10^4 to 10^6 and a Prandtl number Pr = 0.71. The influence of the height of the obstacles (H) will also be treated for values of 10% to 40% of (L) Height of the cavity.

2. Mathematical formulation

Let us consider a square cavity of height L (Fig. 1). This cavity is filled with a fluid having a Prandtl number (Pr = 0.71) initially at the temperature T_c. The horizontal walls of the cavity are assumed to be adiabatic. The vertical walls are maintained at constant temperatures T_f and T_c (hot and cold temperatures respectively), so a horizontal thermal gradient is established. The non-slip condition is applied to all walls of the cavity [8]. The fluid filling the cavity is subjected to the only action of the gravity field (g) directed downwards in a Cartesian coordinate system (o, x, y). The fluid is supposed to be Newtonian and incompressible. The governing equations follow the equation of continuity, the equations of conservation of momentum and energy represented as follows:

$$\frac{\partial u}{\partial x}+\frac{\partial v}{\partial y}=0 \tag{1}$$

$$x \rightarrow \quad U\frac{\partial U}{\partial X}+V\frac{\partial U}{\partial Y}=\frac{\partial P}{\partial X}+Pr\left(\frac{\partial^2 U}{\partial X^2}+\frac{\partial^2 U}{\partial Y^2}\right) \tag{2}$$

$$y \rightarrow \quad U\frac{\partial V}{\partial X}+V\frac{\partial V}{\partial Y}=\frac{\partial P}{\partial Y}+Pr\left(\frac{\partial^2 V}{\partial X^2}+\frac{\partial^2 V}{\partial Y^2}\right)+Pr.Ra.\theta \tag{3}$$

$$u\frac{\partial \theta}{\partial x}+v\frac{\partial \theta}{\partial y}=\left(\frac{\partial^2 \theta}{\partial x^2}+\frac{\partial^2 \theta}{\partial y^2}\right) \tag{4}$$

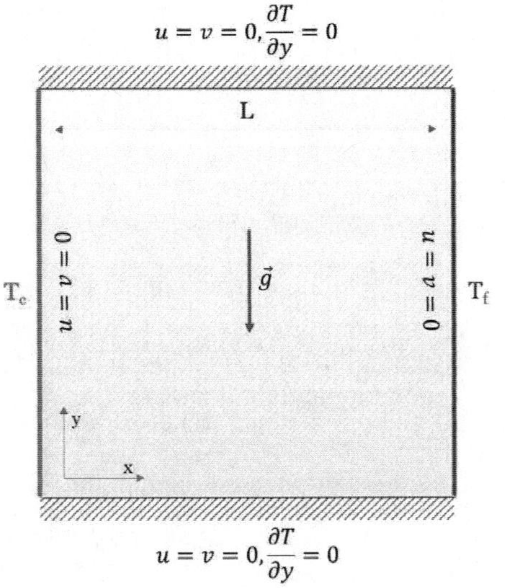

Fig. 1: Schematic representation for the physical model.

- **Boundary conditions of the problem**
- Hot wall: $x = 0$ and $0 \leq y \leq 1$ $U = 0; V = 0; T = T_c$
- Cold wall: $x = 1$ and $0 \leq y \leq 1$ $U = 0; V = 0; T = T_f$
- Bottom and top walls: $y = 0,\ y = 1$ and $0 \leq x \leq 1$ $U = 0; V = 0; \frac{\partial T}{\partial y} = 0$

3. Operated method

In this work we performed the Lattice Boltzmann Method to solve the natural convection problem. This method is based on kinetic of gazes. The governing equation will be discretized as follow: The kinetic equation for the distribution function (temperature distribution, species distributions. $f_k(x,t)$ can be written as:

$$\frac{\partial f_k(x,t)}{\partial t}+c_k\frac{\partial f_k(x,t)}{\partial x}=\Omega_k \tag{5}$$

The left-hand side terms represent the streaming process, where the distribution of stream function along with the lattice link with velocity: $c_k = \frac{\Delta x}{\Delta t}$

The right-hand term Ω_k represents the rate of change of distribution function f_k in the collision process. BGK approximation for the collision operator can be approximated as [9]–[11].

$$\Omega_k = \frac{-1}{\tau}\left[f_k(x,t) - f_k^{eq}(x,t) \right] \tag{6}$$

The term τ represents a relaxation time toward the equilibrium distribution (f_k^{eq}), which is related to the diffusion coefficient on the macroscopic scale. The relation will be discussed latter on.
As usual, the LBM consists of two steps, collisions and streaming
The collisions step without forcing function is:

$$f_k(x,y,t+\Delta t) = f_k(x,y,t)[1-\omega] + \omega f_k^{eq}(x,y,t) \tag{7}$$

k = 0; . . .; 8:
The streaming step is:

$$f_k(x+\Delta x, y+\Delta y, t+\Delta t) = f_k(x,y,t+\Delta t) \tag{8}$$

For the temperature field:

$$g_k(x+\Delta x, t+\Delta t) - g_k(x,t) = -\frac{1}{\tau_\alpha}\left[g_k(x,t) - g_k^{eq}(x,t) \right] \tag{9}$$

Where τ_α is the relaxation time. f_k^{eq} and g_k^{eq} are the local equilibrium distribution function.
g_k^{eq} is defined as follow:

$$g_k^{eq} = \omega_k \phi(x,t)\left(1 + \frac{c_k.\vec{u}}{c_s^2}\right) \tag{10}$$

where $\vec{u}$ is advection velocity vector, and c_k is unit vector along the streaming direction. The weight factors are $\omega_0 = 4/9$. $\omega_1 = \omega_2 = \omega_3 = \omega_4 = 1/9$, and $\omega_5 = \omega_6 = \omega_7 = \omega_8 = 1/36$.
Where: $c_s = c_k/\sqrt{3}$ is the lattice speed of sound.
The dependent variable $\phi(x,t)$ in Eq. 10, can be related to the distribution function g_k, as $\phi(x,t) = \sum_{k=1}^{8} g_k(x,t)$

4. Validation of calculation code

In the first, to validate our calculation code made with LBM, we studied natural convection in a differentially heated closed square cavity,the two horizontal walls of the top and bottom of which are considered adiabatic. This type of configuration is associated with a classic problem intensively studied and used to check the accuracy of numerical codes (Fig. 1). It is emphasized that the right-side wall of the cavity is subjected to a cold temperature and that of the left to a hot temperature, which produces between the two faces, a horizontal temperature gradient ($\Delta T = T_c - T_f$). We compare the results of the streamlines and isotherms deduced with those of the literature. A 100x100 network is used. The number of Prandtl Pr is fixed at 0.71 (air), and the number of Rayleigh Ra, defined by the preceding Eq varies from 10^3 to 10^6.

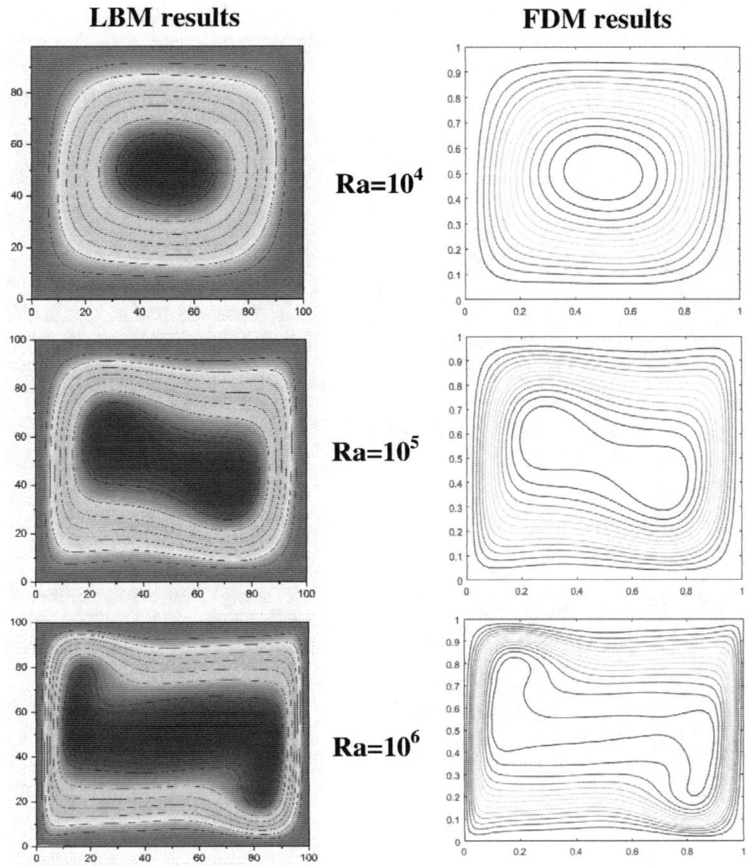

Fig.2: Comparison between the results of stream function from LBM and FDM for different Rayleigh numbers.

This section is reserved for a comparative study between the results of the LBM and the finite difference method to see the concordance rate between the results of the two methods and to try to validate the present calculation code. To do this, we have studied the problem of natural convection in a closed square cavity filled with air, differentially heated, whose two horizontal top and bottom walls are considered adiabatic. This type of configuration is associated with a classical problem that has been intensively studied and used for the verification of the accuracy of calculation codes. It is pointed out that the right and left side walls are maintained at fixed cold and hot temperatures respectively. The validation of the calculation code was performed for the test case (unobstructed square cavity), with a Prandtl number (Pr) set to 0.71 (air case), and the Rayleigh number (Ra) between 10^4 and 10^6. Figures 2 and 3 clearly show that the results of the stream and isothermal lines obtained are similar to those found by the finite difference method[12], [13], which leads us to admit the present computation code to study other future applications.

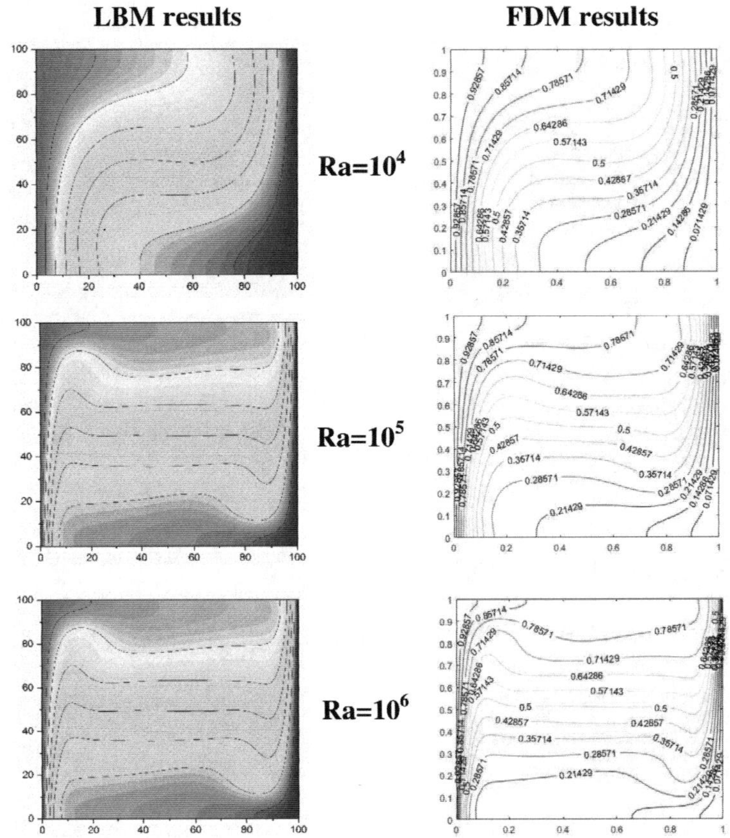

Fig.3: Comparison of isotherms from LBM and FDM for different Rayleigh numbers.

5. Results and discussion

After validation of the calculation code by the comparative approach, this sub-section will be devoted to the study of the influence of the two adiabatic obstacles height, which varies between 0.1L and 0.4L on the convective flow within the cavity. This study is made for a Rayleigh range between 10^4 and 10^6. In what follows, we will show the results obtained for the two thermal and dynamic fields, these results will be accompanied by a discussion on the effect of the adiabatic obstacles on these two fields.

- ***Dynamic field***

Figures 4-6, respectively, show the stream functions for different obstacle heights at different Rayleigh numbers between 10^4 and 10^6.

For obstacles with a height H=0.1L, their presence does not affect changes in the flow structure, and the resulting configuration tends to be an obstacle-free case. This means that the phenomenon of heat transfer for this first case can remain without apparent influence.

IOP Conf. Series: Materials Science and Engineering **948** (2020) 012004 doi:10.1088/1757-899X/948/1/012004

For a height H=0.2L, there is a slight compression towards the centre of the cavity. This slightly influences the structure of the flow for a Rayleigh number equal to Ra $=10^4$. As the Rayleigh number increases, the influence of the compression on the flow structure increases, which becomes more and more apparent, illustrated by a break in the central stream line.

The increase in the height of the obstacles from H=0.3Lto H=0.4L induces a break in the unicellular structure of the flow to form two cells. The shape of the stream lines shows that they are stretched towards the upper part on the left side and the other towards the lower part on the right side. Furthermore, it can be concluded from these findings that for higher heights H=0.2L, we are faced with a construction of two zones of stream lines of different intensity, from which the stream lines on the left side are of a much higher intensity than the contours on the right side. This influence on mechanical profiling, from the case of convection in the presence of obstacles, becomes more and more intense by increasing the number of Rayleigh.

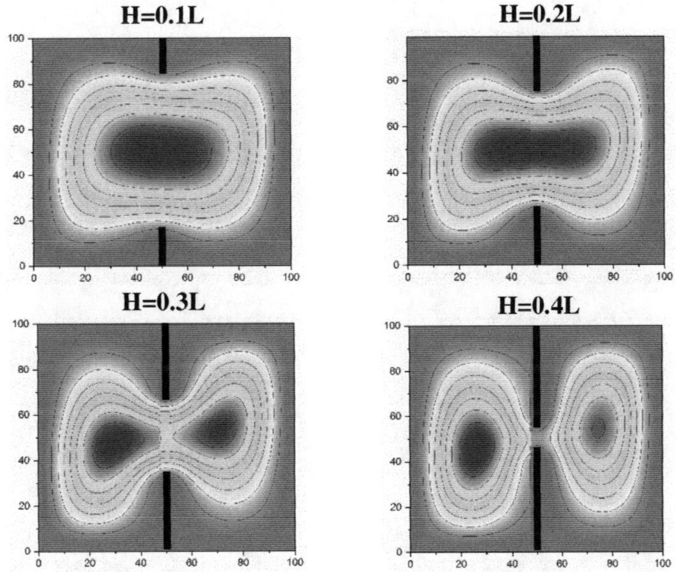

Fig.4: Stream functions for different obstacle heights at Ra=10^4.

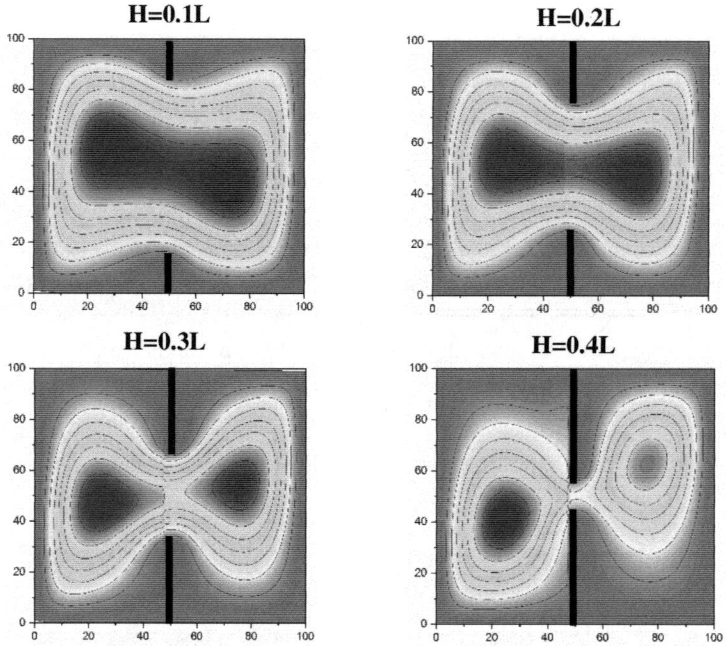

Fig.5:Stream functions for different obstacle heights at Ra=10^5.

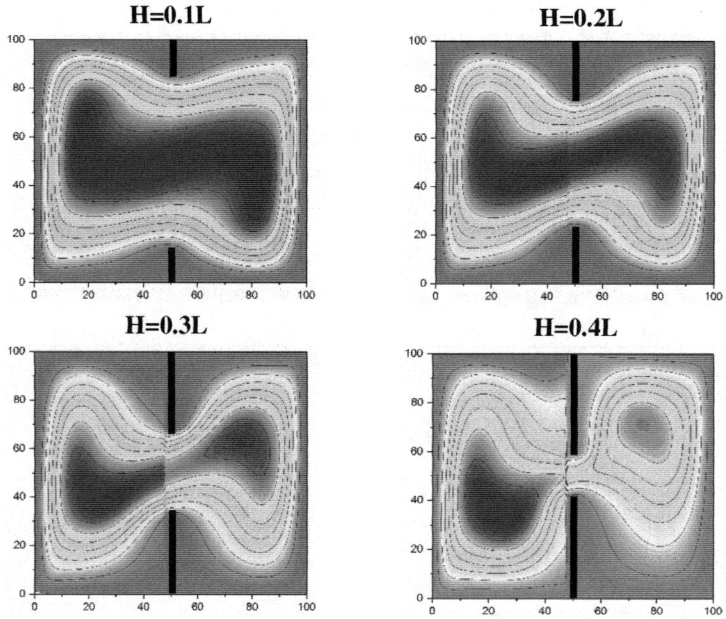

Fig.6: Stream functions for different obstacle heights at Ra=10^6.

- ***Thermal field***

The thermal field, described by the isothermal lines, is represented respectively on figures 7 to 9, and this for different obstacle heights from H=0.1L to H=0.4L and at different Rayleigh numbers between 10^4 and 10^6. The shape of the isotherms for a height H=0.1L is similar to those obtained for a square cavity without obstacles, the only difference being the adiabatic obstacles, in which the isothermal lines join these obstacles forming a mass of parallel lines along with its thickness. The presence of obstacles with a height of H=0.2L slightly changes the shape of the isotherms which stratify in the centre of the cavity near the obstacles. This stratification becomes more preponderant with increasing obstacle heights (H=0.3L and H=0.4L) and makes these layers more twisted away from these isothermal blocks, this effect becomes even more apparent with increasing Rayleigh numbers. It is also noticed that the increase in the height of the adiabatic obstacles can divide our cavity into two zones, a high-temperature zone near the warm wall and a cold temperature zone near the cooler wall, this remark is valid for all Rayleigh numbers, which may explain the use of such a process in thermal insulation applications.

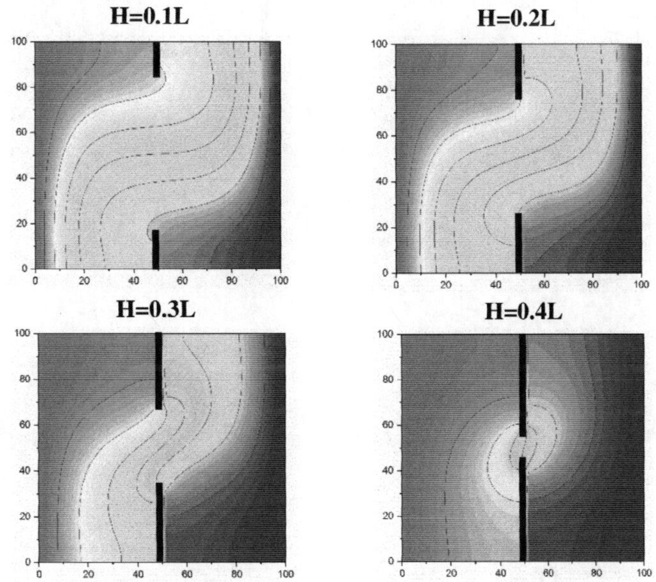

Fig.7: Isotherms for different obstacle heights at Ra=10^4.

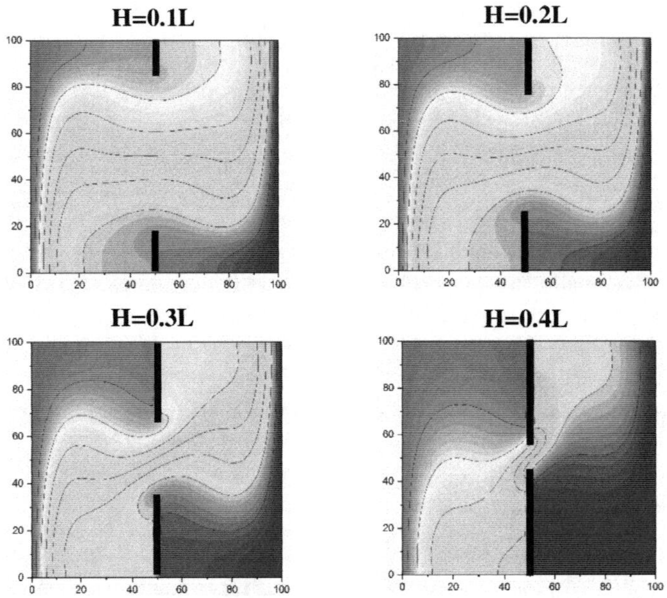

Fig.8: Isotherms for different obstacle heights at Ra=10^5.

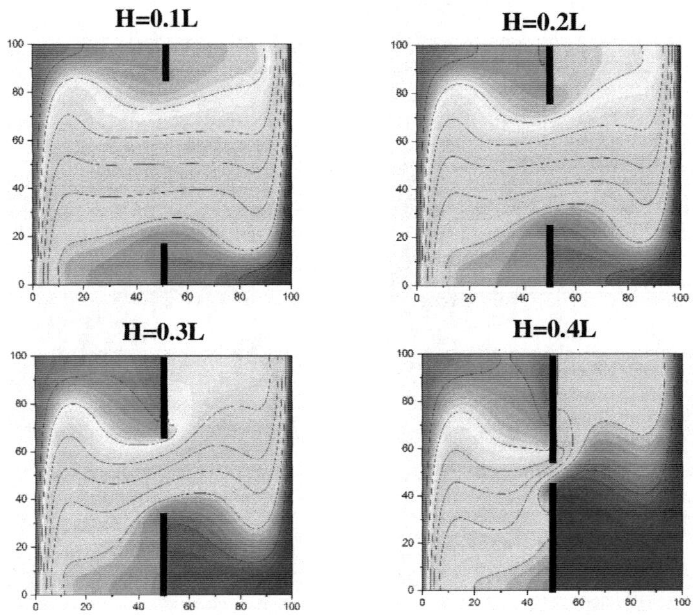

Fig.9: Isotherms for different obstacle heights at Ra=10^6.

- ***Heat transfer Rate***

To concretize this result, and to try to quantify the rate of heat transfer for the case of a cavity with obstacles, as well as the effect of their presence on the diffusion of heat, it was possible to evaluate the average Nusselt number for all the previously reported cases. The results thus found are shown in Fig.10, which shows the evolution of mean Nusselt for each height of obstacle and different Rayleigh numbers. It is found that the increase in obstacle height leads to a decrease in the Nusselt number, and consequently to degradation of heat transfer for our configuration. This decrease becomes more and more important for heights higher than H=0.2L, and this, by increasing the number of Rayleigh. Quantifying the rate of heat transfer employing the average Nusselt number made it possible to conclude the harmfulness of this obstacle arrangement in our study configuration.

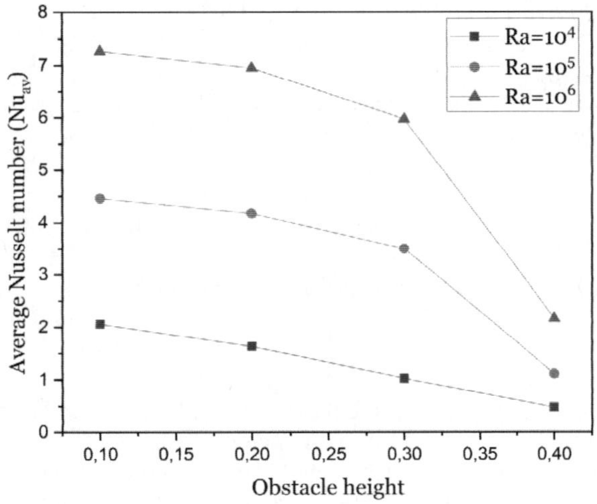

Fig.10: The evolution of the average Nusselt number for different Rayleigh numbers and different obstacle heights.

6. Conclusion

The numerical study of the effect of two obstacles attached to the mid-adiabatic walls of a differentially heated square cavity on the fluid flow and heat transfer by natural convection was examined. A qualitative comparison of the results obtained by the LBM with the results of the finite difference method led us to validate our computational code, in order to conduct several numerical simulations based on the said method, to explain and solve this problem. The results thus obtained, allowed us to conclude that the increase in the height of the obstacles induces a decrease in convective heat transfer and to divide our configuration into two zones, one heated and the other cooled, which showed the feasibility of the application in thermal insulation processes. A second conclusion, concerning the effect of the variation in Rayleigh number on convective heat transfer, it was deduced that the increase in Ra, led to an increase in the average Nusselt number, subsequently, a gain in terms of heat transfer rate will apply by this growth.

REFERENCES

[1] M. Yoshino, Y.-H. Hotta, T. Hirozane, and M. Endo, "A numerical method for incompressible non-Newtonian fluid flows based on the lattice Boltzmann method," *J. Nonnewton. Fluid Mech.*, vol. 147, no. 1–2, pp. 69–78, 2007.

[2] U. Ghia, K. N. Ghia, and C. T. Shin, "High-Re solutions for incompressible flow using the Navier-Stokes equations and a multigrid method," *J. Comput. Phys.*, vol. 48, no. 3, pp. 387–411, 1982.

[3] A. Boualit, N. Zeraibi, T. Chergui, M. Lebbi, L. Boutina, and S. Laouar, "Natural convection investigation in square cavity filled with nanofluid using dispersion model," *Int. J. Hydrogen Energy*, vol. 42, no. 13, pp. 8611–8623, 2017.

[4] A. Bendaraa, M. M. Charafi, and A. Hasnaoui, "Numerical study of natural convection in a differentially heated square cavity filled with nanofluid in the presence of fins attached to walls in different locations," *Phys. Fluids*, vol. 31, no. 5, p. 052003, 2019.

[5] G. D. Mallinson and G. de Vahl Davis, "The method of the false transient for the solution of coupled elliptic equations," *J. Comput. Phys.*, vol. 12, no. 4, pp. 435–461, 1973.

[6] M. Jami, A. Mezrhab, and H. Naji, "Numerical study of natural convection in a square cavity containing a cylinder using the lattice Boltzmann method," *Eng. Comput.*, 2008.

[7] G. de Vahl Davis, "Natural convection of air in a square cavity: a bench mark numerical solution," *Int. J. Numer. methods fluids*, vol. 3, no. 3, pp. 249–264, 1983.

[8] E. Bilgen, "Natural convection in cavities with a thin fin on the hot wall," *Int. J. Heat Mass Transf.*, vol. 48, no. 17, pp. 3493–3505, 2005.

[9] A. A. Mohamad, *Lattice Boltzmann Method*, vol. 70. Springer, 2011.

[10] H. N. Dixit and V. Babu, "Simulation of high Rayleigh number natural convection in a square cavity using the lattice Boltzmann method," *Int. J. Heat Mass Transf.*, vol. 49, no. 3–4, pp. 727–739, 2006.

[11] A. A. Mohamad and A. Kuzmin, "A critical evaluation of force term in lattice Boltzmann method, natural convection problem," *Int. J. Heat Mass Transf.*, vol. 53, no. 5–6, pp. 990–996, 2010.

[12] A. Bendaraa, M. M. Charafi, and A. Hasnaoui, "Numerical modeling of natural convection in horizontal and inclined square cavities filled with nanofluid in the presence of magnetic field," *Eur. Phys. J. Plus*, vol. 134, no. 9, p. 468, 2019.

[13] M. M. Charafi, A. Bendaraa, and A. Hasnaoui, "Numerical modelling of natural convection in a square cavity: Effect of nanofluid volume fraction and inclination," in *MATEC Web of Conferences*, 2018, vol. 241, p. 1006.

IOP Conf. Series: Materials Science and Engineering **948** (2020) 012005 doi:10.1088/1757-899X/948/1/012005

Ab initio study of electron and phonon coupling in silicon

Mustapha El Mliles[1,2] & Youssef El Kouari[1] & Abdelowahed Hajjaji[2,*]

[1] Faculty of Sciences and Techniques, Hassan II Casablanca University, Morocco
[2] National School of Applied Sciences, Chouaib Doukkali University, Morocco
*E-mail: hajjaji.a@ucd.ac.ma

Abstract: We have studied electron-phonon coupling in silicon and its relation to hot carriers thermalization. The calculation of electron-phonon and phonon-electron interactions in silicon was obtained from first-principles density functional theory and many-body perturbation theory. Electron and phonon relaxation times were calculated subsequently. Moreover, the results showed that silicon exhibits a relatively wide phononic band gap along the symmetry line **X-W**, which prevents the Klemens mechanism leading to a phonon bottleneck effect. As demonstrated, phonon-electron interactions can play an important role in mitigating the electron thermalization along this symmetry line. This work shade light on the importance of including the phonon- electron interactions in the ab initio approach for future semiconductor materials engineering.

1. Introduction

Semiconductor materials used in photovoltaic (PV) cells have the property of releasing their electrons under the influence of solar energy. In this process, photons strike electrons and create an electron-hole pair. Afterward, the electron (hole) undergoes an inelastic scattering with phonons and relaxes to the conduction (valence) band edge, releasing energy as heat to the crystal lattice [1, 2] in a process called thermalization. Therefore, solar energy is partially converted into heat, causing efficiency degradation due to the increase in cell temperature [3].

Thermalization dynamics are very important to understand the efficiency degradation in semiconductors and to advance material technologies in this field. Electron-phonon (e-ph) interactions are the origin of the electron thermalization, where the electron creates a phonon at the cost of some of its energy [4]. A semiclassical approach using Monte Carlo simulation was performed to describe the e-ph interactions [5], however, this approach has used some empirical parameters to fit experimental data.

A study was made to understand deeply these e-ph interactions by Bernardi et al. [6, 7], who developed an accurate ab initio approach based on the density functional theory and many-body perturbation theory to calculate the electron relaxation time, and applied it to silicon and gallium arsenide semiconductors. Computed electron relaxation times were in agreement with the experimental data.

The previous studies [6, 7] did not take into consideration ph-e interactions that can play an important role in thermalization dynamics. This work is an attempt to fill this lack, by including ph-e interactions in the ab initio approach used by Bernardi et al. [6, 7], and apply it to investigate the relaxation time of electrons and phonons in silicon (Si).

2. Computational methods

Ab initio calculations were performed on a unit cell Si with a lattice parameter of 5°A. Electronic band structure calculations were carried out within density functional theory (DFT) [8] using the Quantum ESPRESSO code [9], with a local density approximation (LDA) [10] for the exchange-correlation functional [11]. Norm-conserving pseudopotentials were used [12], and the kinetic energy cutoff was set to 80 Ry for wave functions. Phonon calculations were performed within density functional perturbation theory (DFPT) [13]. The EPW code [14] was used to calculate the electron and phonon self-energies, arising from electron-phonon interaction.

The carrier relaxation time is determined by:

$$\tau = \frac{[\mathrm{Im}(\Sigma)]^{-1}\hbar}{2} \quad (1)$$

Where Im(Σ) is the imaginary part of the carrier self-energy.

The electron self-energy is calculated as:

$$\Sigma_{n,k}^{e-ph} = \sum_{qv} \omega_q |g_{mn}^{v}(k,q)|^2 \left[\frac{n(\omega_{qv}) + f(\epsilon_{m,k+q})}{\epsilon_{n,k} - \epsilon_{m,k+q} - \omega_{qv} - i\eta} + \frac{n(\omega_{qv}) + 1 - f(\epsilon_{m,k+q})}{\epsilon_{n,k} - \epsilon_{m,k+q} + \omega_{qv} - i\eta}\right] \quad (2)$$

And the electron relaxation time describing the e-ph interaction is given by:

$$\tau_{n,k}^{e-ph} = \frac{\left[\mathrm{Im}\left(\Sigma_{n,k}^{e-ph}\right)\right]^{-1}\hbar}{2} \quad (3)$$

On the other hand, the phonon self-energy is calculated as:

$$\Sigma_{mn,k}^{ph-e} = \sum_{mn,k} \omega_k |g_{mn}^{v}(k,q)|^2 \frac{f(\epsilon_{n,k}) - f(\epsilon_{n,k+q})}{\epsilon_{n,k} - \epsilon_{n,k+q} - \omega_{qv} + i\eta} \quad (4)$$

And the phonon relaxation time describing the ph-e interaction is given by:

$$\tau_{qv}^{ph-e} = \frac{\left[\mathrm{Im}\left(\Sigma_{qv}^{ph-e}\right)\right]^{-1}\hbar}{2} \quad (5)$$

In Eq. 2 and Eq. 4 the electronphonon matrix elements are given by:

$$g_{mn}^{v}(k,q) = \langle \psi_{m,k+q} | \partial_{qv} V | \psi_{n,k} \rangle \quad (6)$$

With $\psi_{n,k}$ the electronic wave function for band n, wave-vector **k** and eigenvalue $\epsilon_{n,k}$, $\partial_{qv}V$ the derivative of the self-consistent potential associated with a phonon of wave-vector **q**, branch index ν and frequency ω_{qv} . $f(\epsilon_{n,k})$ and $f(\epsilon_{m,k+q})$ are Fermi occupations, and $\omega_{k,q}$ are weights of k-points and q-points, respectively.

The electronic states were calculated self consistently using DFT over a uniform grid of 6×6×6 k-points, then the vibrational states using DFPT over a uniform grid of 6×6×6 q-point grids. Subsequently, the EPW framework [14] transforms the quantities needed to evaluate the e-ph self-energy into the Wannier representation [15, 16] on a finer grid of 20×20 ×20 k and q points. Finally, the carrier relaxation time was calculated using Eq. 3 and Eq. 5.

IOP Conf. Series: Materials Science and Engineering **948** (2020) 012005 doi:10.1088/1757-899X/948/1/012005

In this work, electron-electron (e-e) interactions are neglected, due to their slight contribution to the electron relaxation time compared to e-ph interactions. This assumption is justified based on the results obtained by Bernardi et al. [6]. They showed that the phonon emission processes from e-ph interactions dominate the electron relaxation time.

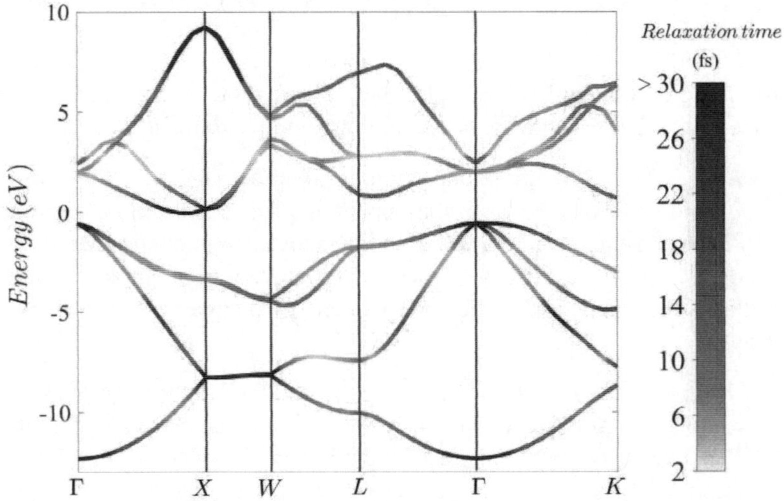

Figure 1 The band structure and the electron relaxation time $\tau_{n,k}$ of Si.

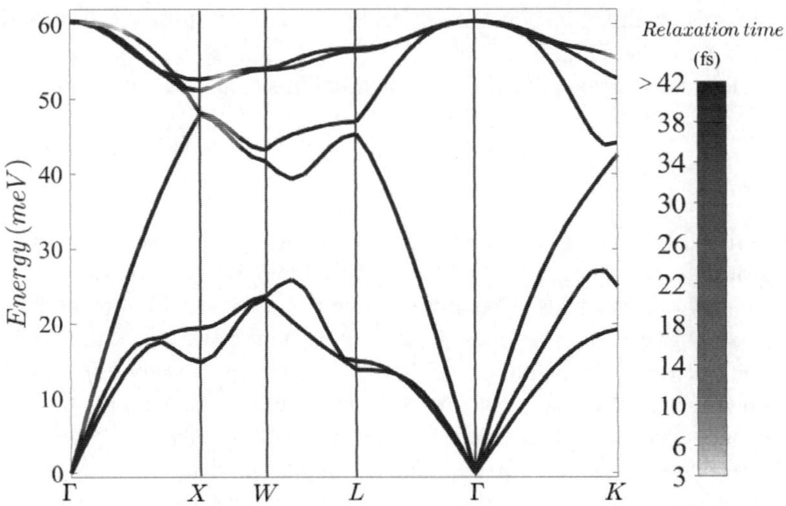

Figure 2 The phonon dispersion and the phonon relaxation time τ_{qv} of Si.

3. Results and discussion

The electronic band structure with the electron relaxation time in Si along high-symmetry directions with Fermi level set at 0 eV are depicted in Fig 1. As observed, the electron relaxation time depends highly on the band and k directions. Also, the symmetry line **X-W** exhibits the longer overall relaxation time compared to the other symmetry lines. This is related primarily to the few energy bands and the separation between them along this direction, which restricts the electron possible states transitions leading to phonons emission.

Therefore, the rate of e-ph interactions along the **X-W** direction is relatively weaker resulting in longer relaxation time of the order of ~ 10fs.

The phonon dispersion with the phonon relaxation time in Si along high-symmetry directions is shown in Fig 2. The dispersion relations result in two modes of phonons, specifically, the acoustic and optical modes corresponding to a zero and non-zero frequency at the Brillouin zone center, respectively. We observed that the phonon lifetime in Si is uniform and exhibits a weak dependence on q directions, except at the optical branch of the Brillouin zone center and the symmetry line **X-W** where the phonon lifetime drops dramatically.

This result can be explained by the phonon bottleneck effect [17], which occurs whenever a large phononic band gap exists between the optical and acoustical modes, preventing optical phonons (OP) from decaying into two lower energy acoustical phonons (AP) (OP → 2AP) in a process known as Klemens decay [18]. The conservation of energy will prevent the Klemens mechanism, as long as the phononic gap energy is greater than the highest acoustic phonon energy.

The phononic band gap is nearly absent in Si (Fig. 2), and the Klemens mechanism is strongly present. Except along the **X-W** direction, where we observed the presence of a relatively wide phononic band gap between optical and acoustical branches, hence slowing the decay of hot phonons. Consequently, the ph-e interactions will take place, increasing the probability of electrons reabsorbing hot phonons.

Hot phonons along the **X-W** direction with a lifetime of the order of ~ 10fs can contribute effectively to the reheating of electrons by optical phonon absorption. As discussed, the **X-W** direction minimizes the difference between electron and phonon lifetime, making the extraction of electrons appreciable within silicon thin films along this direction.

4. Conclusion

Thermalization dynamics were investigated in silicon, through the calculation of electron and phonon relaxation times along high symmetry directions. The electronic band structure in silicon depicts few energy bands relatively separated along the **X-W** direction. Besides, the phonon dispersion reveals the presence of the phonon bottleneck effect along this direction. As a result, the electron and phonon relaxation times are comparable along the **X-W** direction (~ 10fs), allowing the reabsorption of hot phonons by electrons. These results suggest that including phonon-electron interactions in the ab initio analysis allows for a better understanding of thermalization dynamics in silicon and emphasizes the significant role of phonon-electron interactions in engineering semiconductor materials.

IOP Conf. Series: Materials Science and Engineering **948** (2020) 012005 doi:10.1088/1757-899X/948/1/012005

References

[1]A.P. Kirk, M.V. Fischetti, Physical Review B 86, 165206 (2012)
[2]A. Polman, H.A. Atwater, Nature materials 11, 174 (2012)
[3]M.A. Green, Physica E: Low-dimensional Systems and Nanostructures 14, 65 (2002)
[4]M. Bernardi, The European Physical Journal B 89, 239 (2016)
[5]M.V. Fischetti, S.E. Laux, Physical Review B 38, 9721 (1988)
[6]M. Bernardi, D. Vigil-Fowler, J. Lischner, J.B. Neaton, S.G. Louie, Physical review letters 112, 257402 (2014)
[7]M. Bernardi, D. Vigil-Fowler, C.S. Ong, J.B. Neaton, S.G. Louie, Proceedings of the National Academy of Sciences 112, 5291 (2015)
[8]K. Hohenberg, Phys. Rev. 136, B864 (1964)
[9]P. Giannozzi, S. Baroni, N. Bonini, M. Calandra, R. Car, C. Cavazzoni, D. Ceresoli, G.L. Chiarotti, M. Cococcioni, I. Dabo et al., Journal of physics: Condensed matter 21, 395502 (2009)
[10]S. Kohn, Phys. Rev. 140, A1133 (1965)
[11]Y. Zhao, N.E. Schultz, D.G. Truhlar, Exchange-correlation functional with broad accuracy for metallic and nonmetallic compounds, kinetics, and noncovalent interactions (2005)
[12]D. Hamann, Physical Review B 88, 085117 (2013)
[13]S. Baroni, S. De Gironcoli, A. Dal Corso, P. Giannozzi, Reviews of Modern Physics 73, 515 (2001)
[14]J. Noffsinger, F. Giustino, B.D. Malone, C.H. Park, S.G. Louie, M.L. Cohen, Computer Physics Communications 181, 2140 (2010)
[15]N. Marzari, A.A. Mostofi, J.R. Yates, I. Souza, D. Vanderbilt, Reviews of Modern Physics 84, 1419 (2012)
[16]A.A. Mostofi, J.R. Yates, Y.S. Lee, I. Souza, D. Vanderbilt, N. Marzari, Computer physics communications 178, 685 (2008)
[17]G. Conibeer, N. Ekins-Daukes, J.F. Guillemoles, D. K˝onig, E.C. Cho, C.W. Jiang, S. Shrestha, M. Green, Solar Energy Materials and Solar Cells 93, 713 (2009)
[18]P. Klemens, Physical Review 148, 845 (1966)

Design and performance analysis of Rectenna Circuit

K. El Batal,[1,5*], N. Chakhchaoui [2,3], A. Eddiai [1] M. Meddad [4], M. Rguiti [5], M. Mazroui [1], O. Cherkaoui [2]

[1] Laboratory of Physics of Condensed Matter (LPMC), Faculty of Sciences Ben M'Sik, Hassan II University, Casablanca, Morocco

[2] REMTEX Laboratory, Higher School of Textile and Clothing Industries (ESITH), Casablanca, Morocco

[3] BGIM Laboratory, Higher Normal School (ENS), Hassan II University, Casablanca, Morocco

[4] LAS Laboratory of Setif, Mohamed el Bachir el Ibrahimi BBA University, Algeria

[5] Univ. Valenciennes, EA 2443 - LMCPA - Laboratoire des Matériaux Céramiques et Procédés Associés, F-59313 Valenciennes, France.

* khadija.elbatal-etu@etu.univh2c.ma

Abstract: Radio Frequency Energy harvesting is an extremely important research subject, linked to sustainability, and could be a potential solution to conventional energy supplies. The main idea is to develop battery-free wireless sensors able to capture the available energy into the mentioned bandwidth. In this paper, a 1.8–2.4 GHz wideband rectenna is designed for radio frequency (RF) energy harvesting in the context of wireless sensor nodes (WSN). For this purpose, the agilent Advanced Design System (ADS) was used. This article presents the design and simulation of a high output voltage rectifier. First, we designed a rectifier based on Metelics MSS20-141 Schottky diode using a series topology to convert electromagnetic energy into DC. Then, a low-pass filter was implemented to filter out undesirable harmonics generated by the non-linear Schottky diode. Simulation results show that load resistance, input frequency and input power are important parameters to evaluate the performances of RF energy harvester. Also, the three RF bands, GSM1800, UMTS and Wi-Fi, are considered. Therefore, the designed system has an average RF-to-DC conversion of 18 % for an incident power of -10 dBm for a variation of 20 to 1200 Ω.

Keywords: RF energy harvesting, Schottky diode, Rectenna, RF– DC conversion efficiency.

1. Introduction

Wireless sensor networks (WSN) are mainly powered by batteries. however, Batteries have a short lifespan despite substantial progress. which motivates people to develop alternative methods of powering these objects. One exciting approach is to harvest energy from the ambient environment of microsystems. Besides using thermal gradients [1], mechanical vibrations [2-6], or solar energy[7-8] as sources of energy. Recently, electromagnetic waves have attracted attention as an important source of energy. Unlike the majority of energy sources, electromagnetic or RF energy sources are permanently available. The downside of RF sources, while pervasive in our everyday lives, is that the power density they can offer is small relative to other source [9-12].

A rectifying antenna (rectenna) is a key component of the microwave power transfer system .The principal drawback of a rectenna circuit is that it is designed for a very specific operating point. Generally, the key problem when designing rectennas is optimizing the electrical power supplied to the load or the conversion efficiency RF-to-dc [13].

IOP Conf. Series: Materials Science and Engineering **948** (2020) 012006 doi:10.1088/1757-899X/948/1/012006

This paper proposes a novel reconfigurable electromagnetic harvesting device that is ensuring the best possible energy conversion efficiency over a very wide range of input power levels.
The purpose of this work is to design, simulate, and characterize an innovative high voltage rectifier circuit. A series topology is used and a low-pass filter is placed at the output of the circuit to provide a stable DC output voltage. Furthemore, the rectenna was designed and simulated using Agilent's Advanced Design System (ADS). And Harmonic balance method has been employed to take into account the rectifier's non - linear characteristics.

2. Rectenna Design

The objective of the RF energy harvester, also known as Rectenna consists of converting the electromagnetic energy contained in ambient RF sources into continuous dc electrical voltage. The rectenna is generally composed of a receiving antenna and an impedance matching network followed by a rectifier, a dc filter and a load containing the corresponding impedance of the sensor to be supplied. The developed circuit is in series topology, the series design of the rectifier was established for simple adaptation purposes. Firstly, the proposed rectenna is adapted to three RF bands, GSM1800, UMTS and Wi-Fi. Matching circuit is important to ensure maximum power transfer from the antenna to the rectifier circuit. It is then followed by a Schottky diode positioned in series and a low-pass filter. The diode chosen for our circuit is Metelics MSS20-141 diode. Moreover, The output dc filter consists of a 3.3 nH inductance and a 68 pF capacitor [9]. With an incident power of -10dBm, these values were calculated by simulation to achieve a higher output voltage. Moreover, the impedance of the R_{Ω} antenna is set at 50 and the load is a resistance of 430Ω. We have simulated the behaviour of the rectenna circuit under Harmonic balance. The design is displayed in figure 1.

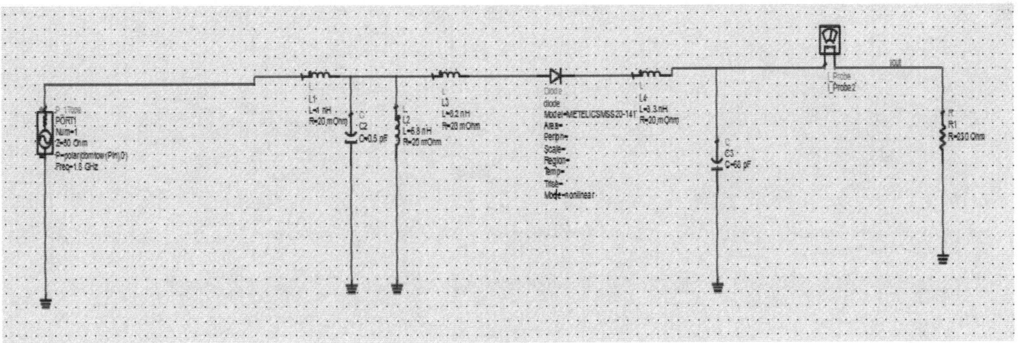

Figure 1. Schematics of the proposed Rectenna circuit simulated under ADS.

3. Simulated Results of Rectenna

The schematic of the rectenna simulated under ADS with the Harmonic Balance mode. Subsequently, we handle the variation of the output voltage depending on the incident power and sweep the input power from -10dBm to 30dBm using "Power Sweep Simulation" for F= 1.8Ghz. The curve shows the increase of the output voltage as a function of power up to the maximum value of 2.9V.Thus, important value is obtained with only one rectification diode, It is clear from the plots that the rectifier's performance has improved significantly.

IOP Conf. Series: Materials Science and Engineering **948** (2020) 012006 doi:10.1088/1757-899X/948/1/012006

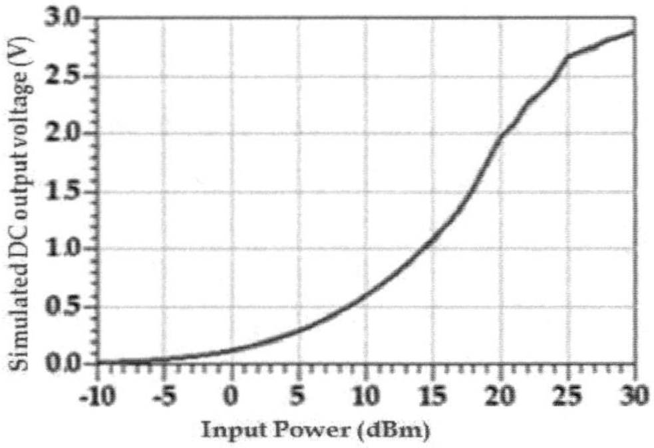

Figure 2. Output DC voltage versus input power for 1.8 GHz.

The conversion efficiency as a function of the load value R_L.for an input power of 20 dBm is shown in Figure 3. The adaptation network was designed in such a way as to transfer the maximum power to the load. And the load of the energy harvester represents the equivalent impedance of the sensor to be supplied. For the three frequencies, GSM, UMTS and Wi-Fi bands, we notice an increase in the maximum voltage approximately equal to 3V at R=1200 Ω for F=2.4 Ghz.

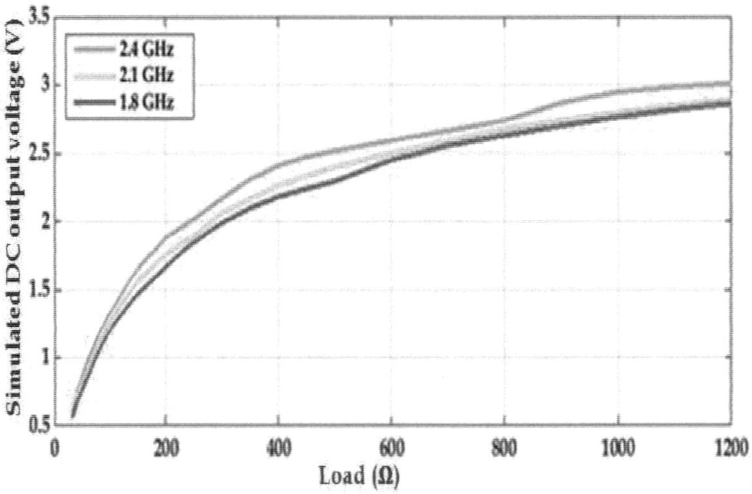

Figure 3. Output DC voltage as a function of the loadR_L. With Pin: 210 dBm. For 1.9, 2.1 and 2.4 GHz.

To achieve an optimal energy harvesting system, the rectenna should have a high RF-DC conversion efficiency, which is defined as:

$$(\%) = \frac{P_{DC}}{P_{RF}} \times 100\% = \frac{V^2{}_{DCout}/R_L}{P_{RF}} \times 100\% \quad (1)$$

Where the P_{DC} is the output DC power harvested, P_{RF} the input RF power and R_L is the load to supply. The efficiency of the rectenna varies with change in the input frequencies as depicts in Figure 4.The effects of the R_L on the RF-to-DC conversion efficiency are evaluated for three frequencies, 1.8, 2.1 and 2.1 GHz, at 20dBm input power level. It can be seen that when the load resistance varies within the 20 to 170 Ω ranges, the rectifier has the highest efficiency at the three frequency bands.

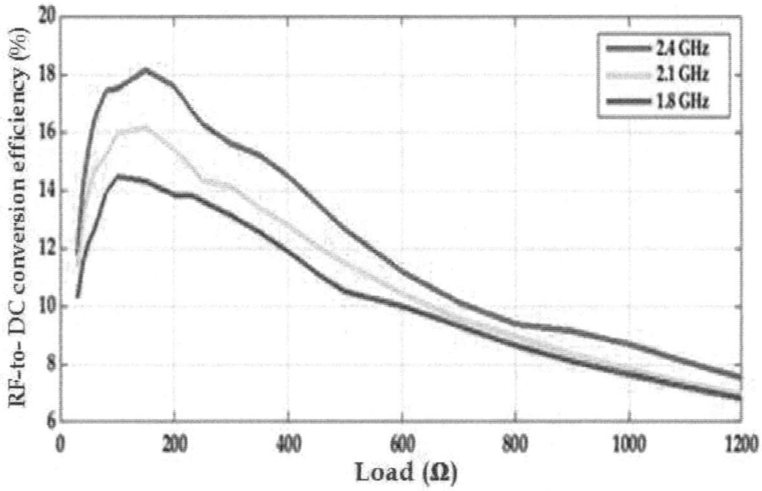

Figure 4. Comparison of conversion efficiency versus load resistance for 1.9, 2.1 and 2.4 GHz.

4. Conclusion

A high-voltage rectifier was presented to address the needs of applications involving wireless power transmission. The rectenna is one of the key components of RF energy harvesting system. In this paper, rectenna designed for GSM,UMTS and Wi-Fi bands has been presented and discussed. We have used Metelics MSS20-141 diode as a series-connected rectifier device. At the circuit output, low-pass filters were used to increase rectifier efficiency by effectively removing undesired RF components. Simulations were performed using Advanced Design System and the results obtained indicate that the proposed harvesting device is capable of recovering the available RF energy by using opportunities sources and providing wireless sensor nodes.

References

[1] Gusarov, B., Gusarova, E., Viala, B., Gimeno, L., Boisseau, S., Cugat, O., Louison, B. (2016). Thermal energy harvesting by piezoelectric PVDF polymer coupled with shape memory alloy. Sensor. Actuat: A-Phys. **243**, 175-181

[2] Meddad M, Eddiai A, Farhan R, Benahadouga S, Mazroui M, Rguiti M, Superl and Microst 2019 127 86-92

[3] Chakhchaoui N, Ennamiri H, Hajjaji A, Eddiai A, Meddad M, and Boughaleb Y 2017 Theoretical modeling of piezoelectric energy harvesting in the system using technical textile as a support Polymers for Advanced Technologies 28(9) 1170-1178

[4] Farhan R, Eddiai A., Meddad M, Mazroui M, and Guyomar D 2019 Electromechanical losses evaluation by energy-efficient method using the electrostrictive composites: experiments and modeling Smart Materials and Structures 28(3) 035024

[5] Eddiai A, Meddad M, Farhan R, Mazroui M, Rguiti M, Guyomar D Superl and Microst 2019 127 20-26

[6] Chakhchaoui N, Jaouani H, Ennamiri H, Eddiai A, Hajjaji A, Meddad M, and Boughaleb Y 2019 Modeling and analysis of the effect of substrate on the flexible piezoelectric films for kinetic energy harvesting from textiles Journal of Composite Materials 53(24) 3349-3361.

[7] Lekbir, A., Meddad, M., Benhadouga, S., Khenfer, R. (2019). Higher-efficiency for combined photovoltaic thermoelectric solar power generation. Int. J. Green. Energy. **16**(5), 371 377

[8] López-Lapeña, O., Penella, M. T., Gasulla, M.: A new MPPT method for low-power solar energy harvesting. IEEE. T. Ind. Electron. **57**(9), 3129-3138 (2009)

[9].Kuhn, V. (2015). Capture opportuniste d'énergie micro-onde pour l'autonomie des objets communicants (Doctoral dissertation).

[10].Pavone, D., Buonanno, A., D'Urso, M., & Della Corte, F. (2012). Design considerations for radio frequency energy harvesting devices. Progress In Electromagnetics Research, 45, 19-35.

[11].Kuhn, V., Seguin, F., Lahuec, C., & Person, C. (2015, September). Matching network improvement for RF energy harvesters in Body Sensor Area Network context. In *2015 European Microwave Conference (EuMC)* (pp. 303-306). IEEE.

[12].Kuhn, V., Seguin, F., Lahuec, C., & Person, C. (2016). Enhancing RF-to-DC conversion efficiency of wideband RF energy harvesters using multi-tone optimization technique. *International Journal of Microwave and Wireless Technologies*, *8*(2), 143.

[13].Okba, A. (2017). *Conception et réalisation de rectennas utilisées pour la récupération d'énergie électromagnétique pour l'alimentation de réseaux de capteurs sans fils* (Doctoral dissertation, Université Paul Sabatier-Toulouse III).

IOP Conf. Series: Materials Science and Engineering **948** (2020) 012007 doi:10.1088/1757-899X/948/1/012007

PSPICE compact model for power MOSFET based on manufacturer datasheet

Abdelghafour Galadi

National School of Applied Sciences of Safi, Cadi Ayyad University, Marrakech, Morocco

Email: agaladi@uca.ma

Abstract. In this paper, large signal model for power MOSFET devices is presented. The proposed model includes quasi-saturation effect and describes accurately the electrical behavior of the power MOSFET devices. The large signal model elements will be provided based on the device structure. Furthermore, the model parameters are extracted from measurements considering the voltages depending effect of the nonlinear gate-source, gate-drain and drain-source interelectrode capacitances. Excellent agreements will be shown between the simulated and the datasheet data. Finally, a description of the model will be provided along with the parameter extraction procedure.

1. Introduction

In power electronic circuits, power MOSFET device is widely used thanks to its high input impedance, high switching speed and thermal stability. To sustain high voltage in the off state, the power MOSFET structure contains a thickness and lowly doped drift layer, see Figure 1. In the on state, the extended-drain drift layer turns into high resistance in case of high voltage power MOSFETs [1]. Consequently, the standard power MOSFETs are limited to low voltage applications. The power MOSFETs present two important effects: the quasi-saturation effect [2]–[4] and the capacitive gate-drain effect. Quasi-saturation is observed in output characteristics of power MOSFET as the compression of the drain current curves at high gate voltages. At quasi-saturation regime, the drain current becomes less sensitive to the increase of gate voltage. This phenomenon is due to the carrier velocity saturation in the drift layer [2]–[4]. In other word, the intrinsic accumulation regions between the P-sources introduce gate-drift capacitance limiting the dynamic performance of the power MOSFETs. Therefore, the modeling of the power MOSFET must take into consideration the peculiar characteristics of the power MOSFET structure.

The intrinsic drift layer and the mobility degradation, in accumulation and inversion layers at the surface of silicon [5], make the modeling of power MOSFET more complex. SPICE power MOSFET models must describe accurately the electrical behavior of the device in all bias conditions. In addition, the model must be simple, fast and accurate without any problem of convergence. In literature, various models for power MOSFET were proposed [6]–[16]. These models use a subcircuit consisting of an equivalent circuit of the power MOSFET structure. The subcircuit combines a traditional MOSFET transistor, representing the channel region, with resistances, capacitors and dependent/independent current/voltage sources to model the electrical behavior of the power MOSFET. However, most of these models need proprietary tools to set their parameters or uses constant capacitance to model the gate-source capacitance C_{GS}.

IOP Conf. Series: Materials Science and Engineering **948** (2020) 012007 doi:10.1088/1757-899X/948/1/012007

A compact improved dynamic model for power MOSFET is presented in this paper. The proposed model is simple and takes into consideration the quasi-saturation effect and the nonlinear interelectrode capacitances. In this model, C_{GS} is modeled by using the Meyer's simplified capacitance model. The model parameter extraction procedure uses the datasheet data to determine the model parameters. The model features excellent agreement between the simulated and datasheet curves in all regions of operation of the power MOSFET.
In this work, the device samples used to validate the model are the STP36NF06L N-channel standard power MOSFET and the HUF76419S3ST_F085 N-channel trench power MOSFET [17].
For this purpose, section 2 describes the equivalent circuit model with the model's equations. Section 3 is devoted to the procedure of parameter extraction and section 4 describes the static and dynamic simulation results obtained with the model compared to measurements.

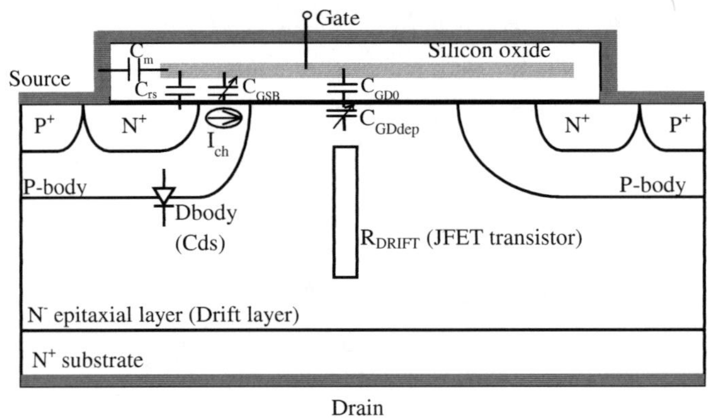

Figure 1. Conventional power MOSFET structure with origin of its subcircuit elements.

2. Model description
Figure 1 shows the power MOSFET structure with the origin of its equivalent circuit. The proposed model is shown in Figure 2. This model uses standard MOSFET to model the channel region in series with JFET to model the drift region [18]. One "effective" gate voltage is added to equivalent circuit to take into account the small variation of the drain current with the gate voltage at the quasi-saturation regime [18]. The Dbody diode represents the intrinsic P-source/N^--drift/N^+-substrate diode. This diode is very useful because it will be used to take into consideration the breakdown voltage of the power MOSFET by the "BV" SPICE diode parameter, the drain-source interelectrode capacitance C_{DS} and the forward behavior in the case of negative drain-source voltage. The parasitic gate-source capacitance C_{GS} is divided to two capacitances in parallel:
1) The constant capacitance C_{rs} equivalent to the overlap of the gate electrode over the N^+ source region plus the capacitance between the gate and source metallizations C_m.
2) The nonlinear channel capacitance between the gate and the P-body source C_{GSB}.
The nonlinear Miller gate-drain capacitance C_{GD} is added to complete the dynamic model. As shown in Figure 1, C_{GD} is equivalent to the constant oxide capacitance C_{GD0} in series with the depletion capacitance C_{GDdep}.

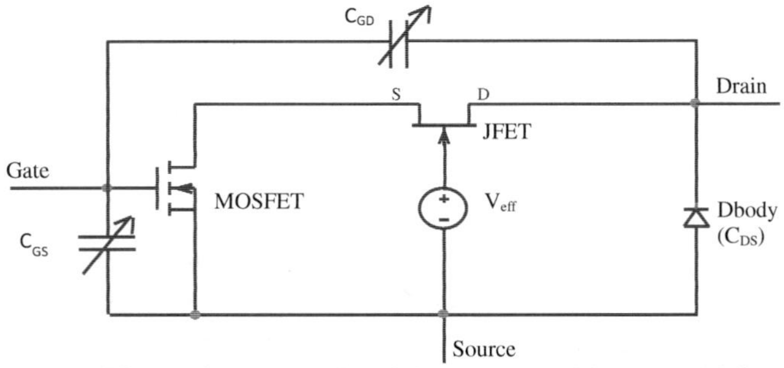

Figure 2. Schematic of the power MOSFET model.

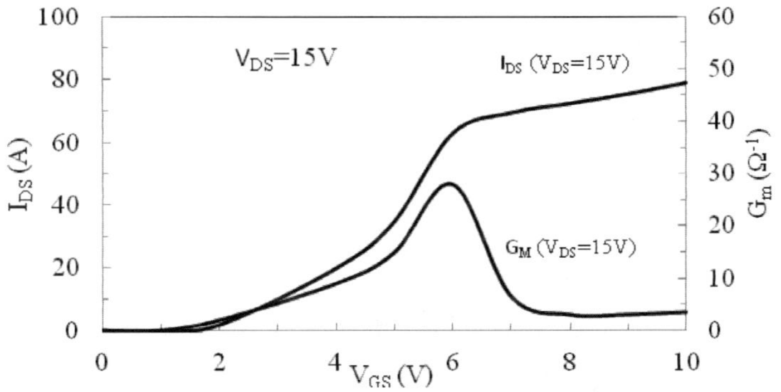

Figure 3. Typical I_{DS}-V_{GS} and G_m-V_{GS} characteristics of power MOSFET showing the quasi-saturation effect.

2.1. Static behavior Modeling

The static model is obtained by the channel MOSFET, the JFET transistor, the "effective" gate voltage V_{eff} and the Dbody diode, as shown in Figure 2. In this model, the drain current depends on a critical gate voltage V_{GSC} which is extracted from the I_{DS}-V_{GS} curve as shown in Figure 3. V_{GSC} is the value of V_{GS} at the peak value of the transconductance parameter G_m. For $V_{GS} \leq V_{GSC}$ the drain current is governed by the channel MOS transistor while for $V_{GS} > V_{GSC}$ the drain current is governed by the JFET transistor [18]. The expression of the channel drain current is based on the nth power law MOSFET model [19], [20]:

$$I_{DS} = \begin{cases} 0 & \text{if} \quad V_{GS} \leq V_{TM} \\ I_{DSAT}\left(1+\lambda_M V_{DS}\right)\left(2-\dfrac{V_{DS}}{V_{DSAT}}\right)\dfrac{V_{DS}}{V_{DSAT}} & \text{if} \quad V_{DS} < V_{DSAT} \\ I_{DSAT}\left(1+\lambda_M V_{DS}\right) & \text{if} \quad V_{DS} \geq V_{DSAT} \end{cases} \tag{1}$$

$$I_{DSAT} = B\left(V_{GS} - V_{TM}\right)^n \tag{2}$$

$$V_{DSAT} = K\left(V_{GS} - V_{TM}\right)^m \tag{3}$$

where V_{GS}, V_{DS} and V_{DSAT} are gate-source, drain-source and drain-source saturation voltage, respectively. V_{TM} is the threshold voltage and I_{DSAT} is the saturation drain current. K and m model parameters control the linear region while B and n control the saturation region of operation in output characteristics. The parameter λ_M is added to take into consideration the channel length modulation.
The JFET drain current is giving by the following equations:

$$I_{DS} = \beta V_{DS}\left(2\left(V_{eff} - V_p\right) - V_{DS}\right), \qquad V_{DS} < V_{GS} - V_p \tag{4}$$

$$I_{DS} = \beta\left(V_{eff} - V_p\right)^2, \qquad V_{DS} \geq V_{GS} - V_p \tag{5}$$

β (BETA) and V_p are the JFET transconductance coefficient (A/V^2) and the pinch-off voltage, respectively. The two JFET SPICE model parameters β and V_p can be extracted using drain current equation for $V_{GS} = V_{GSC}$ in linear and saturation regions in the quasi-saturation regime assuming the JFET gate-source voltage $V_{GSJ} = V_{eff} = 0$ [18].
Finally, the "effective" gate voltage V_{eff} is chosen as a linear function of gate voltage [18]:

$$V_{eff} = gV_{GS} + V0 \tag{6}$$

The equation parameters g and V0 can be extracted using the measurements drain current at quasi-saturation regime when $V_{eff} = 0$.

2.2. *Dynamic behavior Modeling*

To model the dynamic behavior of the power MOSFET, nonlinear interelectrode capacitances are added to the static model as shown in Figure 2. The parasitic interelectrode capacitances of the power MOSFET structure are divided into three nonlinear capacitances: C_{GS}, C_{GD} and C_{DS}.

In this paper, C_{GS} will be estimated by considering its value from blocking condition to conducting condition using the gate charge measurement. The estimated value of C_{GS} will be inspired by the Meyer's simplified capacitance model [21]. On the other hand, C_{GD} and C_{DS} will be extracted from the measured device interelectrode capacitances (C_{rss} and C_{oss}).

2.2.1. *Gate-source capacitance*

C_{GS} is the sum of two capacitances: 1) constant capacitance C_{GS0} ($C_{GS0} = C_{rs} + C_m$) and 2) nonlinear channel gate-source capacitance C_{GSB}, see Figure 1. C_{GSB} is a MOS type capacitance with the channel P substrate short-circuited to N^+ source. Therefore, the modeling of C_{GS} will be used by a modified Meyer's simplified capacitance model [21] depending on both gate and drain voltages.

In the OFF state, C_{GS} is equal to C_{GS0}. In linear region of operation, C_{GS} will be considered equal to the sum of C_{GS0} and the total channel capacitance C_{GSB}. In the saturation region, C_{GS} will be estimated by using the Meyer's simplified capacitance model, i.e. $C_{GS0} + (2/3C_{GSB})$ [21].

In linear region, we choose to add the entire channel capacitance C_{GSB} to the gate-source capacitance because, unlike the conventional MOSFET, in power MOSFET structure the N-drain is linked to the N^+ drain of the power MOSFET, as shown in Figure 1.

As a resume, the nonlinear gate-source capacitance model can be summarized as follows:

$$C_{GS} = \begin{cases} C_{GS0} & \text{if} \quad V_{GS} \leq V_{TM} \\ C_{GS0} + C_{GSB} & \text{if} \quad V_{DS} < V_{DSAT} \\ C_{GS0} + \frac{2}{3}C_{GSB} & \text{if} \quad V_{DS} \geq V_{DSAT} \end{cases} \tag{7}$$

C_{GS0} is the gate-source capacitance in the off state of the power MOSFET device for positive gate voltages. Therefore, this capacitance will be extracted from V_{GS} curve in the gate charge measurement for $V_{GS} < V_{TM}$. C_{GSB} is the capacitance of the channel region between the gate and the P-body source. Thus, C_{GSB} will be extracted from V_{GS} curve in the gate charge measurement at the linear region of operation.

IOP Conf. Series: Materials Science and Engineering **948** (2020) 012007 doi:10.1088/1757-899X/948/1/012007

In this paper, the nonlinear C_{GS} capacitance is modeled by using an ABM module with replacing this capacitance by a controlled current source:

$$i_{GS} = C_{GS} \times ddt\,(V_{GS}) \tag{8}$$

2.2.2. Gate-drain capacitance

The gate-drain Miller capacitance C_{GD} is a key parameter of the SPICE models for power MOSFETs. This Miller capacitance provides a feedback loop between the output and the input of the circuit limiting the frequency response of the transistor. As it can be seen in Figure 1, unlike the conventional MOS capacitance, C_{GD} of the power MOSFET is a MOS capacitance with the substrate N region connected to drain of the device. Therefore, C_{GD} will be modeled as a MOS capacitance depending on both V_{GS} and V_{DS} voltages (V_{DG} voltage). In this paper, C_{GD} is modeled by a constant oxide capacitance C_{GD0} when $V_{DG} \leq 0$ and by the series association of C_{GD0} with a depletion capacitance C_{GDdep} when $V_{DG} > 0$, see Figure 1. Then the following expressions are used:

$$C_{GD} = C_{GD0} \quad \text{for} \quad V_{DG} \leq 0 \tag{9}$$

$$C_{GD} = \frac{C_{GD0}}{\left(1 + \frac{V_{DG}}{V_{JG}}\right)^{MG}} \quad \text{for} \quad V_{DG} > 0 \tag{10}$$

where C_{GD0}, corresponding to the gate-drain oxide capacitance, is a constant capacitance. MG is the gate-drain grading coefficient and V_{JG} is the gate-drain junction potential. These parameters will be extracted by using the provided datasheet reverse transfer capacitance C_{rss}.

The nonlinear behavior of C_{GD} is estimated by using the transition capacitance of the SPICE diode model [22]. In this paper, C_{GD} is represented by a controlled current source formalism based on the ABM module of SPICE using equation 8 with replacing i_{GS} by i_{GD}, C_{GS} by C_{GD} and V_{GS} by V_{GD}.

2.2.3. Drain-source capacitance

As it can be seen in Figure1, C_{DS} is a drain-source junction capacitance. The non-linear behavior of C_{DS} as a direct function of V_{DS} is described by equation 11: as it can be seen in Figure 2, C_{DS} is modeled by the transition capacitance of the SPICE diode model ("Dbody" diode). This diode will also takes into account the breakdown voltage of the power MOSFET and the forward biased behavior.

$$C_{DS} = \frac{C_{DS0}}{\left[1 + \frac{V_{DS}}{V_{JD}}\right]^{MD}} \tag{11}$$

C_{DS0} is the zero drain-source bias capacitance, MD is the drain-source grading coefficient and V_{JD} is the drain-source junction potential. C_{DS0}, MD and V_{JD} are parameters of the SPICE Dbody diode model.

3. Parameter extraction

The critical gate-source voltage V_{GSC} parameter is extracted from the transfer characteristic at the peak value of the transconductance G_m as shown in Figure 3.

IOP Conf. Series: Materials Science and Engineering **948** (2020) 012007 doi:10.1088/1757-899X/948/1/012007

3.1. Static model parameter extraction

The channel MOSFET model parameters are extracted and specified from measured output characteristics for $V_{GS} < V_{GSC}$ with respect to the extraction procedure introduced in [18]. For $V_{GS} > V_{GSC}$, the JFET and the "effective" gate voltage parameters are extracted from the output characteristics at the quasi-saturation regime. The detailed extraction model parameters procedure can be found in [18].

3.2. Dynamic model parameter extraction

3.2.1. C_{GD} and C_{DS} capacitances

The model parameters of these two capacitances will be extracted from the measured C_{rss} and C_{oss} capacitances: $C_{GD} = C_{rss}$ and $C_{DS} = C_{oss}-C_{rss}$. The zero-bias junction capacitance (C_{GD0} and C_{DS0}), the grading junction coefficient (MG and MD) and the junction potential (V_{JG} and V_{JD}) parameters can be extracted by using the commonly used methods described in [22]: the three-point CV method or the linear regression with fixed diffusion potential method.

3.2.2. C_{GS} capacitance

For the gate-source capacitance two parameters need to be calculated: C_{GS0} which is the gate-source capacitance in off state and C_{GSB} which is the total channel capacitance when the transistor is in the linear region of operation. These two capacitances are extracted from the gate charge V_{GS}-Q_G measurement as shown in Figure 4. Gate charge measurement gives the total amount of charge needed to turn on the power MOSFET from the blocking state. In gate charge measurement, a constant independent current source is applied at the gate terminal to charge the gate capacitance. $1/C_{GS0}$ is equal to the slope of the V_{GS}-Q_G curve when V_{GS} is inferior to V_{TM}. C_{GSB} is calculated from the slope of the V_{GS}-Q_G curve when the transistor enters in linear region, as shown in Figure 4.

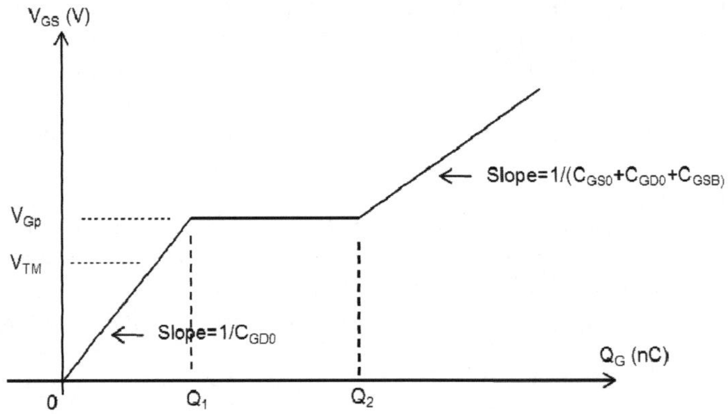

Figure 4. Extraction procedure for the gate-source capacitance.

4. Results and discussion

4.1. Static validation

In [18], it was reported that the static model provides an accurate representation of the static behavior of the power MOSFETs. Figure 5 shows I_{DS}-V_{GS} simulated and datasheet transfer characteristics of the standard power MOSFET. At the quasi-saturation regime, a good agreement between the simulated and the provided datasheet is observed. Unlike the standard power MOSFET models, the error percent introduced by the proposed model remains practically constant with increase of gate-source voltage V_{GS}.

IOP Conf. Series: Materials Science and Engineering **948** (2020) 012007 doi:10.1088/1757-899X/948/1/012007

4.2. Dynamic validation

The datasheet and the model simulated static reverse and output capacitances for the standard and the trench power MOSFETs are shown in Figures 6 and 7 respectively. Good agreement is observed, however discrepancy is observed in the transition region between low and high V_{DS} voltages. This divergence is due to the non-uniform doping channel region in VDMOS device which make the determination of C_{GD} and C_{DS} grading coefficients more complex [15].

The proposed interelectrode capacitance model gives excellent agreement in case of the standard and the trench devices compared to the datasheet curves.

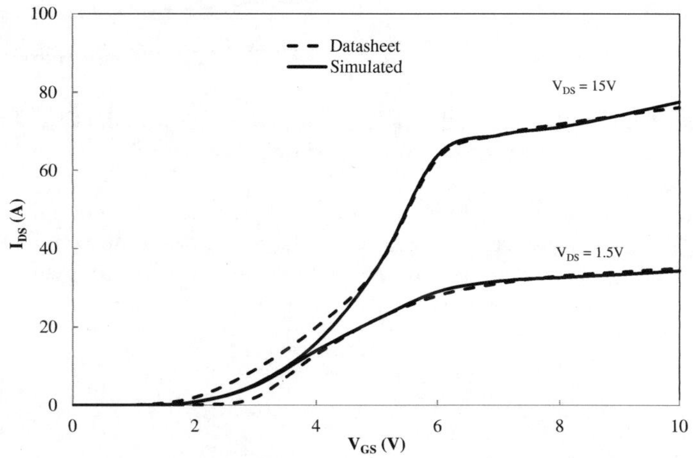

Figure 5. Simulated and datasheet I_{DS}-V_{GS} characteristics of the N-channel standard power MOSFET transistor for V_{DS} = 1.5 V and V_{DS} = 15 V.

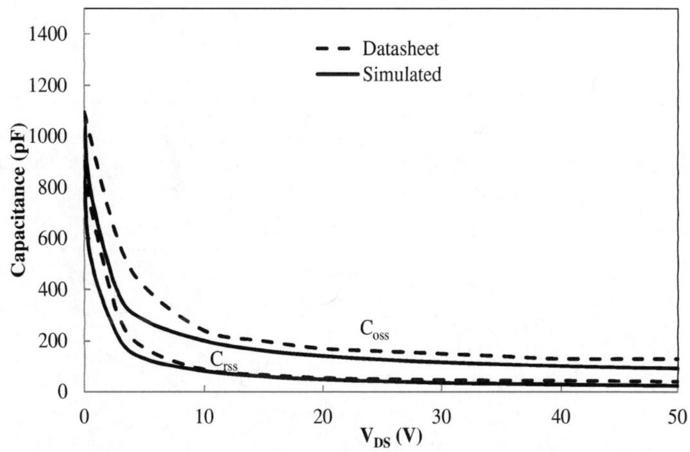

Figure 6. Device datasheet (dashed) and SPICE simulated (solid) C_{rss} and C_{oss} of the N-channel standard power MOSFET transistor (V_{GS} = 0V).

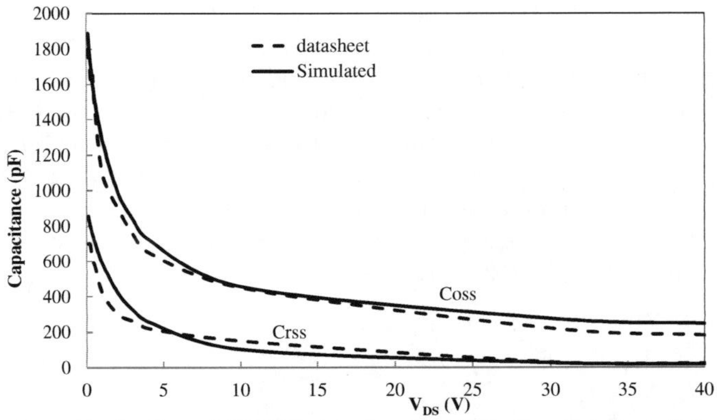

Figure 7. Device datasheet (dashed) and SPICE simulated (solid) C_{rss} and C_{oss} of the N-channel trench power MOSFET transistor (V_{GS} = 0V).

On the other hand, the dynamic validation is made by the simulated and datasheet gate charge test for the standard (Figure 8) and the trench (Figure 9). As it can be seen, excellent agreement between simulated and measured gate charge test with an average error percentage less than 10% is observed.

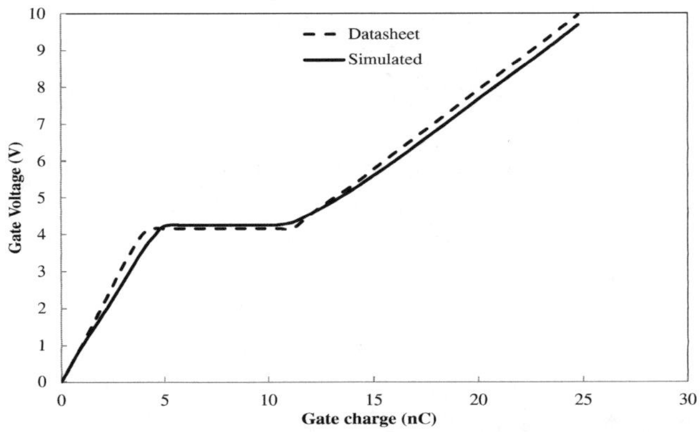

Figure 8. Device datasheet (dashed) and SPICE simulated (solid) gate charge results of the N-channel standard power MOSFET transistor.

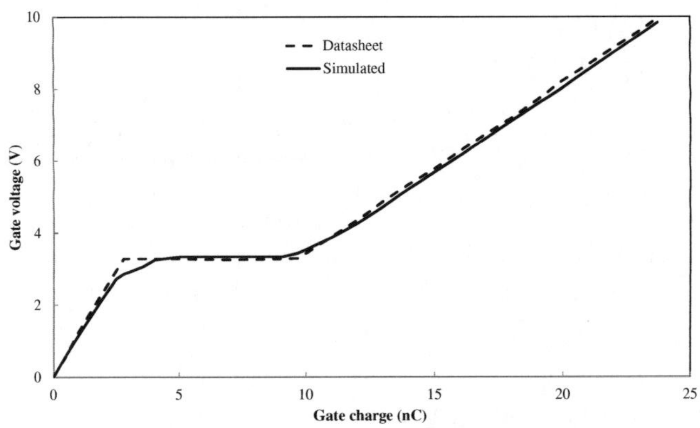

Figure 9. Device datasheet (dashed) and SPICE simulated (solid) gate charge results of the N-channel trench power MOSFET transistor.

5. Conclusion

An enhanced equivalent circuit model for power MOSFET devices has been presented. The model takes into consideration the nonlinear interelectrode capacitances. The model parameters were extracted from device datasheet informations. Comparison between PSPICE model simulations and datasheet values allow the validation of the model for the trench and standard power MOSFET structures. Future direction of this work can be devoted to give a unified model of all power MOSFET structures in order to capture the switching behavior of the power MOSFET in simulation of power electronic converters.

References

[1] Baliga B J 2010 *Advanced Power MOSFET Concepts* (Springer), chapter 2, pp. 23-61.
[2] Darwish M N 1986 *IEEE Trans. Electron Devices* ed. 33, no. 11, pp. 1710–1716.
[3] Kreuzer C H, Krischke N and Nance P 1996 *Proc. of Int. Electron Devices Meeting* pp. 489–492.
[4] Wang L, Wang J, Gao C, Hu J, Li P, Li W and Yang S H Y 2009 *IEEE Trans. Electron Devices* vol. 56, no. 3, pp 492–498.
[5] Sun S C and Plummer J D 1980 *IEEE J. of Solid-State Circuits* vol. 15, pp. 562-573.
[6] Mukunoki Y, Nakamura Y, Horigushi T, Kinouchi S, Nakayama Y, Terashima T, Kuzumoto M and Akagi H *IEEE Trans. Electron Devices* vol. 16, issue 11, pp. 4339-4345.
[7] Johannesson D and Nawaz M 2016 *Microelectronics J.* vol. 53, pp. 167–176.
[8] Fu R, Grekov A, Hudings J, Mantooth A and Santi E *2012 IEEE Trans. Ind. Appl.* vol. 48, no. 1, pp. 181-190.
[9] Mudholkar M, Ahmed S, Ericson M N, Frank S S, Britton C L and Manthoot H A 2014 *IEEE Trans. Power Electron.* vol. 29, no. 5, pp. 2220–2228.
[10] Arribas A P, Shang F, Krishnamurthy M and Shenai K 2015 *IEEE Trans. Electron Devices* vol. 62, no. 2, pp. 449-457.
[11] Phankong N, Yanagi T and Hikihara T 2011 *Proc. IEEE Eur. Conf. Power Electron. Appl.* pp. 1–8.
[12] Chauhan Y S *et al.* 2006 *Proc. of the 7th Int. Symposium on Quality Electronic Design (ISQED '06)* pp. 109-114.
[13] Aarts Annemarie C T and Kloosterman Willy J 2006 *IEEE Trans. Electron Devices* vol. 53, no. 4, pp. 897–902.
[14] Mantooth A, Peng K, Santi E and Hudgins J L 2015 *IEEE Trans. Electron Devices*, vol. 62, no. 2, pp. 423–433.
[15] Yao W, Gildenblat G, McAndrew C C and Cassagnes A 2012 *IEEE Trans. Electron Devices* vol. 59, no. 3, pp. 542–550.
[16] Kraus R and Castellazzi A 2016 *IEEE Trans. Electron Devices* vol. 31, no. 8, pp. 5863–5870.
[17] "N-channel power MOSFET", STP36NF06L ST datasheet, 2006.
"N-channel trench power MOSFET", HUF76419S3ST_F085 Fairchild semiconductor datasheet, 2013.
[18] Galadi A 2016 *J. of Computational Electron.* vol. 15, issue 2, pp. 619-626.
[19] Sakurai T and Newton A R 1991 *IEEE Trans. Electron Devices* vol. 38, no. 4, pp. 887-894.
[20] Sakurai T and Newton A R *IEEE J. of Solid State Circuits* vol. 26, no. 2, pp. 122-131.
[21] Bhattacharyya A B 2009 *Compact MOSFET Models for VLSI Design* (John Wiley, chapter 6, pp. 289–335).
[22] Russel H T 2001 *Rectifier Applications Handbook* (Motorola, chapter 3, pp. 47–71).

IOP Conf. Series: Materials Science and Engineering **948** (2020) 012008 doi:10.1088/1757-899X/948/1/012008

Global MPP Tracking for Photovoltaic Panel under Shading Conditions using Robust Numerical Algorithm

Oumaima Mesbahi [1,2,3], Khalid Kandoussi [1], Rabie El otmani [1], Abdeloawahed Hajjaji [1], Fernando M. Janeiro [3,4], Mouhaydine Tlemçani [2,3*].

[1]Science Engineer Laboratory for Energy (LabSIPE), National School of Applied Sciences, University of Chouaib Doukkali, Morocco
[2]ICT of Évora,Portugal
[3]Universidade de Évora, Portugal
[4]Instituto de Telecomunicações, Portugal

*tlem@uevora.pt

Abstract. This paper presents a new numerical approach for the modelling of the PV panels under partial shading conditions. It also solves the major drawback of the PV systems, which is the tracking of maximum power point under partial shading conditions where the P-V characteristic has a more complex structure with many local maximums on the graph. This paper aims to resolve this phenomenon using Newton method in order to find the global maximum power point.

1. Introduction

The use of renewable sources to produce energy is a promising option for energy supply that meets the world's growing demand [1], with benefits such as abundance, the absence of pollution and the availability on several places of the terrestrial globe.
Solar energy [2, 3] is a good choice for electric power generation using as a technology the photovoltaic (PV) panels. However, the efficiency of the chain of conversion of this energy is relatively low [4, 5], due to the effect of the shadows on the performance of the photovoltaic panels. This shadow can be created by clouds, neighbouring buildings, trees or sometimes even dust. Where the panels depend heavily on the environmental conditions such as the ambient temperature and the solar radiation [5, 6], these two are mainly affected by the shading. Causing not only a loss in the produced power, but also generate multiple summits in the power-voltage (P-V) and current-voltage (I-V) characteristics of the PV array.
To extract the maximum power produced by a photovoltaic array, there have been proposed in the literature many classical conventional methods (Perturb & Observe, Incremental conductance) [7-12]. Many studies have proved that these methods are known to be very effective for tracking the maximum power point (MPP) which changes according to the solar radiation. Nevertheless, these classical methods rely on the fact that the PV array is exposed to a uniform solar radiation, where the P-V characteristic has only one maximum power point. Whereas if there is a non-uniform radiation applied to the PV array, the P-V characteristic will have a more complex curve resulting in many local maxima. This leads to the failure of those methods on tracking the global maximum power point [13-15]. Many researches have been conducted in order to improve the performance of these methods

under partial shading conditions [15-20], either by adding a modification to the algorithms or by applying a variable step to the classical methods. In spite of this, the application of these new approaches in the real time remains difficult due to the fact that these methods have limitations on tracking the maximum power point under rapidly changing atmospheric conditions, especially if non-uniform.

In order to bridge over this inconvenience and to achieve a better maximum power point tracking, this paper presents a new, rapid and simple method for detecting the global power point under non-uniform illumination based on a numerical model analysis of the PV array with Newton method.

2. Modelling of the photovoltaic panel behaviour

2.1. The equivalent equation of the PV cell

The PV cell converts the sunlight radiation to electricity. In the literature the photovoltaic cell can be represented by the equivalent circuit shown in figure 1. This circuit consists of a diode which represents the PN junction of the cell, and a constant current source whose current amplitude depends on the intensity of the radiation. There is also the material resistivity which is represented by R_p, parallel resistance, characterizing the leakage current at the surface of the cell due to the impurities near the PN junction. The serial resistance R_s.

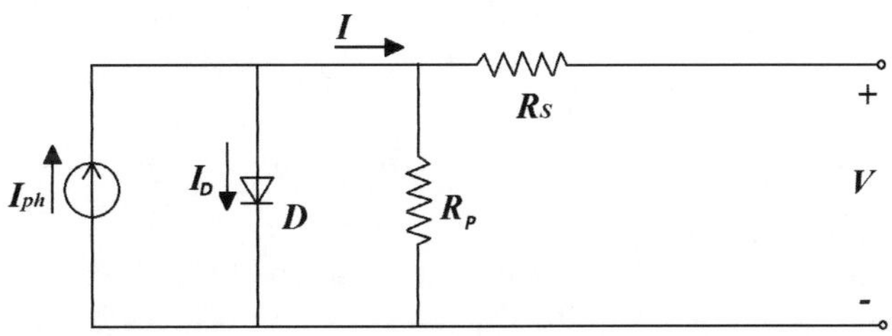

Figure 1. Equivalent circuit of a photovoltaic cell

The mathematical model of the PV cell can be obtained by applying the Kirchhoff's circuit laws to the circuit above resulting into

$$I = I_{ph} - I_D - \frac{V + IR_s}{R_p} \tag{1}$$

with

$$I_D = I_S \left[\exp\left(\frac{V + IR_s}{V_{th}} \right) - 1 \right] \tag{2}$$

and

$$V_{th} = \frac{nkT}{q} \tag{3}$$

where I_{ph} is the photocurrent, I_D is the current through the diode, I_S is the saturation current of the diode, q stands for the electronic charge, k is Boltzmann constant, n is the junction ideality factor, T is the junction temperature , I is the current supplied by the cell and V is the output voltage of the cell. From (1) to (3) results that the current is

$$I = I_{ph} - I_S \left[\exp\left(\frac{V + IR_s}{V_{th}} \right) - 1 \right] - \frac{V + IR_s}{R_p} . \tag{4}$$

Resistance R_p is generally neglected and the photocurrent is approximately equal to the short circuit current $I_{sc} = I_{ph}$. Also, when there is an open circuit

$$V_{oc} = V_{th} \ln\left(\frac{I_{sc}}{I_S} \right) \tag{5}$$

and the saturation current can be expressed as

$$I_S = I_{sc} \exp\left(-\frac{V_{oc}}{V_{th}} \right) . \tag{6}$$

So, equation (4) for the supplied current becomes

$$I = I_{sc} \left[1 - \exp\left(\frac{V - V_{oc} + IR_s}{V_{th}} \right) \right] . \tag{7}$$

2.2. Modeling of photovoltaic panel

To obtain a more accurate modelling of PV cell, the impact of the solar radiation and the ambient temperature should also be considered [5, 6]. Normally, the cell parameters are provided by the manufacturers under standard test conditions (STC), which are 25 °C and 1000 W/m^2. For this, the different components of the equation (7) are written as a function of the available radiation G and present temperature T according to following equations

$$I_{sc} = I_{sc,0} \frac{G}{G_0} \left[1 + k_0 \left(T - T_0 \right) \right] \tag{8}$$

$$V_{oc} = V_{oc,0} + C_3 \left(T + C_2 G - T_0 \right) \tag{9}$$

where G_0 and T_0 are the standard test conditions of irradiance and temperature and $I_{sc,0}$ and $V_{oc,0}$ are the measured current and voltage under these conditions. The coefficients k_0 and C_3 are the temperature coefficient of the photocurrent and voltage respectively, and C_2 is the radiation coefficient of the open circuit voltage V_{oc}. By establishing these two equations, it is assured that the variation of the radiation and the ambient temperature is taking into consideration.

The PV model is an assembly of many PV cells which can be connected into series or parallel or mixed configurations and are accompanied in many cases by a bypass diode, as can be in figure 2.

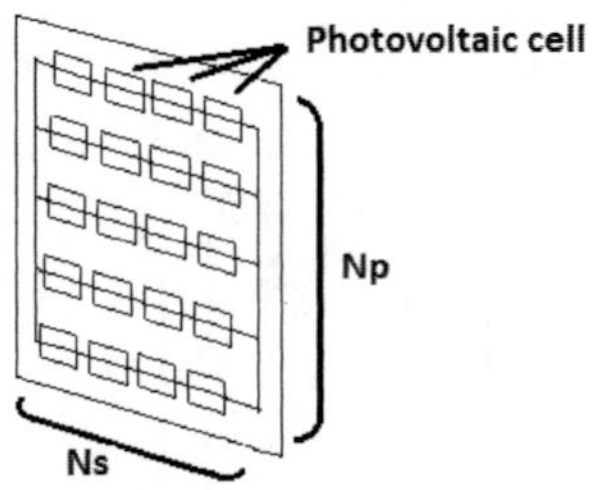

Figure 2. Photovoltaic panel

Figure 2 represents the configuration of a photovoltaic panel where N_p and N_s are the number of cells connected in parallel and series configuration. The PV panel current is then given by:

$$I_{pv} = N_p I_{sc}\left[1-\exp\left(\frac{V_{pv} - N_s V_{oc} + I_{pv}\frac{N_s}{N_p}R_s}{V_{th,pv}}\right)\right]. \qquad (10)$$

This equation is an implicit function which depends on the overall short circuit current and on the open circuit voltage of the panel along with the equivalent series resistance.
For the modelling of the PV panel shown in figure 3, MATLAB Simulink has been used, through a combination between Simscape and Simulink libraries. MATLAB Simulink allowed the simulation of the behaviour of the panel and characterize its nonlinear P / V and I / V curves to better understand the effect of radiation and temperature on the output current and voltage.

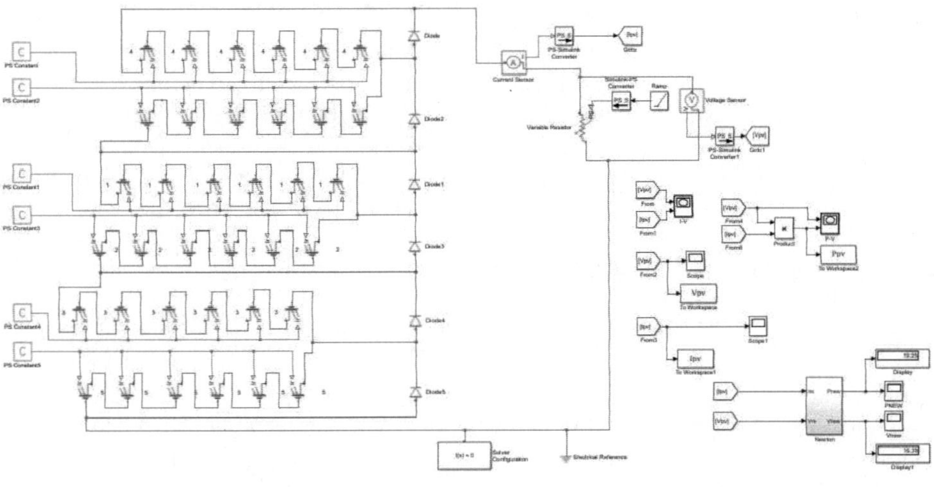

Figure 3. MATLAB interface model for a PV panel

2.3. The effect of the shading on the photovoltaic panel

When a cell is shaded, it behaves as an overload, dissipating the electrical energy by Joule effect. This leads to temperature increases and generates the so-called hot spots, thus helping to create an open circuit in the whole of the photovoltaic module system. Therefore, the detection of the maximum power of the photovoltaic panel under shaded conditions in real time is vital.

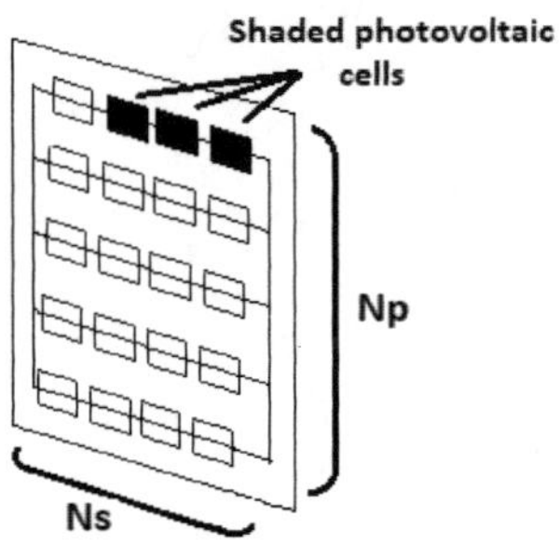

Figure 4. PV panel structure with different illuminations

In order to overcome this inconvenience and to achieve a better maximum power point tracking (MPPT) process, this paper presents a new, fast and simple method for MPPT under non-uniform illumination based on a numerical model analysis of the PV array using Newton method.

3. Newton method for shaded photovoltaic panel

3.1. Newton numerical approach

In numerical analysis, the Newton method, in its simplest application, is the most well-known and efficient algorithm to find the precise numerical approximation to the roots of a real valued function

$$f(x)=0. \tag{11}$$

The method calculates the tangent at the point x_0 (the function derivative) which is the initial value chosen near the solution. The next value to use is the intersection between the tangent and the x-axis. It is an iterative process which will stop when the convergence criteria are achieved. The Newton method can be used in the case of continuous and differential functions in a certain interval $[a,b]$. The differentiability condition assures the presence of a linear tangent for the function. The continuity condition assures that there are no jumps in the function. The graph slope is obtained by

$$f'(x_n)=\frac{\Delta y}{\Delta x}=\frac{f(x_n)}{x_n-x_{n+1}} \tag{12}$$

are if the terms are reorganized, equation (12) reduces to

$$x_{n+1}=x_n-\frac{f(x_n)}{f'(x_n)} \tag{13}$$

The Newton method can also be used to find local minimums and maximums. When a minimum or a maximum is achieved, the solution x^* is the stationary point of the $f(x)$ function (x^* is the solution to $f'(x)$). Newton method steps can be resumed as follows:

- Chose an initial approximation x_0;
- Calculate the value of the function at this point $f(x_0)$;
- Calculate the function derivative $f'(x)$ f' and evaluate its value at $f'(x_0)$;

- The approximation x_1 is obtained with $x_1 = x_0 - \frac{f(x_0)}{f'(x_0)}$;
- Repeat the process till finding the solution x^* using $|xn+1 - xn| < \varepsilon$ as a stopping criteria.

3.2. Application of the Newton method to the MPPT

As a first step the Newton iterative method has been applied for uniform illumination where there is the presence of only one maximum power point, so if the PV array is operating with a a voltage lower than the maximum V_m, then $dP/dV \neq 0$ since $dP/dV = 0$ only occurs at the maximum power point. This paper seeks to find the solution of $dP/dV = 0$ by the Newton iterative method in order to detect the global maximum power point rapidly with high precision.

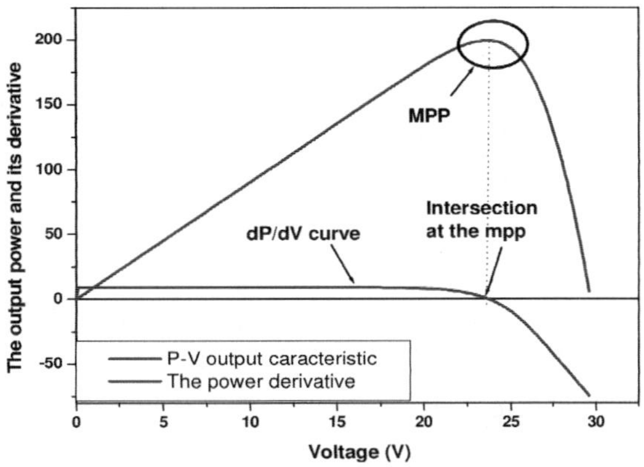

Figure 5. Mathematical relations of Output Power and Output Voltage

According to what has been described, the Newton method is useful in finding the roots of any non-linear function. As can be seen from figure 8, the roots of the derivative of the power curve correspond to the point of MPP, which can be obtained through Newton's method. For this reason, the function of the power derivative is taken as the objective function for the Newton method, where the power curve has been approximated by the quadratic form

$$q(V) = P(V_k) + (V - V_k)P'(V_k) + \frac{(V - V_k)^2}{2} P''(V_k) \tag{14}$$

To maximize this expression, its derivative is taken to be zero

$$q'(V) = 0 = P'(V_k) + (V - V_k)P''(V_k) \tag{15}$$

resulting into

$$V_{k+1} = V_k - \frac{P'(V_k)}{P''(V_k)} \tag{16}$$

with

$$P'(V_k) = \frac{P(V_k) - P(V_{k-1})}{V_k - V_{k-1}} \quad \text{and} \quad P''(V_k) = \frac{P'(V_k) - P'(V_{k-1})}{V_k - V_{k-1}}. \tag{17}$$

3.3. Simulation and results

A 100W PV array was chosen for modelling in MATLAB. The electrical characteristics specifications of the used panel are shown in table 1.

Table 1. Electrical characteristics of the PV panel

Parameter	Variable	Value
Maximum power	P_m	100 W
Voltage at Pmax	V_m	17.6 V
Current at Pmax	I_m	5.71 A
Short circuit current	I_{sc}	6.4 A
Open circuit voltage	V_{oc}	21 V

3.3.1. Uniform illuminations

The MATLAB model simulation results are represented in the figure 3, through the P-V and I-V characteristics obtained with uniform illumination and temperature on the overall panel corresponding to the Standard Test Conditions ($T = 25°C$, $G = 1000W/m^2$).

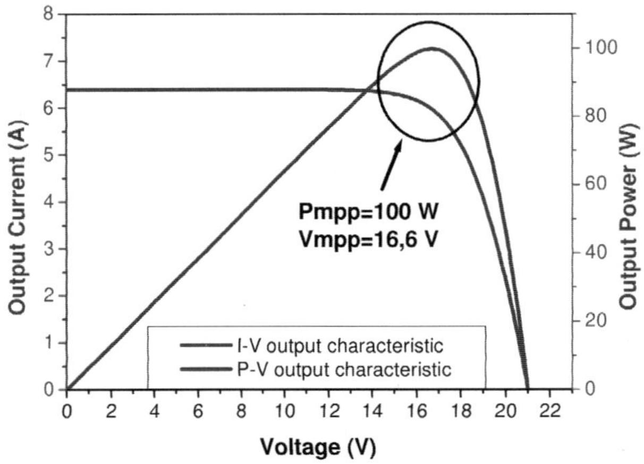

Figure 6. PV panel simulation characteristics under STC

Figure 6 shows the characteristics of the PV panel obtained by the MATLAB modelling, where the blue curve represents the I-V characteristic and the violet represents the P-V characteristic. Table 2 shows the values of the voltage and the output power of the MPP obtained by the Newton method.

Table 2. Newton method results for MPP

Newton method results (uniform illuminations)	
Voltage at MPP (Newton)	16.354 V
Power at MPP (Newton)	99.765 W

Observation of the curves on figure 6 and table 2, it was found that the P-V and I-V characteristics of the simulation were closely matched with the characteristics found by the manufacturer under the same conditions. And as can be seen, the P-V characteristic presents a peak where the maximum power is. The detection of this output power was obtained by applying the Newton method which showed that the maximum power is 99.765 W as presented in table 2. Therefore, the use of Newton's method has been verified and validated for finding the maximum point under uniform illumination and temperature.

3.4. Non-Uniform illuminations

Before applying the Newton method to non-uniform illumination condition, a mathematical model was required to establish the equation of the PV system under partial shadowing conditions. A wide study has been undertaken to set up a mathematical model of the PV panel under partial shading conditions, which helped test the behaviour of a PV array under non-uniform illuminations. The same parameters shown in table 1 were applied to the mathematical model and the global current calculated is obtained by the sum of all the output currents with a variation of the irradiance that has been made in order to achieve the impact of the shading on the photovoltaic panel.

Using a PV array of dimensions 1×36, receiving four different levels of radiation which are: 500 W/m^2, 1000 W/m^2, 100 W/m^2 and 1300 W/m^2. The same process of the Newton method has been applied in this case where at first the four peaks of the P-V characteristics have been detected, thus the global maximum power point has been found.

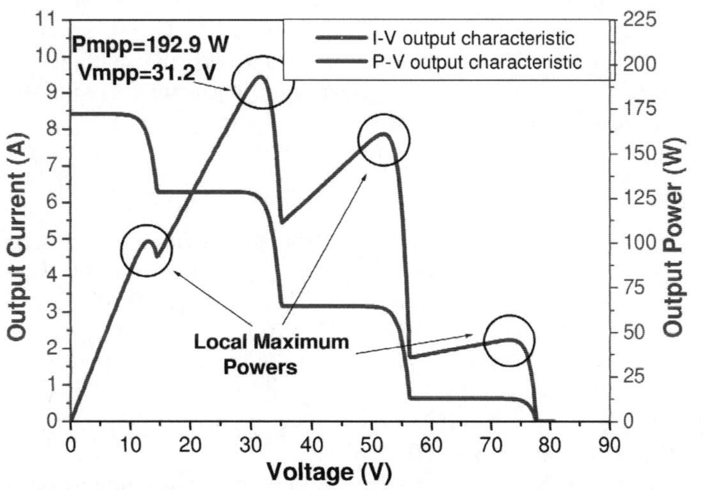

Figure 7. The output I-V and P-V characteristics of PV array under different illuminations

Figure 7 shows the MATLAB simulated characteristics of the previously described PV array receiving four different levels of illumination. As can be seen, the overall PV characteristic presents multiple peaks which are not well-handled by the conventional MPPT methods.

Table 3. Voltage and power at MPP of a PV panel under different illuminations obtained by Newton's method

Newton method results (non-uniform illuminations)	
Voltage at MPP	31.9 V
Power at MPP	191.6681W

IOP Conf. Series: Materials Science and Engineering **948** (2020) 012008 doi:10.1088/1757-899X/948/1/012008

As can be seen in figure 7 and table 3 the overall P-V characteristic has a very complex curve with a maximum power point at 192.9 W and 31.25 V, accompanied by three more local maximum powers. In this case, the downside of the conventional methods [13, 14] is evident since they will only follow the first local maximum power point. Table 3 shows the maximum power point coordinates obtained by Newton method indicating that the maximum power point is 191.67 W at a voltage of 31.9 V, which correspond to the global MPP coordinates given by the figure 7. This proves the accuracy and the precise convergence of the iterative Newton process. Thus, under partial shading conditions, finding the global power point for a PV array can be accomplished by applying the Newton iterative method.

4. Conclusion

This paper firstly presented the modelling of a PV panel under the MATLAB interface based on a numerical approach using the equations that describe the behaviour of the PV panel. Also, it brings out the major phenomena facing this technology which is the partial shading, where the conventional popular MPPT methods, namely the Perturb and Observe and Incremental conductance methods, cannot perform MPPT under these conditions due to multiple local maximums that are exhibited on the P-V characteristic. Although some researchers have worked on improving these methods for MPPT under partial shading conditions, they still have some drawbacks. In order to overcome this inconvenience, this paper, focus mainly on developing an easy and fast numerical approach to track the maximum power point under partial shading conditions using the Newton method to solve the equation of the PV system under these conditions, where a MATLAB based PV model is presented with the aim to detect the global maximum power point under partial shading conditions. According to the results, it has been shown that this method has good performance with small errors under non-uniform illumination and with fast convergence.

References

[1] F. Blaabjerg and D. M. Ionel, "Renewable Energy Devices and Systems – State-of-the-Art Technology, Research and Development, Challenges and Future Trends," *Electric Power Components and Systems,* vol. 43, pp. 1319-1328, 2015/07/21 2015.

[2] J. T. Bialasiewicz, "Renewable Energy Systems With Photovoltaic Power Generators: Operation and Modeling," *IEEE Transactions on Industrial Electronics,* vol. 55, pp. 2752-2758, 2008.

[3] U. Gangopadhyay, S. Jana, and S. Das, "State of Art of Solar Photovoltaic Technology," *Conference Papers in Energy,* vol. 2013, p. 9, 2013.

[4] J. Khan and M. H. Arsalan, "Solar power technologies for sustainable electricity generation – A review," *Renewable and Sustainable Energy Reviews,* vol. 55, pp. 414-425, 2016/03/01/ 2016.

[5] A. Vasić, B. Lončar, M. Vujisić, K. Stanković, and P. Osmokrović, "Aging of the photovoltaic solar cells," in *Microelectronics Proceedings (MIEL), 2010 27th International Conference on,* 2010, pp. 487-490.

[6] A. D. Hansen, P. E. Sørensen, L. H. Hansen, and H. W. Bindner, "Models for a stand-alone PV system," 8755027741, 2001.

[7] D. Hohm and M. E. Ropp, "Comparative study of maximum power point tracking algorithms," *Progress in photovoltaics: Research and Applications,* vol. 11, pp. 47-62, 2003.

[8] Rashel, M. R., Albino, A., Tlemcani, M., Gonçalves, T. C., & Rifath, J. (2016, November). MATLAB Simulink modeling of photovoltaic cells for understanding shadow effect. In 2016 IEEE International Conference on Renewable Energy Research and Applications (ICRERA) (pp. 747-750). IEEE.

[9] A. Patel, V. Kumar, and Y. Kumar, "Perturb and observe maximum power point tracking for photovoltaic cell," *Innovative Systems Design and Engineering,* vol. 4, pp. 9-15, 2013.

[10] M. R. Rashel, A. Albino, M. Tlemçani and T. C. F. Goncalves, "Sensitivity analysis of parameters of a photovoltaic cell under different conditions," in 10th International Conference on Software, Knowledge, Information, Management & Applications, China, 2016

[11] R. I. Putri, S. Wibowo, and M. Rifa'i, "Maximum Power Point Tracking for Photovoltaic Using Incremental Conductance Method," *Energy Procedia,* vol. 68, pp. 22-30, 2015/04/01 2015.

[12] X. Weidong and W. G. Dunford, "A modified adaptive hill climbing MPPT method for photovoltaic power systems," in *2004 IEEE 35th Annual Power Electronics Specialists Conference (IEEE Cat. No.04CH37551)*, 2004, pp. 1957-1963 Vol.3.

[13] K. Ishaque and Z. Salam, "A review of maximum power point tracking techniques of PV system for uniform insolation and partial shading condition," *Renewable and Sustainable Energy Reviews,* vol. 19, pp. 475-488, 3// 2013.

[14] K. Ishaque, Z. Salam, and G. Lauss, "The performance of perturb and observe and incremental conductance maximum power point tracking method under dynamic weather conditions," *Applied Energy,* vol. 119, pp. 228-236, 4/15/ 2014.

[15] E. Karatepe, T. Hiyama, M. Boztepe, and M. Çolak, "Voltage based power compensation system for photovoltaic generation system under partially shaded insolation conditions," *Energy Conversion and Management,* vol. 49, pp. 2307-2316, 8// 2008.

[16] Y. E. Abu Eldahab, N. H. Saad, and A. Zekry, "Enhancing the tracking techniques for the global maximum power point under partial shading conditions," *Renewable and Sustainable Energy Reviews,* vol. 73, pp. 1173-1183, 6// 2017.

[17] J. Ahmed and Z. Salam, "An improved perturb and observe (P&O) maximum power point tracking (MPPT) algorithm for higher efficiency," *Applied Energy,* vol. 150, pp. 97-108, 7/15/ 2015.

[18] Y. H. Ji, D. Y. Jung, J. G. Kim, J. H. Kim, T. W. Lee, and C. Y. Won, "A Real Maximum Power Point Tracking Method for Mismatching Compensation in PV Array Under Partially Shaded Conditions," *IEEE Transactions on Power Electronics,* vol. 26, pp. 1001-1009, 2011.

[19] D. Sera, T. Kerekes, R. Teodorescu, and F. Blaabjerg, "Improved MPPT algorithms for rapidly changing environmental conditions," in *Power Electronics and Motion Control Conference, 2006. EPE-PEMC 2006. 12th International*, 2006, pp. 1614-1619.

[20] J. Zhao, X. Zhou, Z. Gao, Y. Ma, and Z. Qin, "A novel global maximum power point tracking strategy (GMPPT) based on optimal current control for photovoltaic systems adaptive to variable environmental and partial shading conditions," *Solar Energy,* vol. 144, pp. 767-779, 3/1/ 2017.

New analytical expressions of photovoltaic solar module physical parameters: Effects of module temperature and incident solar irradiance

H El Achouby, M Zaimi, A Ibral, O Zegoudi and E M Assaid*

Electronics and Optics of Semiconductor Nanostructures and Sustainable Energy Team, Laboratory of Instrumentation of Measure and Control, Department of Physics, Faculty of Sciences, University Chouaib Doukkali, P. O. Box 20, El Jadida, Morocco

E-mail: assaid.e@ucd.ac.ma

Abstract. In this work, we present three methods to extract model physical parameters of a photovoltaic solar module modelled by a single-diode electronic circuit. The model physical parameters we are looking for are photocurrent I_{ph}, saturation current I_s, ideality factor η, series resistance R_s and shunt conductance G_p. The first method is based on exact solution of module characteristic equation, the second one uses exact expression of dynamic conductance and the third one makes use of exact expression of the integral of translated current-voltage characteristics (I-I_{sc}). In all these extraction methods, we achieve numerical fitting of experimental data of output current, differential conductance and area under translated characteristics with suitable analytical expressions to get model physical parameters at standard test conditions (STC). Then, we assume these values as initial conditions and derive exact analytical expressions reporting effects of module temperature and incident solar irradiance on photocurrent, saturation current and ideality factor using temperature coefficients provided as well as irradiance coefficients extracted from module manufacturer datasheet. We examine new analytical expressions of model physical parameters in the case of Shell SQ80 photovoltaic solar module at arbitrary operating conditions of module temperature and solar irradiance. We evidence good agreement between experimental characteristics and forecasted analytical characteristics generated by new analytical expressions.

1. Introduction

Today, global energy demand is steadily increasing cause energy needs in developed economies such as USA, Europe, Japan and newly emerging countries such as India, China, Brazil, Russia and South Africa grow continuously. Unfortunately, global fossil origin fuel reserves have reached the phase of overexploitation and even in some cases depletion. In addition, geopolitical instabilities in Middle East and Arabian Persian Gulf about nuclear program of Iran make the supply of world oil market uncertain in medium term. Otherwise, heavy and unreasonable burning of fossil origin fuels such as oil, natural gas and coal leads to a regular increase of water vapor, methane, carbon dioxide and nitrogen dioxide contents in the atmosphere. Because of anthropogenic emissions coming from deforestation and combustion of fossil fuels, our planet Earth is facing major environmental challenges such as air pollution and global warming. This causes a general climate disorder engendering melting of polar ice, rising of sea levels, severe droughts and catastrophic floods. Nowadays, the use of clean and sustainable energies is the alternative solution capable of giving adequate answers to all these crucial questions. Solar energy as a source of clean and sustainable energy available everywhere is able to meet the growing energy needs of all humanity. It is free from all geopolitical issues. It is also independent of the issue of transportation, supply since we harvest sunlight and produce electricity locally. With its components: solar thermal, concentrated solar power and solar photovoltaic, Sun is hope for all inhabitants of our planet.

In the present paper, we focus on solar photovoltaic topics. More precisely, our interest will be in all devices that convert incident solar irradiation into electricity such as photovoltaic solar cells, modules and arrays. The issue we are tackling in this work is the investigation of effects of module

IOP Conf. Series: Materials Science and Engineering **948** (2020) 012009 doi:10.1088/1757-899X/948/1/012009

temperature and incident solar irradiance on model physical parameters of photovoltaic solar module. Indeed, module temperature as well as incident solar irradiance vary throughout a sunny day. Temperature and irradiance are maximum at solar midday and minimum at sunrise and sunset [1]. Thus, it is very important to predict evolving of model physical parameters throughout a sunny day in order to guess variations of current-voltage characteristics and to predict accurately maximum power point coordinates, optimal load resistance value, fill factor and module efficiency.

In the literature, majority of papers dealing with this topic consider photocurrent as a bilinear function of T and G [2–4], saturation current as exponential function of T and quasi-constant function of G [2,5,6], ideality factor often as independent function on T and G [4,5] and rarely as a linear function of T [2,7]. Recently, El Achouby et al. [8,9] considered series resistance and parallel conductance as independent functions on solar irradiance and module temperature, determined numerically and analytically variations of photocurrent, saturation current and ideality factor versus G and T and gave sets of I=f(V) and P=f(V) characteristics for G=1000W/m² and different values of module temperature and T=25°C and different values of solar irradiance. Zaimi et al. [10] considered all model-physical parameters as dependent functions on solar irradiance and module temperature and determined numerically their values for solar module operating under non-standard conditions. In the present work, we consider that photocurrent, saturation current and ideality factor are all dependent on T and G. We assume that series resistance and shunt conductance are independent on T and G.

After general introduction object of this section, we lay in the second section the foundations of our work. We present three different methods to extract physical parameters of equivalent electronic circuit modelling behavior of a photovoltaic solar module built by series mounting of a given number of solar cells. We derive new analytical expressions describing variations of model physical parameters versus module temperature and incident solar irradiance. In the third section, we investigate the case of Shell SQ80 solar module and present results of our achievements. We extract its physical parameters at STC. We determine analytical expressions of photocurrent, saturation current and ideality factor at given temperature and irradiance. We carry out an estimate of accuracy for each model physical parameter extraction method by comparing optimized characteristics to experimental characteristics at standard test conditions (STC) (T=25°C, G=1000W/m^2). Then we achieve an evaluation of precision for analytical expressions modelling variation of physical parameters against temperature and irradiance by comparing predicted characteristics to experimental characteristics at arbitrary operating conditions of temperature and irradiance. In the last section, we close our paper with general conclusion.

2. Theoretical framework

2.1. *Photovoltaic solar module fundamental equations*

In this paper, we model photovoltaic solar module by single-diode equivalent electronic circuit with five physical parameters (see figure 1):

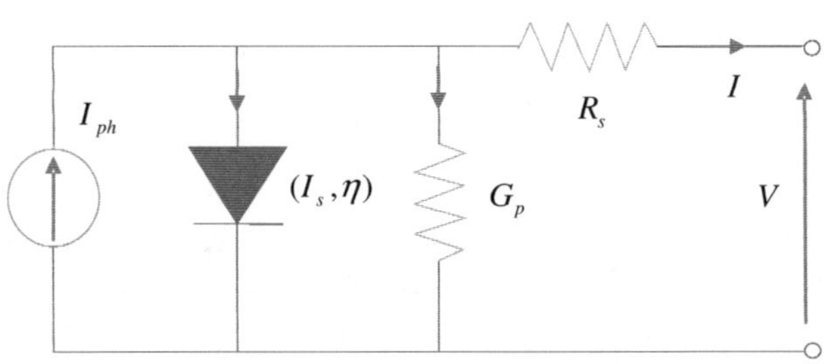

Figure 1. Electronic circuit modelling a photovoltaic solar module [11]

The characteristic equation of single-diode equivalent electronic circuit writes [11–15]:

$$I = I_{ph} - I_s\left[\exp\left(\frac{V + I\,R_s}{N_s\,\eta\,V_{th}}\right) - 1\right] - G_p\left(V + I\,R_s\right) \tag{2.1}$$

Where I_{ph} is the photocurrent, I_s is the diode saturation current, η is the ideality factor, R_s is the series resistance and $G_p=1/R_{sh}$ is the shunt conductance, it is the reciprocal of parallel resistance R_{sh}. $V_{th} = k_B T/q$ is the thermal voltage, it is the energy quanta corresponding to thermodynamic agitation within semiconductor. k_B is Boltzmann constant, q is elementary charge, N_s is the number of cells mounted in series to build photovoltaic solar module and T is the module temperature. In the following, we give module key equations at cardinal points.

At open-circuit point, photovoltaic solar module voltage V_{oc} corresponding to an output current $I=0$ is expressed as follows:

$$0 = I_{ph} - I_s\left[\exp\left(\frac{V_{oc}}{N_s\,\eta V_{th}}\right) - 1\right] - G_p V_{oc} \tag{2.2}$$

At short-circuit point, photovoltaic solar module current I_{sc} corresponding to an output voltage $V=0$ is given by:

$$I_{sc} = I_{ph} - I_s\left[\exp\left(\frac{R_s I_{sc}}{N_s\,\eta V_{th}}\right) - 1\right] - G_p R_s I_{sc} \tag{2.3}$$

At maximum power point, photovoltaic solar module current I_{mpp} and voltage V_{mpp} are linked via the following equation:

$$I_{mpp} = I_{ph} - I_s\left[\exp\left(\frac{V_{mpp} + R_s I_{mpp}}{N_s \eta\, V_{th}}\right) - 1\right] - G_p\left(V_{mpp} + R_s I_{mpp}\right) \tag{2.4}$$

2.2. Numerical extraction of model physical parameters

In this subsection, we use three different methods to determine physical parameters of single-diode equivalent electronic circuit describing behavior of photovoltaic solar module operating at standard test conditions (STC) (G=1000 W/m^2, T=25ºC). The methods of extraction are fully analytic. They rely on mathematical calculations where no approximation is used. They do not use initial guess values of model physical parameters to trigger iterations.

According to the first method, we determine analytical solution of photovoltaic module characteristic equation (see equation (2.1)), giving output current versus output voltage and model physical parameters:

$$I(V) = -\frac{\eta N_s V_{th}}{R_s} W\left(\frac{R_s I_s}{\eta N_s V_{th}\left(1 + G_p R_s\right)} \exp\left(\frac{R_s(I_{ph} + I_s) + V}{\eta N_s V_{th}\left(1 + G_p R_s\right)}\right)\right) + \frac{I_{ph} + I_s - G_p V}{1 + G_p R_s} \tag{2.5}$$

W(x) is LambertW function, it is the analytic solution of transcendental equation $w(x)e^{w(x)}=x$. In order to extract five model physical parameters according to the present method, we achieve a numerical fitting of *I=f(V)* experimental characteristics to equation (2.5) via Statistics [NonlinearFit] (SNF) package [16] of Maple computer algebra software [17].

Following the second method, we derive an analytical expression of module differential conductance $G_{dyn}=dI/dV$ from equation (2.1) [18–21]:

$$G_{dyn}(I,V)=-\left(\frac{I-I_{ph}-I_s+G_pV+G_pI\,R_s-G_pN_s\eta V_{th}}{\left(I-I_{ph}-I_s+G_pV+G_pI\,R_s-G_pN_s\eta V_{th}\right)R_s-N_s\eta V_{th}}\right) \tag{2.6}$$

Which leads to an expression linking output current I and output voltage V to the five physical parameters. Then we achieve a two-dimensional fitting of differential conductance experimental data to analytical equation (2.6) via Statistics [NonlinearFit] package of Maple computer algebra software.

According to the third method, we calculate Co-Content function *CC (I, V)* which is defined as the area under $(I(V)-I_{sc})$ translated characteristics: $CC(I,V)=\int_0^V (I(V)-I_{sc})dV$ [22,23]:

$$\begin{aligned}CC(I,V)=&\left(a_6+\frac{(a_4\,a_5\,a_6)}{a_1}-a_4\,a_5\right)V+\left(\frac{a_5}{2}+\frac{a_4\,a_5^2}{2\,a_1}\right)V^2\\&-\left(\frac{a_4\,a_5}{a_1}\right)V\left(I-I_{sc}\right)+\left(a_4-\frac{a_4\,a_6}{a_1}\right)\left(I-I_{sc}\right)+\left(\frac{a_4}{2\,a_1}\right)\left(I-I_{sc}\right)^2\end{aligned} \tag{2.7}$$

Where

$$a_1=-\frac{N_s\eta V_{th}}{R_s}\quad;\quad a_2=\frac{R_s\,I_s}{N_s\eta V_{th}(1+G_pR_s)}\quad;\quad a_3=R_s(I_{ph}+I_s)$$

$$a_4=N_s\eta V_{th}(1+G_pR_s)\quad;\quad a_5=-\frac{G_p}{1+G_pR_s}\quad;\quad a_6=\frac{I_{ph}+I_s}{1+G_pR_s}-I_{sc}$$

To determine five model physical parameters, we carry out two-dimensional fitting of Co-Content function *CC (I, V)* experimental values to analytical equation (2.7) using Statistics [NonlinearFit] package of Maple computer algebra software.

2.3. *Effects of module temperature and incident solar irradiance on model physical parameters*

Model physical parameters of photovoltaic solar module are very sensitive to variations of temperature T and global incident solar irradiance G. Thus, it is crucial to be able to guess physical parameters evolving versus T and G variables in order to foresee changes occurring in current-voltage characteristics and to guess its maximum power point (P_{mpp}), optimal load resistance (R_{mpp}), Fill Factor and module efficiency.

Thereafter, and for the purpose of simplicity, we assume that series resistance R_s and parallel conductance G_p are independent of T and G. Therefore, we consider that changes of T and G throughout a sunny day only affect photocurrent I_{ph}, saturation current I_s and ideality factor η.

In the majority of works concerned with this topic, scholars assume that open-circuit voltage V_{oc} and short-circuit current I_{sc} vary linearly with temperature from standard temperature T_{STC} equal to 25°C and at fixed global irradiance equal to 1000W/m^2 [1, 2, 4-6]:

$$V_{oc}(T)=V_{oc}(T_{STC})+K_v\left(T-T_{STC}\right) \tag{2.8}$$

$$I_{sc}(T)=I_{sc}(T_{STC})+K_i\left(T-T_{STC}\right) \tag{2.9}$$

Where $V_{oc}(T_{STC})$ and $I_{sc}(T_{STC})$ are open-circuit voltage and short-circuit current for a given illumination and for the reference temperature T_{STC} equal to 25°C. K_v is the temperature coefficient of open-circuit voltage (V/°C). K_i is the temperature coefficient of short-circuit current (A/°C).

In this study, we suggest for the first time in the literature two affine expressions to model effect of global incident irradiance on open-circuit voltage and short-circuit current:

$$V_{oc}(G)=V_{oc}(G_{STC})+\alpha_v(G-G_{STC}) \tag{2.10}$$

$$I_{sc}(G)=I_{sc}(G_{STC})+\alpha_i(G-G_{STC}) \tag{2.11}$$

Where $V_{oc}(G_{STC})$ and $I_{sc}(G_{STC})$ are open-circuit voltage and short-circuit current for a given module temperature and under reference irradiance G_{STC} equal to 1000Wm^{-2}. $\alpha_v=dV_{oc}/dG$ is the irradiance coefficient of open-circuit voltage (Vm^2/W). $\alpha_i=dI_{sc}/dG$ is the irradiance coefficient of short-circuit current (Am^2/W). In the present work, α_v and α_i are extracted from module manufacturer datasheet.

In the case where T and G vary at the same time, we model open-circuit voltage and short-circuit current by two-dimensional affine equations:

$$V_{oc}(T,G)=V_{oc}(T_{STC},G_{STC})+K_v(T-T_{STC})+\alpha_v(G-G_{STC}) \tag{2.12}$$

$$I_{sc}(T,G)=I_{sc}(T_{STC},G_{STC})+K_i(T-T_{STC})+\alpha_i(G-G_{STC}) \tag{2.13}$$

In all references in literature devoted to this subject, ideality factor is considered as constant, photocurrent I_{ph} is explicitly dependent on module temperature and solar irradiance while saturation current I_s is explicitly dependent on temperature and implicitly dependent on irradiance. At STC, the saturation current I_s is given by [5,24,25]:

$$I_s(T)=DT^3\exp\left(\frac{-qE_g}{\eta_{STC}k_BT}\right) \tag{2.14}$$

Where E_g is the semiconductor band gap energy equal to 1.12eV for crystalline silicon at 25°C. D is a real constant, which depends on diffusion process within the module. At STC, the analytical expression of D is determined from equation (2.14):

$$D=\frac{I_s(T_{STC})}{T_{STC}^3\exp\left(-qE_g/\eta_{STC}k_BT_{STC}\right)} \tag{2.15}$$

In the present paper, we consider that ideality factor is dependent on both module temperature and incident solar irradiance. We then generalize the writing of equation (2.14) to arbitrary operating conditions of T and G. The new expression of saturation current becomes:

$$I_s(T,G)=DT^3\exp\left(\frac{-qE_g}{\eta(T,G)k_BT}\right) \tag{2.16}$$

2.4. New analytical expressions modelling effects of module temperature and incident solar irradiance on physical parameters

2.4.1. Expression of photocurrent I_{ph}

By neglecting excess minority carrier's diffusion current within the module in equation (2.3), we lead to the following analytical expression of photocurrent I_{ph} versus module temperature and incident solar irradiance:

$$I_{ph}(T,G)=I_{ph}(T_{STC},G_{STC})+K_i\left(1+G_pR_s\right)(T-T_{STC})+\alpha_i\left(1+G_pR_s\right)(G-G_{STC}) \tag{2.17}$$

2.4.2. Expression of ideality factor η

In equation (2.2), we neglect thermal agitation current with regard to minority carrier's diffusion current and lead to an equation containing ideality factor and quantities depending on both module temperature and incident solar irradiance:

$$0 = I_{ph}(T,G) - I_s(T,G)\exp\left(\frac{V_{oc}(T,G)}{\eta(T,G)N_sV_{th}(T)}\right) - G_pV_{oc}(T,G) \tag{2.18}$$

Analytical resolution of equation (2.18) leads to exact expression of ideality factor versus module temperature and incident solar irradiance [8]:

$$\eta(T,G) = \frac{N_sE_g - V_{oc}(T,G)}{\ln\left(-DT^3/\left(G_pV_{oc}(T,G) - I_{ph}(T,G)\right)\right)N_sV_{th}} \tag{2.19}$$

2.4.3. Expression of saturation current I_s

By plugging equation (2.19) into equation (2.16), we obtain a new analytical expression of saturation current against module temperature and incident solar irradiance:

$$I_s(T,G) = DT^3\exp\left(\frac{-E_g\left(\ln\left(-DT^3/\left(G_pV_{oc}(T,G) - I_{ph}(T,G)\right)\right)N_s\right)}{N_sE_g - V_{oc}(T,G)}\right) \tag{2.20}$$

3. Results and discussions

In Table I, we give all technical settings of Shell SQ80 photovoltaic solar module as given in manufacturer datasheet.

Table 1. Shell SQ80 photovoltaic solar module technical settings at Standard Test Conditions (STC) as given by manufacturer datasheet [26].

Settings	N_s	I_{sc} (A)	V_{oc} (V)	I_{mpp} (A)	V_{mpp} (V)	K_i (mA/°C)	K_v (mV/°C)
Shell SQ80	36	4.85	21.8	4.58	17.5	1.4	-81

3.1. Experimental and optimized current-voltage characteristics of shell SQ80 solar module at standard test conditions

We achieved extraction of model physical parameters of Shell SQ80 photovoltaic solar module at standard test conditions (G=1000W/m², T=25°C), from current-voltage characteristics available in manufacturer's datasheet via the methods presented above. We gathered model physical parameters values given by means of all methods of extraction in Table II.

Table 2. Model physical parameters of Shell SQ80 solar module at standard test conditions extracted using Methods I, II and III.

Parameters	**Estimated values**		
	Method I	**Method II**	**Method III**
$\mathbf{I_{ph}}$ (A)	4.8500056431	4.8500165894	4.8500146358
$\mathbf{I_s}$ (10^{-8} A)	2.0462947584	2.5215848057	2.0324189480
η	1.2243484559	1.2343158965	1.2212970838
$\mathbf{R_s}$ (Ω)	0.3399880892	0.3193999322	0.3365411991
$\mathbf{G_p}$ (Ω^{-1})	0.0000253586	0.0000273558	0.0002278270

In Figure 2, we plot experimental current-voltage characteristics as well as optimized characteristics obtained via all methods of extraction at standard test conditions (STC) corresponding to a module temperature equal to 25°C and a solar spectrum AM1.5G which is equivalent to an incident solar irradiance equal to 1000W/m².

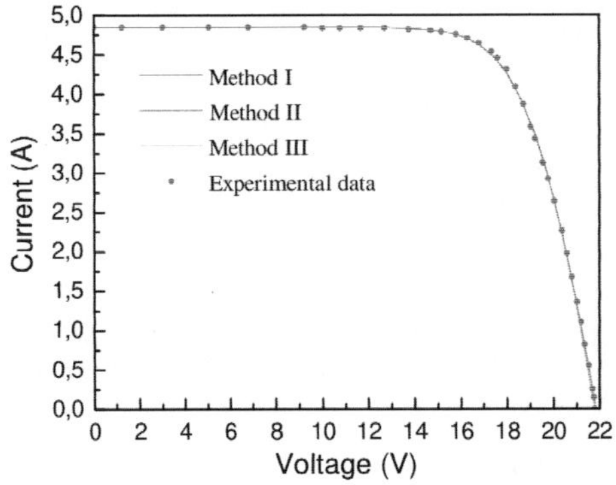

Figure 2. Experimental and optimized *I=f(V)* characteristics of Shell SQ80 solar module at standard test conditions (G=1000W/m², T=25°C).

To do assessment of analytical expressions predicting values of model physical parameters I_{ph}, I_s and η for module temperature and incident solar irradiance different from those corresponding to standard test conditions, we draw current-voltage and power-voltage characteristics for T=25°C, G=200, 400, 600, 800, 1000 Wm^{-2}, and for G=1000Wm^{-2}, T=20, 30, 40, 50, 60°C. To do so, we use equations (2.17), (2.19) and (2.20) while keeping R_s and G_p constant. Then we compare obtained curves to experimental data available in manufacturer datasheet and corresponding to identical operating conditions.

*3.2. Influence of incident solar irradiance*In Figures 3, we plot current-voltage *I=f(V)* (a) and power-voltage *P=f(V)* (b) characteristics of Shell SQ80 solar module for different irradiance intensities G and a module temperature T equal to 25°C. Dots correspond to experimental values coming from datasheet and solid lines correspond to predicted characteristics obtained assuming R_s and G_p constant and using analytical expressions of output current (equation 2.5), photocurrent $I_{ph}(T,G)$ (equation 2.17), ideality factor $\eta(T,G)$ (equation 2.19) and saturation current $I_s(T,G)$ (equation 2.20). We note that there is a very good agreement between forecasted analytic characteristics and experimental characteristics.

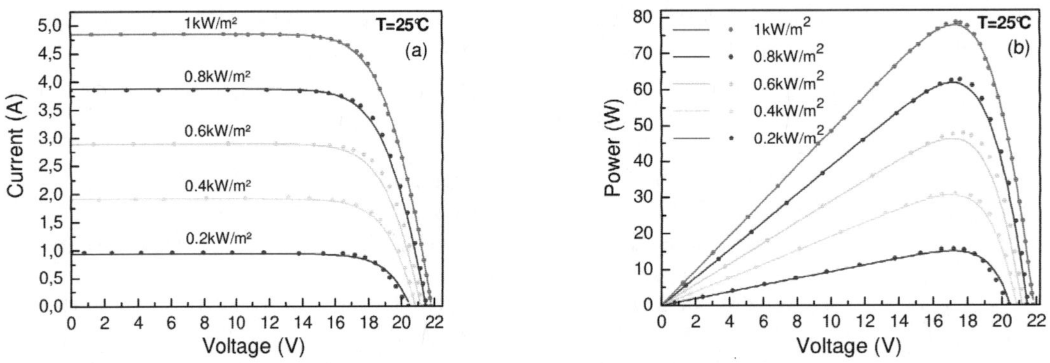

Figure 3. Experimental (dots) and predicted (solid lines) *I=f(V)* (a) and *P=f(V)* (b) characteristics of Shell SQ80 photovoltaic solar module at reference module temperature and different irradiance values.

3.3. Influence of module temperature

In Figures 4, current-voltage $I=f(V)$ (a) and power-voltage $P=f(V)$ (b) characteristics of Shell SQ80 photovoltaic solar module are drawn for different module temperatures T and a given intensity of incident solar irradiance G equal to 1000 Wm^{-2}. Experimental values from datasheet correspond to dots and forecasted characteristics obtained assuming R_s and G_p constant, and using analytical expressions of output current (equation 2.5), photocurrent $I_{ph}(T,G)$ (equation 2.17), ideality factor $\eta(T,G)$ (equation 2.19) and saturation current $I_s(T,G)$ (equation 2.20) correspond to solid lines. One can notice that experimental data and foreseen characteristics are in good agreement.

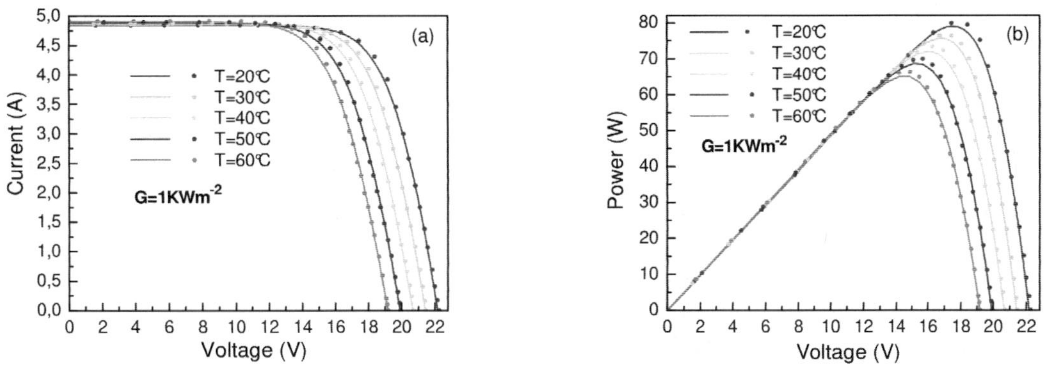

Figure 4. Experimental (dots) and forecasted (solid lines) $I=f(V)$ (a) and $P=f(V)$ (b) characteristics of Shell SQ80 photovoltaic solar module at reference incident solar irradiance and different module temperature values.

3.4. Accuracy and error calculation

We define absolute error $\varepsilon_{absolute}$ as the difference between experimental and computational values of output current:

$$\varepsilon_{absolute} = \left| I_{exp} - I_{cal} \right| \tag{3.1}$$

We may do a finer analysis of error by calculating normalized root mean squared error (NRMSE). We define NRMSE as square root of mean quadratic errors to mean experimental values ratio [27]:

$$NRMSE = \sqrt{\frac{1}{N}\sum_{i=1}^{N}\left(I_i^{Exp} - I_i^{cal}\right)^2} \Big/ \frac{1}{N}\sum_{i=1}^{N} I_i^{Exp} \tag{3.2}$$

At the beginning, we evaluate accuracy of physical parameters extraction methods in case of Shell SQ80 photovoltaic solar module functioning under standard test conditions. In Figure 5, we plot absolute error of output current versus output voltage for all methods of extraction presented above. We remark that absolute error is negligible for output voltage values lying between 0 and V_{mpp} which correspond to output current values lying between I_{sc} and I_{mpp}. For output voltage values lying between V_{mpp} and V_{oc}, output current values decreases exponentially between I_{mpp} and 0 and absolute error increases while remaining below 0.11 A. On the other hand, the reader can easily remark that *SNF* method achieving numerical fit of $I=f(V)$ experimental characteristics to equation (2.5) (method I) ensures best accuracy while accuracies of methods achieving numerical fit of G_{dyn} (method II) and $CC(I,V)$ (method III) are approximately the same.

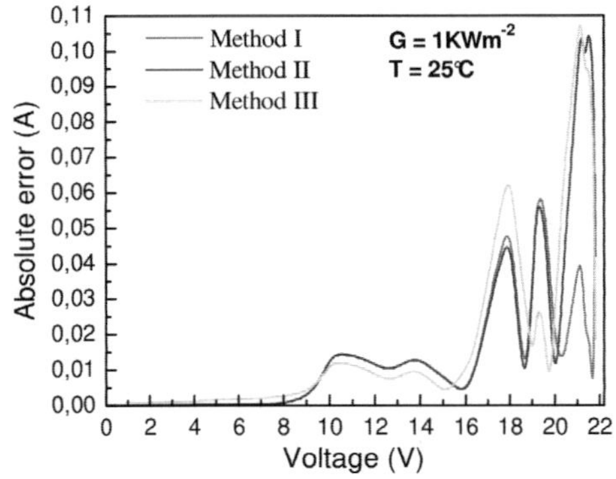

Figure 5. Absolute error of output current against output voltage for methods of extraction of physical parameters in case of Shell SQ80 photovoltaic solar module operating under standard test conditions.

Now, we tackle evaluation of accuracy of analytical model established to forecast $I=f(V)$ characteristics for Shell SQ80 solar module operating under arbitrary conditions of incident solar irradiance and module temperature. We remind the reader that this analytical model is founded on analytical expressions of I_{ph}, η and I_s while and it considers R_s and G_p constant.

In Figure 6, we present absolute error of output current versus output voltage for different values of irradiance and reference value of module temperature T=25°C (a). We also present absolute error of output current versus output voltage for different values of module temperature and reference value of solar irradiance equal to 1000Wm^{-2} (b).

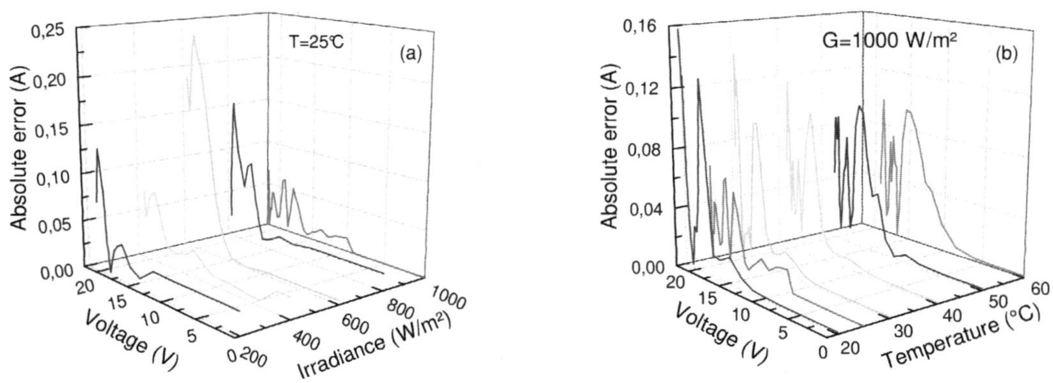

Figure 6. Absolute error of current versus voltage for mathematical model derived to predict $I=f(V)$ characteristics in case of Shell SQ80 photovoltaic solar module for different values of incident irradiance and reference value of module temperature T=25°C (a) and for different values of module temperature and reference value of incident solar irradiance G=1000 Wm^{-2} (b).

In the following histograms, we present NRMSE of mathematical model derived to predict $I=f(V)$ characteristics in case of Shell SQ80 solar module for different solar irradiance values and reference module temperature (a) as well as for different module temperature values and reference solar irradiance (b). Established mathematical model is based on analytical expressions of physical parameters $I_{ph}(T,G)$, $I_s(T,G)$ and $\eta(T,G)$, it uses as boundary conditions the values of photocurrent, saturation current and ideality factor corresponding to STC, determined via the three methods of extraction and gathered in Table II.

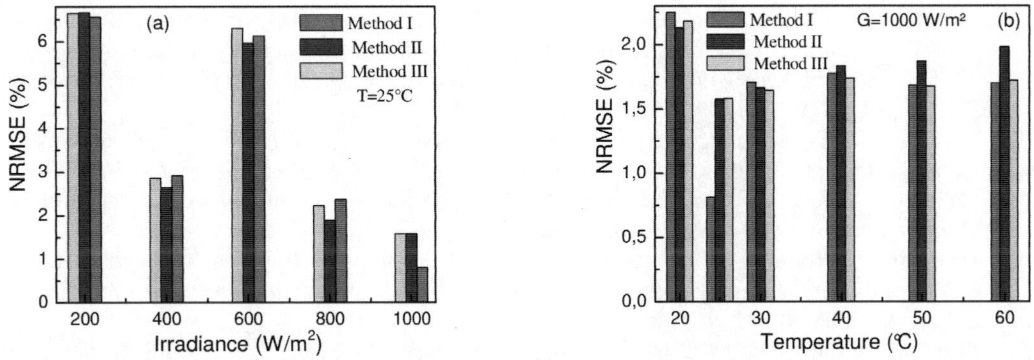

Figure 7. NRMSE of output current corresponding to mathematical model derived to predict current-voltage characteristics in case of Shell SQ80 photovoltaic solar module for different solar irradiance values and reference module temperature (a) and for different module temperature values and reference solar irradiance (b). Mathematical model is founded on analytical expressions of physical parameters $I_{ph}(T,G)$, $I_s(T,G)$ and $\eta(T,G)$ and uses as boundary conditions values of photocurrent, saturation current and ideality factor corresponding to STC and determined via the three methods of extraction.

4. Conclusion

In this study, we present three methods to determine physical parameters of single-diode electronic circuit modelling photovoltaic solar module operating under standard test conditions (STC) (G=1000 Wm^{-2} and T=25°C). The first method extracts physical parameters via one-dimension statistical non-linear fit of experimental current-voltage characteristics to exact solution of characteristic equation. The second and third methods determine physical parameters by means of two-dimension statistical nonlinear fit of experimental values of differential conductance and co-content function to respective analytic expressions. We show that optimized characteristics generated using physical parameters coming from methods of extraction and experimental characteristics are in good agreement. We assume that series resistance and shunt conductance are independent on module temperature and incident solar irradiance. We consider values of model physical parameters obtained at STC as boundary conditions and establish analytical equations modelling variations of photocurrent, saturation current and ideality factor versus module temperature and incident solar irradiance. We derive mathematical model capable to predict current-voltage characteristics, power-voltage characteristics, maximum power point and fill factor for arbitrary operating conditions of temperature and irradiance. We emphasize that forecasted characteristics coming from mathematical model for given values of module temperature and incident solar irradiance are in good accordance with experimental characteristics.

Acknowledgments
The authors H El Achouby and M Zaimi are grateful to Kingdom of Morocco Ministry of Higher Education and to University Chouaib Doukkali (UCD) for their financial supports.

References

[1] Zaimi M, El Achouby H, Ibral A, Assaid E M, El Maliki M S and Saadani R 2019 Temporal Monitoring of Temperature and Incident Irradiance for Predicting Photovoltaic Solar Module Peak Power and Efficiency Using Analytical Expressions of Model Physical Parameters *In:2018 6th International Renewable and Sustainable Energy Conference (IRSEC)*, (Rabat, Morocco). IEEE, pp. 1-7 doi: 10.1109/IRSEC.2018.8702973

[2] Chin V J, Salam Z and Ishaque K 2015 Cell modelling and model parameters estimation techniques for photovoltaic simulator application: A review *Applied Energy* **154** 500–19

[3] Chatterjee A, Keyhani A and Kapoor D 2011 Identification of Photovoltaic Source Models *IEEE Transactions on Energy Conversion* **26** 883–9

[4] Lineykin S, Averbukh M and Kuperman A 2014 An improved approach to extract the single-diode equivalent circuit parameters of a photovoltaic cell/panel *Renewable and Sustainable Energy Reviews* **30** 282–9

[5] Shongwe S and Hanif M 2015 Comparative Analysis of Different Single-Diode PV Modeling Methods *IEEE Journal of Photovoltaics* **5** 938–46

[6] Sera D, Teodorescu R and Rodriguez P 2007 PV panel model based on datasheet values *2007 IEEE International Symposium on Industrial Electronics* 2007 IEEE International Symposium on Industrial Electronics (Vigo, Spain: IEEE) pp 2392–6

[7] Humada A M, Hojabri M, Mekhilef S and Hamada H M 2016 Solar cell parameters extraction based on single and double-diode models: A review *Renewable and Sustainable Energy Reviews* **56** 494–509

[8] El Achouby H, Zaimi M, Ibral A and Assaid E M 2018 New analytical approach for modelling effects of temperature and irradiance on physical parameters of photovoltaic solar module *Energy Conversion and Management* **177** 258–71

[9] El Achouby H, Zaimi M, Ibral A and Assaid E M 2018 Exact analytical expressions modelling effects of incident irradiance and temperature on physical parameters of photovoltaic solar module *In: 2018 Renewable Energies, Power Systems & Green Inclusive Economy (REPS-GIE)*. IEEE, pp 1–6

[10] Zaimi M, El Achouby H, Ibral A and Assaid E M 2019 Determining combined effects of solar radiation and panel junction temperature on all model-parameters to forecast peak power and photovoltaic yield of solar panel under non-standard conditions *Solar Energy* **191** 341–59

[11] Rauschenbach H S 1980 *Solar cell array design handbook: the principles and technology of photovoltaic energy conversion.* (NewYork: Van Nostrand Reinhold)

[12] Aazou S and Assaid E M 2009 The real photovoltaic solar cell *Maple application center*

[13] Aazou S and Assaid E M 2009 Modelling real photovoltaic solar cell using Maple *2009 International Conference on Microelectronics - ICM* 2009 International Conference on Microelectronics - ICM (Marrakech, Morocco: IEEE) pp 394–7

[14] Aazou S and Assaid E M 2011 Model physical parameters effects on real solar cell characteristics and power curve *Global journal of physical chemistry* **2** 61–7

[15] Aazou S, Ibral A, White M S and Kaltenbrunner M 2013 Organic Bulk Heterojunction Solar Cells Based on P3HT and Anthracene-Containing PPE-PPV: Fabrication, Characterization and Modeling *Journal of Optoelectronics and Advanced Materials* **13** 395–404

[16] Khan S 2017 *Parameter Estimation for Photovoltaic Diodes*

[17] Release Maple V 2015 *Copyright 1981–2015. Waterloo Maple Inc.*

[18] Ouennoughi Z and Chegaar M 1999 A simpler method for extracting solar cell parameters using the conductance method *Solid-State Electronics* **43** 1985–8

[19] Chegaar M, Ouennoughi Z and Hoffmann A 2001 A new method for evaluating illuminated solar cell parameters *Solid-State Electronics* **45** 293–6

[20] Chegaar M, Ouennoughi Z and Guechi F 2004 Extracting dc parameters of solar cells under illumination *Vacuum* **75** 367–72
[21] Bouzidi K, Chegaar M and Aillerie M 2012 Solar Cells Parameters Evaluation from Dark I-V Characteristics *Energy Procedia* **18** 1601–10
[22] Ortiz-conde A, Garciasanchez F and Muci J 2006 New method to extract the model parameters of solar cells from the explicit analytic solutions of their illuminated characteristics *Solar Energy Materials and Solar Cells* **90** 352–61
[23] Ortiz-Conde A and García Sánchez F J 2005 Extraction of non-ideal junction model parameters from the explicit analytic solutions of its I–V characteristics *Solid-State Electronics* **49** 465–72
[24] Gow J A and Manning C D 1999 Development of a photovoltaic array model for use in power-electronics simulation studies *IEE Proceedings - Electric Power Applications* **146** 193–200
[25] De Soto W, Klein S A and Beckman W A 2006 Improvement and validation of a model for photovoltaic array performance *Solar Energy* **80** 78–88
[26] Shell Solar *Photovoltaic solar module SQ80 Datasheet*
[27] Kumar M and Kumar A 2017 An efficient parameters extraction technique of photovoltaic models for performance assessment *Solar Energy* **158** 192–206

IOP Conf. Series: Materials Science and Engineering **948** (2020) 012010 doi:10.1088/1757-899X/948/1/012010

Thermal performance of building materials based on feather and polystyrene wastes

Ouahiba MRAJJI [1,2,5] , Mohamed EL WAZNA [1,2], Salma OUHAIBI [3] , Said SAIR [1], Abdeslam EL BOUARI [1], Omar CHERKAOUI [2], Naoual BELOUAGGADIA[4].

[1] Laboratory of Physic-Chemical of Applied Materials, Sciences Faculty of Ben M'sik, Hassan II University of Casablanca Morocco.
[2] Laboratory REMTEX, Higher School of Textile and clothing Industries, Km 8, Route d'EL JADIDA, Casablanca Morocco.
[3] Laboratory of Engineering and Materials (LIMAT), Faculty of Sciences Ben M'Sik, Hassan II University of Casablanca, Morocco
[4] Laboratory of signals, distributed systems and artificial Intelligence, ENSET, Hassan II University, Mohammedia, Morocco.
[5] Corresponding author
ouahiba.mrajji-etu@etu.univh2c.ma

Abstract. The development of composites using renewable resources has received attention of scientists over the last decade, with the aim of manufacturing environmentally friendly composites, reducing cost and the dependence on petroleum resources. This experimental study emphasizes the development of new cost-efficient composite based on feather/cotton nonwoven and polystyrene waste. The microstructure and thermal properties have been determined in each FPS material. Vacuum molding method was adopted for fabricating of the biocomposites with two contents of polystyrene matrix (30% and 50%). As a result, the morphological analysis revealed better compatibility and adhesion between the nonwoven and the polystyrene matrix. The manufactured composites showed an insulating character with thermal conductivity coefficient value of 35.95 and 33.85 mW / (m.K) respectively for the composite reinforced with 30 and 50% polystyrene. The origin of the variation of this property is due to densities and the percentage of the polystyrene matrix. The composites developed with polystyrene showed very high performance, combined with sufficient durability to be potential candidates for the development of new sustainable materials for the building sector.

Keywords: Polystyrene, thermal conductivity, vacuum molding, cost-efficient, waste management.

1. Introduction:

Currently, the atmospheric concentration of carbon dioxide increase which requires an urgent response through the manufacture and use of more environmentally friendly materials obtained from renewable energy sources that contribute to the reduction of CO_2 emissions [1]. The development of green composites has attracted great interest, these new materials was more usable due to their interesting properties, durable, can be easily disposed without harming the environment, low construction cost and applicable in many consumer industries such as construction sector [2]. The global sector of natural fiber-reinforced polymer composites reached US$2.1 billion in 2010 [3]. This material was developed from different types of natural fibers, such as sisal [4], bamboo [5], feather [6] and wood [7]. The attractive characteristics of natural fibres were their reasonable cost, lightness, renewability, biodegradability and low energy consumption during manufacturing. Today, the solid waste generated by poultry sector also remains one of the most important renewable resources used in the formulation of composite materials [8]. Indeed, 2400,000 tons of feather waste was generated per year in Morocco

[9]. The evaluation of this feather waste, which is fairly rich in 91% keratin in the form of nonwoven textiles, is extremely important to solve the incineration problem and to bring new materials to the economy. The feather fiber has remarkable reinforcing properties, they have been imported in all types of polymers: thermosetting (unsaturated polyesters, epoxy resins), or thermoplastics (PS, HDPE, PP and PLA).In all countries of the world, either was the level of their development, sale of the materials using the expansive polystyrene as packing generates important quantities of nonbiodegradable wastes. Management of these wastes is important. Polystyrene (PS) was one of the most versatile synthetic polymers due to its hardness, water resistance and low cost per unit weight. The properties of natural fibre reinforced polystyrene composite have made them the materials of choice for diverse applications. Eusèbe Agoua et al. [10] studied the thermal conductivity of composites manufactured with wood wastes and expanded polystyrene. The results revealed a good compatibility between the two components. The thermal conductivity achieved for different granulation and glue content have presented insulation character. The wood and polystyrene permits to produce economic and environment composites. Recycling of the feathers wastes and the expansive polystyrene packing to produce other building materials appears like an environmental and economic alternative. In this context, the present work aims to finalize a composite composed of a mixture of feather / cotton nonwoven and glue obtained from polystyrene of recuperation dissolved in diethyl ether. The formulation will be made on a basis allowing to obtain products with various thermal and mechanical characteristics according to their uses.

2. Materials and methods

2.1. Preparation of reinforcement:

The nonwoven webs were prepared according to the needling technique. The feathers fiber was provided by a slaughterhouse in the province of Casablanca, Morocco and the cotton fibers were obtained from a Moroccan manufacture unity in form of staple fiber. The feathers fibers are less crimped; a percentage of cotton waste has been added to ensure the nonwoven formation. The treated fiber was cut into small portion of 10-30 mm of length and inserted with waste cotton (50% in weight) into the industrial manufacturing line (needling DILO DI-LOOM OD-II 10069/2012 machine) which utilizes a mechanical consolidation method to entangle the fibers with each other without the addition of any binder. The carded webs are transferred directly to the bonding stage where they are repeatedly punctured by a battery of needles.

2.2. Preparation of matrix:

Polystyrene is obtained from the protective packaging waste of several companies; diethyl ether was purchased from Loba Chemie. The matrix is prepared by dissolution of PS waste in diethyl ether.

$$\frac{m_{Polystyrene}}{m_{Diethyl\ Ether}} = 3 \qquad \text{eq.1}$$

Where, $m_{Diethyl\ Ether}$ is the mass of diethyl ether and $m_{polystyrene}$ is the mass of the polystyrene PS.

2.3. Composite manufacturing process:

Vacuum molding method was adopted for fabricating the feather/cotton nonwoven reinforced polystyrene composites. Composites with dimension of 200 mm*200mm were manufactured with two content of polystyrene matrix (30% and 50%).The relasing agent was applied firstly, then the matrix was poured into the mat of feather/cotton until it is completely soaked. The mould was closed and the lamination is evacuated under vacuum of 500 bar using the vacuum pump at temperature 40°C for 4hours. The samples were left to harden for 24 h and then removed from the mold. The figure 1 shows a schematic of composite manufacturing process.

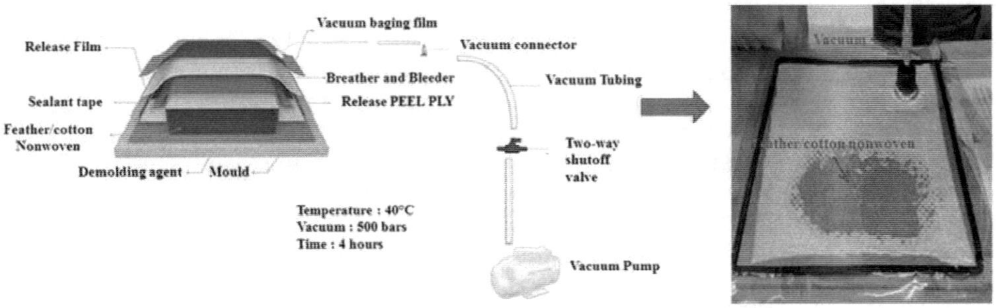

Figure 1. Composite manufacturing process

2.4. Characterization:

The morphology and adhesion between components of composite materials was studied by HIROX SH-4000 M scanning electron microscopy at acceleration voltages between 10 and 20 kV. The thermal conductivity of the manufactured materials was determined using a thermal conductivity analyzer (λ-meter EP500e) in accordance with EN 12667 with a sample size of 200 mm×200 mm at three temperatures 10, 25 and 40 °C[11].

3. Results and discussion

3.1. Morphological analysis

The morphological analysis of composites shows a good compatibility of fiber with polystyrene matrix, the fibres are completely covered with the matrix; forming a homogeneous material with no defects and a good fiber-matrix interface (figure 2). The fibers are well dispersed in polystyrene matrix and do no revealed agglomerates and microvids on the surface of the composite due to the reinforcement structure (nonwoven) and composite processing technique (vacuum molding). The feathers fibers permit an even arrangement due to nonwoven structure and adherence to polystyrene owing to their hydrophobic nature. The adhesion of the nonwoven-matrix is improved with the increase in the PS matrix rate in the composites.

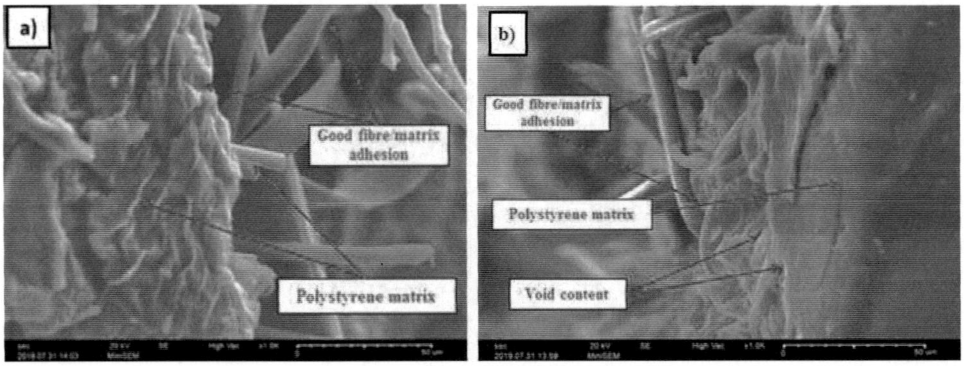

Figure 2. SEM image of feather /cotton nonwoven reinforced 30% polystyrene composite (a) and feather /cotton nonwoven reinforced 50% polystyrene composite (b).

3.2. Thermal conductivity:

The thermal conductivity coefficients identify the insulating nature of the sample; it depends on several parameters such as the morphology of fiber, the density, surface, the void content, matrix and homogeneity of sample. The thermal conductivity values of the various samples are shown in the figure 3. The measurements of the thermal conductivity of the samples are respectively obtained in the margins of 40.25 to 33.85 mW / m.K at 10°C. The best value 33.85mW/m.K is attributed for feather/cotton nonwoven reinforced 50% polystyrene composite. The origin of the variation of this property is due to insulating character of polystyrene leads to a decrease the thermal conductivity. The thermal conductivity of all the samples increases with the increase of temperature; this is explained by the increase of heat conduction through the samples. According to the results composites samples had thermal conductivity λ <60 (mW)/ (K.m) up to 40°C temperature, it can be concluded that these samples are suitable for all applications that require a resistivity against the heat flow.

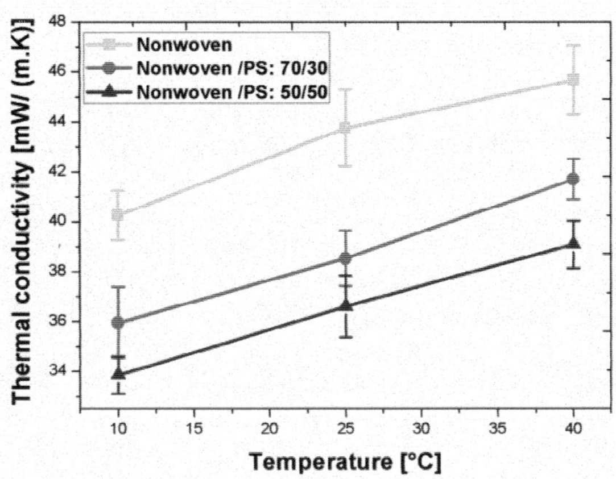

Figure 3. Thermal conductivity at 10, 25 and 40°C

The thermal conductivity results are very satisfactory compared to fiber–reinforced polystyrene composites, the thermal conductivity of polystyrene composites reinforced with natural fiber typically exceed 100 (m.W) / (m.K) [10, 12]. The composites prepared present many advantages such as: Economical by introducing low cost materials and ecological advantage with the use of renewable waste, less energy in the construction, recovery of polystyrene.

4. Conclusion:

In order to promote the use of waste fibers in the thermal insulation of civil and mechanical constructions, a new composite material based on feather/cotton waste and polystyrene is developed. In this study, the feather / cotton nonwoven is used as reinforcement and the polystyrene waste, dissolved in diethyl ether, is used as a matrix. In this study, morphological and thermal measurements were performed to characterize the feather/cotton nonwoven reinforced polystyrene composite. From the above results, the following conclusions have been given:

- Needled nonwoven fabrics have many advantages for use as reinforcement in polymer composites, such as good z-directional strength which greatly reduces delamination problems.
- The results of the thermal conductivity obtained for different samples showed that such materials can be developed to be used for example as insulating material that depend on a resistivity against the heat flow.
- The lowest value of thermal conductivity was observed for feather/cotton nonwoven reinforced with 50% of polystyrene 33.85 at 10°C.

These eco-friendly materials can found applications in various fields such as construction sector due to interesting properties and their good cost-performance ratio.

Reference:

[1] Claramunt J, Fernández-Carrasco L J, Ventura H and Ardanuy M, 2016. Natural fiber nonwoven reinforced cement composites as sustainable materials for building envelopes. Construction and Building Materials, 115, 230-239.
[2] Saheb D N and Jog J P, 1999. Natural fiber polymer composites: a review. Advances in Polymer Technology: Journal of the Polymer Processing Institute, 18(4), 351-363.
[3] FlaxStalk Natural Fiber Solutions, BIO World Congress Montreal, Quebec June 17, 2013, https://www.bio.org/sites/default/files/bio%20world%20congress%20june%2017%202013_joe_hogue.pdf (Accessed on 22 July 2019).
[4] Varghese S, Kuriakose B and Thomas S, 1994. Stress relaxation in short sisal-fiber-reinforced natural rubber composites. Journal of Applied Polymer Science, 53(8), 1051-1060.
[5] Jindal U C, 1986. Development and testing of bamboo-fibres reinforced plastic composites. Journal of composite materials, 20(1), 19-29.
[6] Choudary R B, Krishna N N and Bhargava N R M R, 2018. Study on CFF-Polyester Composites. Materials Today: Proceedings, 5(2), 8514-8522.
[7] Maldas D and Kokta B V, 1991. Influence of maleic anhydride as a coupling agent on the performance of wood fiber-polystyrene composites. Polymer Engineering & Science, 31(18), 1351-1357.
[8] Subramani T, Krishnan S, Ganesan S K and Nagarajan G, 2014. Investigation of mechanical properties in polyester and phenylester composites reinforced with chicken feather fiber. Int J Eng Res Appl, 4(12), 93-104..
[9] Interprofessional Federation of the Poultry Sector (FISA) Statistic, 2017. http://www.fisamaroc.org.ma/index.php?option=com_content&view=article&id=17&Itemid=53 (accessed 5 September 2018).
[10] Agoua E , Allognon-Houessou E, Adjovi E and Togbedji B, 2013. Thermal conductivity of composites made of wastes of wood and expanded polystyrene. Construction and Building Materials, 41, 557-562.
[11] EN 12667:2001. Thermal performance of building materials and products Determination of thermal resistance by means of guarded hot plate and heat flow meter methods. Products of high and medium thermal resistance
[12] Masri T, Ounis H, Sedira L, Kaci A and Benchabane A, 2018. Characterization of new composite material based on date palm leaflets and expanded polystyrene wastes. Construction and Building Materials, 164, 410-418.

IOP Conf. Series: Materials Science and Engineering **948** (2020) 012011 doi:10.1088/1757-899X/948/1/012011

MOLECULAR-DYNAMICS STUDY OF SELF-DIFFUSION: OF THE Au_4;Au_4/Ag(110) SYSTEM

F. Eddiai[1,*], M. Dardouri[2], H. El Azrak[1], A. Hassani, K. Sbiaai [1,], A. Hassnaoui [1]**

1. Laboratory of Materials Science, Milieu and Modulation (LS3M), Univ My. Slimane, *Polydisciplinary Faculty,* 25000 Khouribga, Marocco.
2. Laboratory of the Physics of Condensed Mather (LPMC), Univ. Chouaib Doukkali, El Jadida, Maroc.
* f.eddiai.fpk@gmail.com ** *k.sbiaai@usms.ma*

Abstract: In this work, we will shed light on the results obtained from the molecular dynamics method in the temperature ranging of 300-700K. This investigation concerning the coalescence for the two tetramers islands of system having different forms (SS, TT and NN) deposited at different spacing sites in-channel and cross-channel. The stability of the systems, the diffusion phenomena produced during the dynamics, as well as the diffusion/coalescence lifetime during the evolution of temperature are dependent. The dynamic study shows that between 450K and 650K the homogeneous partial coalescence occurs through the mechanism of the jump. As time goes by increasing the temperature which favors the process of exchange the heterogeneous total coalescence is obtained.

Key words: geometry tetramer 4S, 4N and 4T; island diffusion; evolution dynamic; molecular dynamics simulation; Au_4; Au_4/Ag(110) .

1. Introduction

The epitaxial growth of a bimetallic surface is rich in dynamic and kinetic phenomenon that attracts the attention of the scientific community that address the problems related to surfaces and interfaces during the growth of thin layers [1-5]. This technology is motivated by a wide range of applications, such as catalytic chemistry, mass transport, microelectronic fabrication, magnetic, photonic and photovoltaic devices [6-12]. During the dynamic evolution of the studied systems, different processes and phenomena are produced. Several experimental and theoretical studies have addressed the mechanisms of diffusion of the elementary elements (adatom, dimer ... islets) responsible for the growth of the surfaces which are the key way to understand this phenomenon on the atomic scale [13-14]. Islands stabilities and adatom dynamics are the main building blocks of thin film growth mode [19-21]. All these phenomena occur via several mechanisms leading to three growth modes: layer-by-layer mode, 3D mode and Stranski-Krastanov mode. These processes include the atoms deposited on substrates, the diffusion of adatoms on surface and the processes of jumps. The latter has been demonstrated experimentally in different systems, such as Ir/W(110), Na/Cu(100), Pd/W(211) [22-24]. Diffusion by exchange process has been observed in different systems as an example Ni/Ni(110) [25]. Nucleation, formation of stable islands and diffusion of dimers and trimers are carried out by these processes. The study by Yi Fu et al. pointed out that the most stable cluster structure is the linear form, but that the energy differences between the linear structure and the other structures for Rh tetramers are smaller than those of Ir tetramers [26]. Surface growth begins with the nucleation of adatoms and dimer, which causes the phenomenon of coalescence by the mechanisms of attachment / detachment of atoms on islands to form a large islands [27-32]. The shape and size of the island play a role in the nucleation and growth of homo systems as well as heteroepitaxial systems. The understanding of the diffusion mechanisms and energetic factors induced by the thermodynamic behavior of the cluster and the atoms on the surfaces reveals an important interest for the study of the crystal growth and the formation of the films [8] [33-36]. Yi Fu et al. have indicated that the linear chains reach their stability until the number of atoms reach eight atoms [26].

The dynamic diffusion of a single adatom or group of adatoms (clusters) on metal surfaces is a fertile ground for understanding the mechanism of crystal growth and heterogeneous catalysis at the atomic scale [32] [37-39]. While the details of atomic movements are difficult to observe experimentally because the diffusion process takes place over a short period of time; it is, therefore, difficult to follow the diffusion movements of adatoms on a substrate. Molecular dynamics is a useful method for tracking the trajectory as well as the processes that will take place during the dynamic evolution of epitaxy system ad-species [40]. In stage is devoted to the growth of islands by the phenomenon of coalescence [41] [42]. During this growth phase, the two affected clusters are normally larger. The growth of small islands of Au on silver has been studied theoretically and experimentally [43] [44].

In this work, we studied the phenomenon of coalescence for two tetramer islands $Au_4; Au_4/Ag(110)$ by method molecular dynamics. The main axes of this study focused on the stability of the systems, the diffusion phenomena produced during the dynamics, as well as the diffusion / coalescence lifetime during the evolution of the temperature.

In this paper, we described our systems as well as our simulation method in section 2. Then, we discussed the results obtained during the system investigation $Au_4; Au_4/Ag(110)$ in section 3. Finally, we devoted our attention to a summary in the form of a conclusion.

2. Simulation method

There are many problems in solid state physics requiring the knowledge of the total energy $\boldsymbol{E_{tot}}$ of a multi-site network according to the profiles $(\boldsymbol{\rho}, \boldsymbol{r})$ of occupation of these sites by atoms of the same type or binary. In literatures, numerous empirical and semi-empirical models make it possible to model the total energy of the metallic systems as a function of the constituent atoms. In this work, our systems are modeled by a semi-empirical potential based on the Embedded Atom method (EAM) that has been proposed by Foiles et al. [45- 47] . In the approach in question (EAM), the total energy of the system is given by the sum of two terms, one represents the peer interaction that ensures the body-body repulsion of the atomic nuclei in a given cutoff distance , and the other describes the potential of immersing an atom in an electron density. This total energy is expressed by the following formula :

$$\boldsymbol{E_{tot}} = \sum_i \boldsymbol{F_i}(\boldsymbol{\rho_i}) + \frac{1}{2}\sum_i \sum_{j(\neq i)} \boldsymbol{\varphi_{ij}}(\boldsymbol{r_{ij}})$$

with

$$\boldsymbol{\rho_i} = \sum_{i(j\neq i)} \boldsymbol{\rho_j^{a}}(\boldsymbol{r_{ji}})$$

$\boldsymbol{\rho_i}$ is the sum of the individual electron density provided by the other atoms of system, $\boldsymbol{F_i}(\boldsymbol{\rho_i})$ is the energy needed to integrate the atom i into the local electronic charge density $\boldsymbol{\rho_i}$, $\boldsymbol{\varphi_{ij}}(\boldsymbol{r_{ij}})$ is a pairwise electrostatic interaction between atom i and atom j separated by distance $\boldsymbol{r_{ji}}$ and $\boldsymbol{\rho_j^{a}}$ is the electron density provided by the atom j.

Here, we focused our attention on the diffusion of two islands of type Au_4 (tetramers) on Ag (110) surface $(Au_4; Au_4/Ag(110))$. The motivation of this study is to reveal the atomic processes responsible for the diffusion of the clusters and their coalescences. This study is performed through using molecular dynamics. Simulations were performed for a two-sided free surface substrate containing 5400 atoms for a six-layer symmetry (110) having (30 × 30) atoms in each layer. Here, the x and y axes are located in the plane of the surface, while the z axis is perpendicular to the surface. In the surface plane, the periodic boundary conditions have been made along the x and y directions, but not in the z direction, to ensure the deposition of adatms. Optimization of the geometry is performed at 0K by minimizing their energy and then relaxing the system for 40 µs under the micro-canonical conditions, as shown in the following diagram :

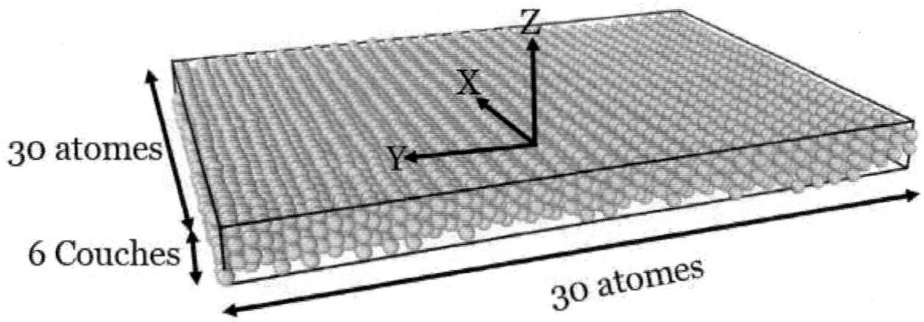

Fig. 1. *Representation of the substrate construction contains three areas; free region (2 layers at the top); Isothermal region (2 layers in the middle) fixed region (2 layers down).*

This part was the subject of an investigation on the dynamics of the two islands (tetramers) on a substrate of an orientation (110), and this for different forms (4S-4S, 4T-4T, 4N-4N) [49] and positions as well as separation distances Fig.2. The study in question was developped in a temperature range from 300K to 700K.

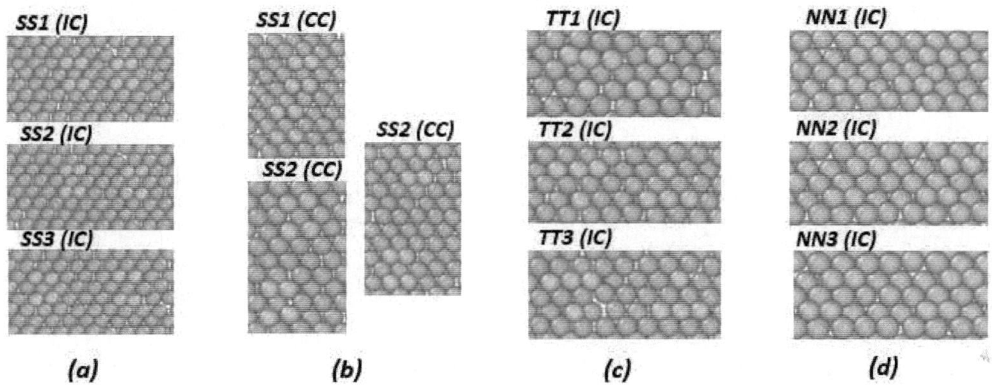

Fig. 2. *Representation of the different structures of the system* $Au_4; Au_4/Ag(110)$*: (a) two square islands separated by n site following the channel (IC) SSn, (b) two separate square islands by n site through the channel (CC) SSn, (c) two islands of form T separated by n site along the channel (IC) TTn (d) two islands of N form separated by n site following the channel (IC) NNn . (N = 1,2)*

3. *Results and discussion*

In what follows, we will scroll through all the results obtained during the static and dynamic study of the coalescence of two tetramers islands $Au_4; Au_4/Ag(110)$for temperatures ranging from 300K up to 700K.

3.1 Static study

In the first place, a study was made on the removal of an adatom attached to an island which can take several sizes from 2 to 16 atoms; our investigation also focused on the influence of island size on the displacement of the torn adatom for the systems (Au / Ag, Cu / Ag, Pt / Ag) at 300K.

Fig. 3b, which represents the displacement of an adatoma attached to islands of different sizes with respect to the substrate adsorption site. It shows that the Au adatoma has a mismatch of the order of 0.24% <1% [44], experienced a shift of 0.11 Å for the sizes of islands of less than 6 atoms, after this shift, the displacement increases with the enlargement of the islands size up to a value of 0.15 Å, beyond the islets a size of 12 atoms this shift is stabilized. However, for the two Pt / Ag and Cu / Ag systems that have an incoherent mesh mismatch greater than 1%, of 4.1% and 11.6% respectively, the adatom displacement attached to the 2 to 12 atom size islands increases. In a linear fashion from 0.23 to 0.35 Å, beyond a block size greater than 12 adatoms the displacement recognized a small increase to a value of 0.36Å. Thus, the cleavage between the adatoms and the substrate produces a displacement of the adatoms relative to the absorption site; this displacement increases with the increase of disagreement for a size of an island less than 12 atoms. However, in the opposite case, when the adatoms have larger sizes compared to that of the substrate, the displacement of the adatoms increases as well and when the sizes of the islands [48].

Figure 3 (a) shows the detachment energy of an adatoma at a cluster as a function of island size. It is shown that the energy of the Au / Ag system decreases exponentially with the increase of the size of the islands up to the size of 12 adatoms, after which it stabilizes towards a value of 0.255 eV, while the two other systems Pt / Ag and Cu / Ag have known values between 0.235 and 0.22 eV depending on the number of adatoms constituting these islands. By analogy, from the representation of the displacement of adatom as a function of the size of the islands, Figure 3 (b) gives the same description to the atomic tearing energies for the three systems as a function of the size of the system.

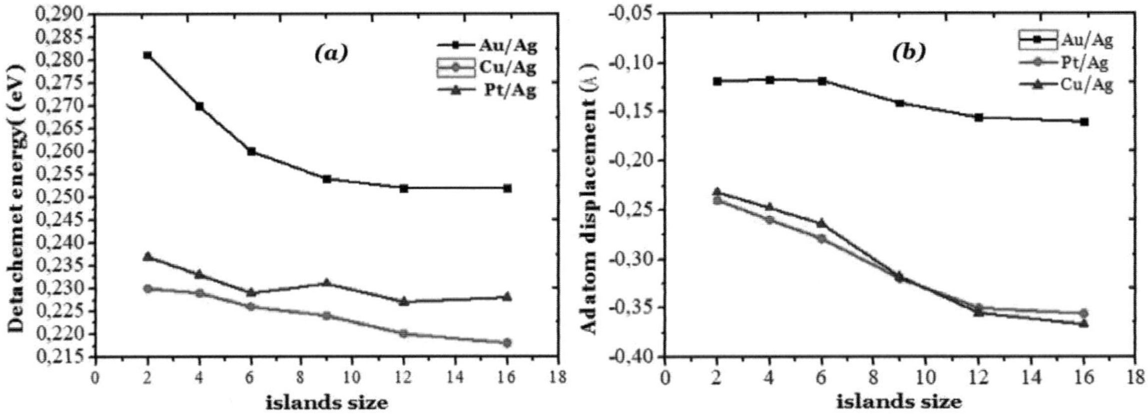

Fig. 3. Representation of the rate of detachment energy of an adatom bound to an island of different size (a), and the displacement of an adatom for different sizes of islands (b), for the systems (Au / Ag, Cu / Ag, Pt / Ag) at 300K

Figure 4 (a) shows the study of the formation energy as a function of the number of separation sites for two tetramers of SS (IC) and SS (CC) form; it is seen that for these two forms of SS , the formation energy is very important for the case of two adjacent islands of in-channel SS (IC) and which are 1.76 eV compared to the case cross-channel SS (CC) which is 1.55 eV. After a separation site, the formation energies coincide in the same values that are close to 1.26 eV fig.4a. On the other hand, for both forms of the TT and NN in-channel (IC) system, the process is reversed. The same energy value is found for the case of two adjacent islands having a value of 1.75 eV, which is explained by the same basic geometry of these two forms. After a separation site, in each case, the formation energy tends to a value of 1.17 eV for TT and 1.24 eV for the form NN. Therefore, the separation of the islands beyond an adsorption site gives a formation energy value depending on the geometric shape of these islands.

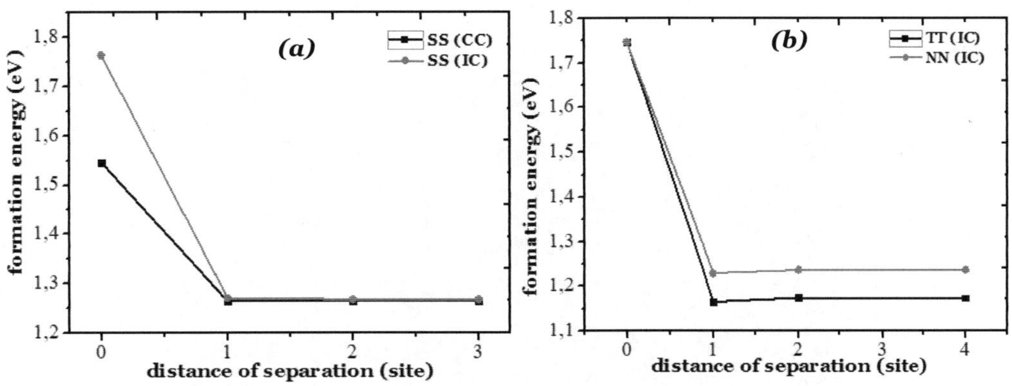

Fig. 4. *the formation energy according to the positions.*

3.2 Dynamic study

In this part, we will focus our attention on the dynamic evolution of the two islands (tetramers) of system$Au_4; Au_4/Ag(110)$, during a duration of 10ns at different temperatures included between 300K and 700K. This study focused on the forms of the cited systems, namely SS (IC), (CC) as well as TT and NN (IC). It is noted that in the range of 300K to 450K the islands of forms SS remains stable along the dynamic evolution, unlike other forms, from where the diffusion is triggered from 400 K, which explains the instability of TT and NN shapes compared to the SS form [26] [49].

Figures 5-8 show the evolution of SS-type islands at different temperatures, summarizing the set of possible training mechanisms for this configuration.

Fig. 5 represents two SS-in-Channel (IC) shaped islands at a temperature of 550 K. The connection was made by the migration of the two atoms from one island to the other successively along the channel fig. 5b-c, thereafter the third atom makes an exchange with an atom of the substrate which belongs to the channel on which the large islet is adsorbed, in order to facilitate the grouping by simple diffusion mechanisms, arriving at a partial homogeneous coalescence (islet) of 6 adatoms). However, for the temperatures of 600K and 650K, the exchange mechanisms become priority to make the connections between the islands in order to have a total heterogeneous coalescence.

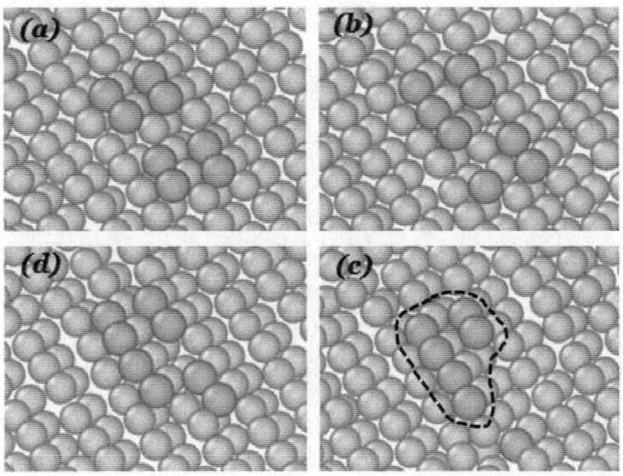

Fig. 5. *Dynamic evolution of two islands of an SS in Channel form at 550 K.*

Figure 6 shows that the two islands of SS system at a temperature of 550 K started with a connection between them through a mechanism of exchange of a dimer (fig. 6b) with the evolution in time this system forms an island with six heterogeneous atoms by two mechanisms of exchange and diffusion of a dimer fig. 6c-d.

IOP Conf. Series: Materials Science and Engineering **948** (2020) 012011 doi:10.1088/1757-899X/948/1/012011

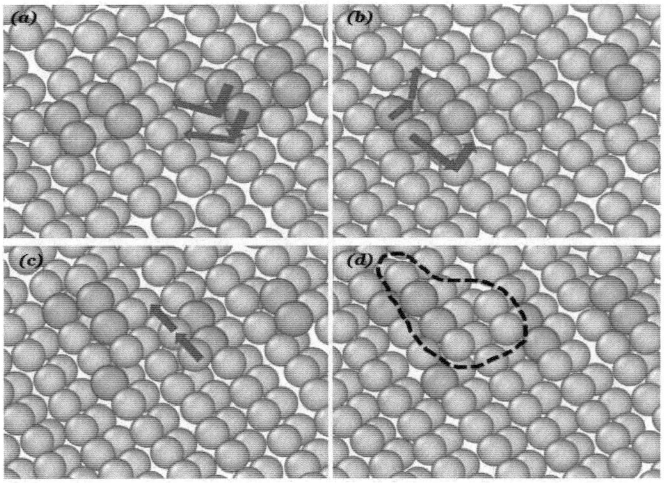

Fig. 6. *Dynamic evolution of two islands of an SS cross Channel form at 550 K.*

In fig.7, two islands having an SS cross-Channel shape at the connected temperature of 600 K are represented. This result shows that two atoms belonging to the same island making an exchange mechanism with two atoms of the substrate forming part of the channel on which the other island is adsorbed. In order to facilitate the grouping by simple diffusion mechanisms fig. 7b. Thus, the coalescence between the 2 islands SS (CC) begins with the mechanisms exchange with the atoms of the channel on which is adsorbed, falling in continuation in the mechanisms of coalescence in the case of SS (IC) systems.

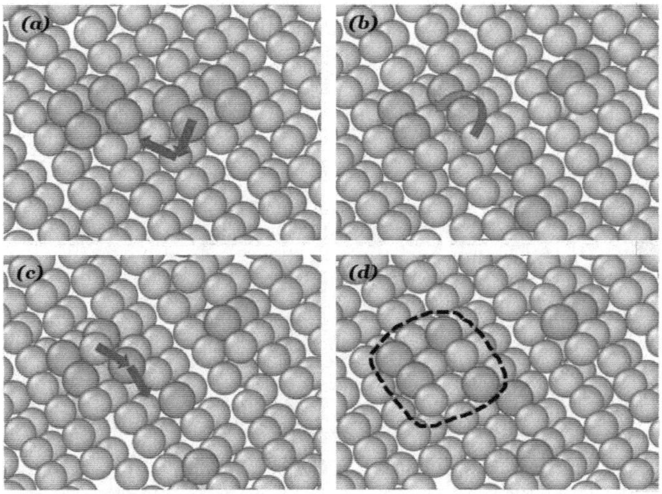

Fig. 7. *Dynamic evolution of two islands for SS cross channel form at 600 K.*

The results of fig. 8 represent the evolution of the SS1-SS1 (IC) structure at 650K. The coalescence between the two islands is done through a multitude of exchange-jump processes in order to arrive at a heterogeneous island of 8 adatoms (figs. (B), (c) and (d))

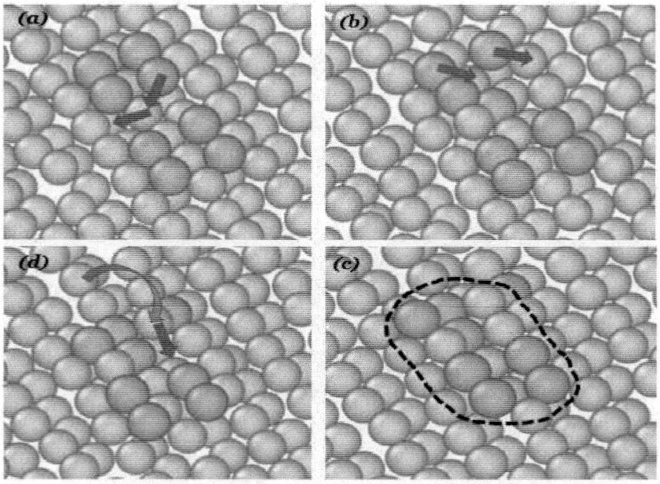

Fig. 8. *Dynamic evolution of two islands of an SS in Channel form at 650 K.*

Figures 9-10 show the evolution of NN and TT type islands at different temperatures, summarizing the set of possible training mechanisms for this configuration. Fig. 9 shows that at a temperature of 500K, the NN islands undergo an exchange, which gives a total heterogeneous coalescence fig.9, b, the evolution over time makes it possible to arrive at an island comprising nine heterogeneous atoms across the planet. Attachment of an adatome fig. 9c-d, for a temperature of 600K, we see that a total heterogeneous coalescence has taken place, with the construction of an island of eight atoms through mechanisms of jumps and exchanges (fig. 10 b-c-d).

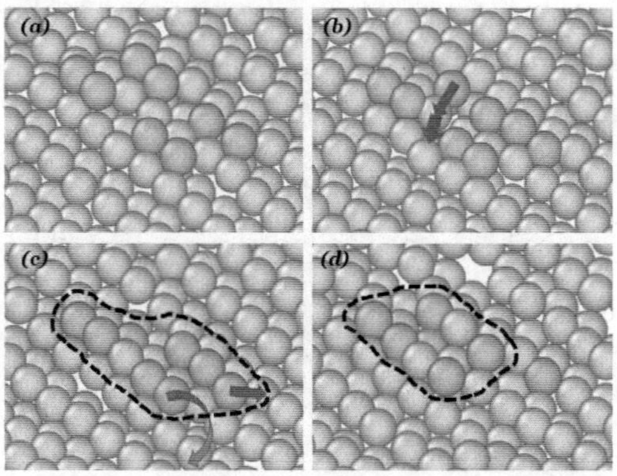

Fig. 9. *Dynamic evolution of two islets for NN in channel form at 500 K*

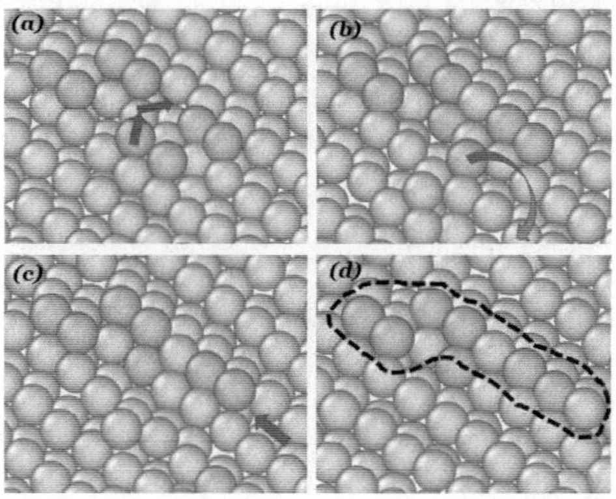

Fig. 10. *Dynamic evolution of two islands for NN in channel form at 600 K.*

Figures 11: b-c shows that at a temperature of 450K each one of two islands T attached to one adatom of the other island T to give a single island formed by two linear tetramers (the most stable form), we can say that we have a homogeneous total coalescence. The evolution of the temperature towards 600K makes it possible to distinguish almost the same behavior with the appearance of a homogeneous partial coalescence; whereas, for 550K fig. 12, through the same diffusion processes the TT form tends towards a heterogeneous partial coalescence fig. 12 c, but over time the island changes its structure fig.12 d

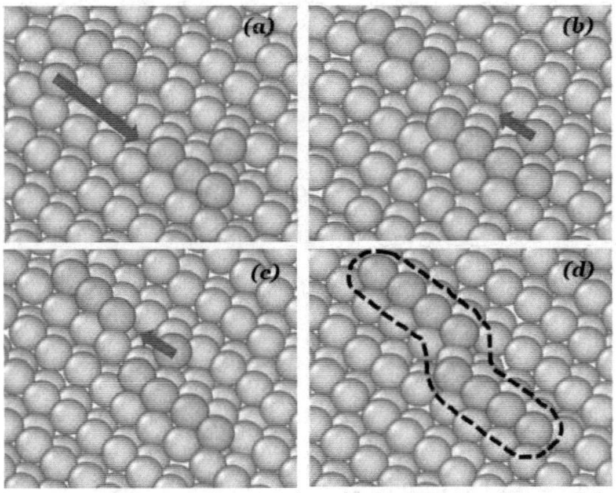

Fig. 11. *Dynamic evolution of two islands for TT in channel form at 450 K..*

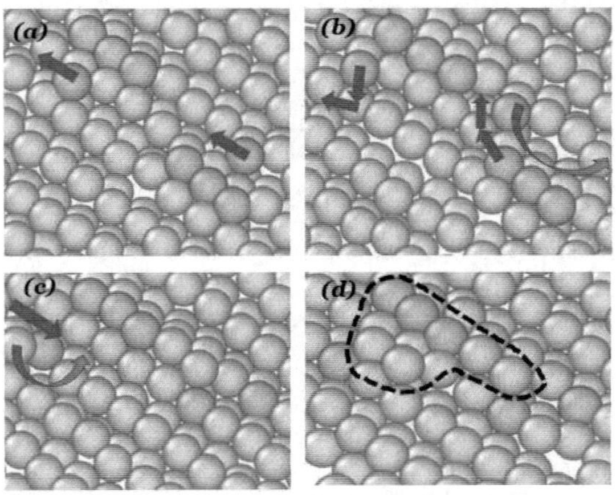

Fig. 12. *Dynamic evolution of two islands for TT in channel form at 550 K*

Fig. 13 (a) shows the evolution of the lifetime of the system of two islands of SS type deposited in-Channel as a function of temperature, the corresponding results showing that the system SS1 diffuses from the temperature 400K; however, both SS2 and SS3 systems withstand a temperature of 450K. However, during the increase of the temperature a decrease in the life of the system is observed, beyond 600K the three systems have a very short life (<20 ps).

The results of fig. 13 (b) show the evolution of the time of coalescence of the two-island SS type system deposited in-Channel as a function of temperature. This appearance shows that the three systems SS1, SS2 and SS3 obey successively coalescing at 400K, 450K and 500K. As the temperature increases, the time of coalescence decreases significantly; but beyond 600K, the coalescence time of the three systems becomes fast, which is reflected by a decrease in the slope of variation of the service life with respect to the temperature.

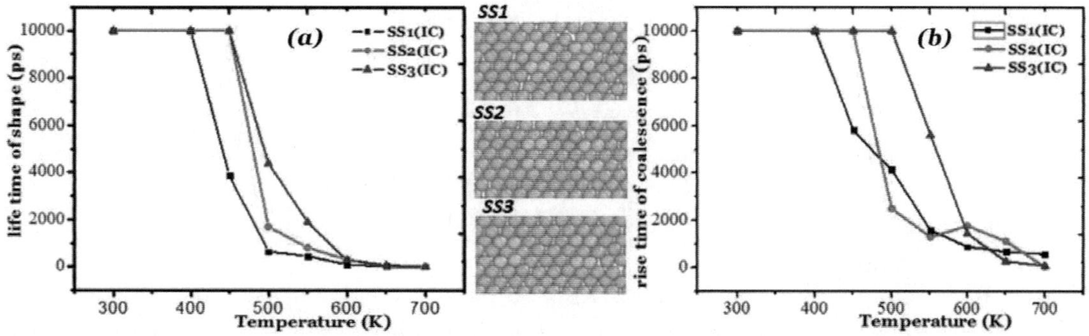

Fig. 13. *(a) Represents the system life of two islets (SS) as a function of temperature, (b) Represents the time of the appearance of coalescence of two islands (S) as a function of temperature for the system* $Au_4; Au_4/Ag(110)$ cross channel

In the case where the two SS islands are deposited in cross-Channel, fig. 14 (a), describes the evolution of the life of this form as a function of temperature. It shows that from a temperature of 400K SS1 system diffuses for a very short time, against both systems SS2 and SS3 withstands a temperature of 450K, as the temperature increases the duration of life decreases after 600K the three systems coalesce quickly.

Figure: 14 (b) represents the evolution of the coalescence time of the system of two islands of SS type deposited in-Channel as a function of the temperature; it is observed that from 500K the three systems SS1, SS2 and SS3 obey a coalescence. As the temperature increases, there is a decrease in the time when diffusion processes occur to coalesce the two islands.

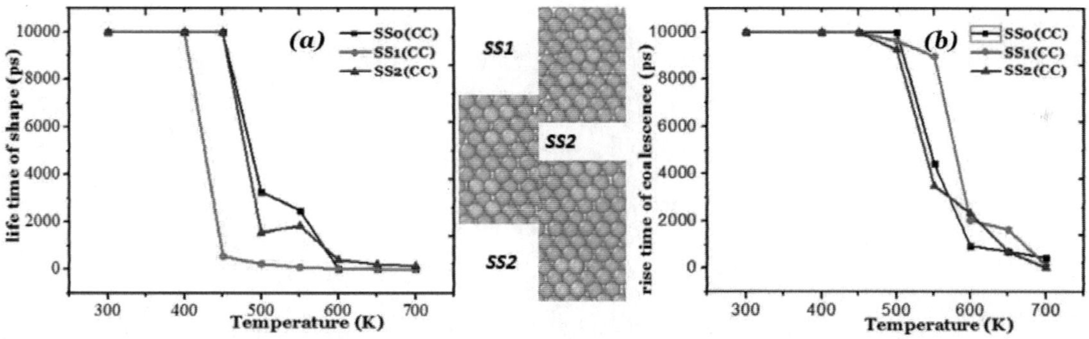

Fig. 14. *(a) Represents the system life of two islets (SS) as a function of temperature, (b) Represents the time of the appearance of coalescence of two islands (S) as a function of temperature for the system* $Au_4; Au_4/Ag(110)$ in channel

Figure: 15 (a) shows the evolution of the service life of the two TT-in-Channel island systems as a function of temperature, it is shown that from a temperature of 400K, the three TT1 systems , TT2 and TT3 diffuse by a sudden fall simultaneously during the increase of temperature, beyond 550K the systems its moments of coalescence becomes small (<60 ps). Fig. 15 (b) shows the evolution of the coalescence time of the two TT-in-Channel island system as a function of temperature. Note that from 400K the three systems TT1, TT2 and TT3 obey a coalescence during an increase in temperature, there is then a decrease in the duration of coalescence.

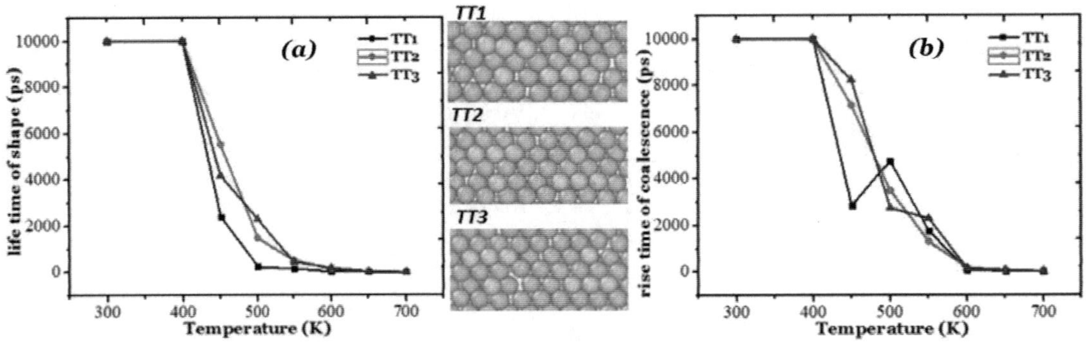

Fig. 15. *(a) Represents the system life of two islets (TT) as a function of temperature, (b) Represents the time of the appearance of coalescence of two islands (TT) as a function of temperature for the system $Au_4; Au_4/Ag(110)$ in channel.*

Figure 16 (a) shows the evolution of the lifetime of the system of two TT-in-Channel islands as a function of temperature, it is shown that starting from a temperature of 400K the three NN1 systems, NN2 and NN3 are spread by a sudden simultaneous fall. During the increase of the temperature, and beyond 550K the systems its moments of coalescence become smaller (<60 ps).

Figure 16 (b) represents the evolution of the time of the coalescence of the system of two islands of NN type in-Channel as a function of the temperature, we note that from 400K the three systems NN1, NN2 and NN3 obey to a coalescence. It is obvious that when the temperature increases a decrease in the duration of the coalescence is observed.

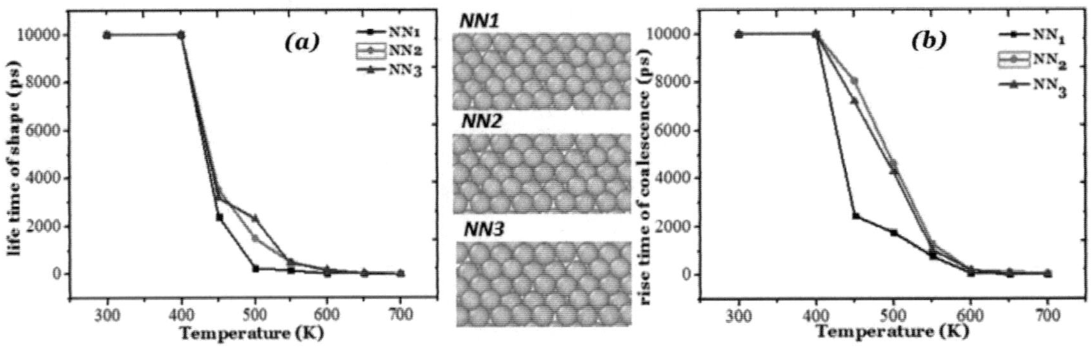

Fig. 16. *(a) Represents the life of each form of two islands (N) as a function of temperature, (b) Represents the birth time of the coalescence of two islands (N) as a function of temperature for ($Au_4; Au_4/Ag(110)$) in cannal.*

Figures 17-18 illustrates a comparison of the duration of diffusion / coalescence versus temperature for the four subject systems of our investigation. It has been shown that for the SS (IC) system, at the presence of a single SS1 separation site, the scattering is triggered at a very high rate at 450K by sweeping a duration of 3800 ps between diffusion and coalescence, the latter decreasing by increasing the temperature, marking periods more or less short since 600K. For both systems SS2 and SS3, the diffusion starts at 500K with different duration of the diffusion until the coalescence having values 2800 and 5700 ps respectively (fig.17-a).

As regards, the Cross-Channel (CC) process for the SS form, it can be seen that, in the presence of two separation sites SS2, the diffusion starts at a temperature of 450K with a diffusion / coalescence duration of 9300 ps. This duration decreases gradually as the temperature increases. For both SS1 and SS3 systems, the periods required to reach coalescence after diffusion are 6900 and 7800 Ps respectively. In the same way, this duration decreases by increasing the temperature (fig.17-b)

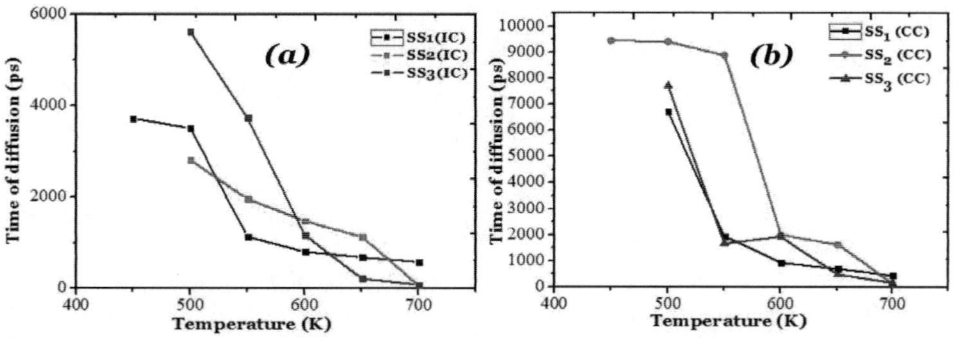

Fig. 17. *(a) Represents the diffusion time of the two island system (SS (IC)) as a function of temperature, (b) Represents the diffusion time of the two island system (SS (CC)) as a function of temperature, , for the system* $Au_4; Au_4/Ag(110)$.

In the TT and NN form systems(fig.18a-b), diffusion occurs at a temperature of 450K for the three separation sites TTn and NNn with n = 1.2 and 3. The diffusion / coalescence time is included in the range 3600-4700 ps for the form TTn, as for the form NNn said duration diffusion / coalescence, fits in the range 2400-4700 ps. This duration decreases by increasing the temperature and marks very small values (<90 ps) beyond 600K.

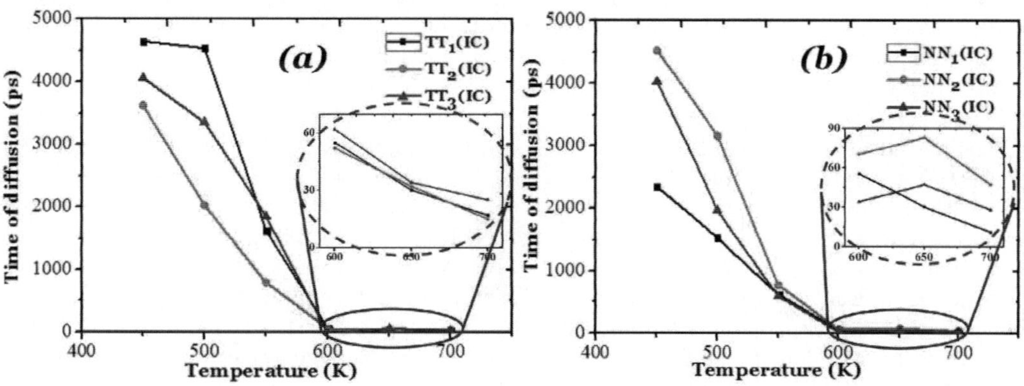

Fig. 18. *(a) Represents the diffusion time of the two island system (TT (IC)) as a function of temperature, (b) Represents the diffusion time of the two island system (NN (IC)) as a function of temperature, for the system* $Au_4; Au_4/Ag(110)$.

It can be concluded, then, that the temperature has a remarkable effect on the duration of diffusion / coalescence. The increase in temperature causes a decrease in the duration in question which intensifies beyond 600K and this for all systems studied. A slightly moderate decrease between 450 and 600K explains the good activity of the coalescence of islands in this temperature range.

4. Conclusion

In the present work, we studied by the molecular dynamics method, the coalescence of two tetrameric islands $Au_4; Au_4/Ag(110)$ of the three forms SS, TT and NN. These islands are deposited at different spacings following and through the channel in a temperature range from 300K to 700K. From the results obtained, it can be concluded that the detachment energy of an adatom of a cluster is decreased by increasing the size of the island. Thus, the adatom displacement attached to islands of varying size from 2 to 12 adatoms increases proportionally with its size in a way that reduces the spacing between the island and the attachment adatom producing an island restriction. It is noted that the separation of islands beyond an absorption site gives a value of formation energy constant and dependent on the geometric shape of these islands. This dynamic study shows that at low temperature (<400K for TT, NN (IC) and <450K for SS (IC), SS (CC)) the system remains stable. On the other hand, in the temperature range between 450 and 650 K, one meets the phenomenology of the homogeneous partial coalescence induced by mechanisms of jumps, thus, the increase of the temperature in this interval favors the mechanisms of exchange which generates a coalescence total heterogeneous.

REFERENCES

[1] C. M. Chang, C. M. Wei, and S. P. Chen, "Self-diffusion of small clusters on fcc metal (111) surfaces," *Phys. Rev. Lett.*, vol. 85, no. 5, p. 1044, 2000.
[2] H. Yang, Q. Sun, Z. Zhang, and Y. Jia, "Upward self-diffusion of adatoms and small clusters on facets of fcc metal (110) surfaces," *Phys. Rev. B - Condens. Matter Mater. Phys.*, vol. 76, no. 11, pp. 1–7, 2007, doi: 10.1103/PhysRevB.76.115417.
[3] M. Dardouri, A. Hassani, A. Hasnoui, and Y. Boughaleb, "Kinetic Monte Carlo Simulation of Au Monolayer Evolution on Cu (110)," *Phys. STATUS SOLIDI B*, vol. 1800404, no. 110, pp. 1–10, 2018, doi: 10.1002/pssb.201800404.
[4] L. Xu, G. Henkelman, C. T. Campbell, and H. Jónsson, "Pd diffusion on MgO(1 0 0): The role of defects and small cluster mobility," *Surf. Sci.*, vol. 600, no. 6, pp. 1351–1362, 2006, doi: 10.1016/j.susc.2006.01.034.
[5] K. Morgenstern, I. Festkörperphysik, A. Oberflächen, and U. Hannover, "Fast scanning tunnelling microscopy as a tool to understand changes on metal surfaces□: from nanostructures to single atoms *," *phys. stat. sol. (b)* 242, No. 4, 773–796 (2005) / DOI 10.1002/pssb.200440002.
[6] M. Valden, X. Lai, and D. W. Goodman, "Onset of catalytic activity of gold clusters on titania with the appearance of nonmetallic properties," *Science*, vol. 281, no. 5383, pp. 1647–1650, 1998, doi: 10.1126/science.281.5383.1647
[7] T. Lee and R. M. Dickson, "Discrete two-terminal single nanocluster quantum optoelectronic logic operations at room temperature," *PNAS*, vol. 100, no. 6, pp. 3043–3046, 2003, doi/10.1073/pnas.0635474100
[8] L. Ventelon and F. Willaime, "Core structure and Peierls potential of screw dislocations in α -Fe from first principles□: cluster versus dipole approaches," *Journal of Computer-Aided Materials Design,* vol.14, no. 2007, pp. 85–94, 2008, doi: 10.1007/s10820-007-9064-y.
[9] P. Taylor, F. Ercolessi, M. Parrinello, and E. Tosatti, "Simulation of gold in the glue model," *philosophical magazine A*, 1988, vol. 58, No. 1,213-226, doi: 10.1080/01418618808205184.
[10] H.-J.Freund, M.Bäumer and H.Kuhlenbeck, "Catalysis and Surface Science□: What Do We Learn from Studies of Oxide- Supported Cluster Model Systems," *Advances in Catalysis,* vol. 45, pp. 333–384, 2000. doi.org/10.1016/S0360-0564(02)45017-1
[11] G. A. Somorjai, "Modern Surface Science and Surface Technologies□: An Introduction," *Chemical Reviews*. vol. 2665, no. 95, pp. 1223–1236, 1996. //doi.org/10.1021/cr950234e.
[12] A. Sabik, C. Zaum, K. Morgenstern, and W. Kamin, "Complex Surface Di ff usion Mechanisms of Cobalt Phthalocyanine Molecules on Ag (100) Graz y, " *J. Am. Chem. Soc*, no. 137, 47, pp.14920–14929, 2015, doi: 10.1021/jacs.5b08001.
[13] Y. Han, C. R. Stoldt, P. A. Thiel, and J. W. Evans, "Ab Initio Thermodynamics and Kinetics for Coalescence of Two-Dimensional Nanoislands and Nanopits on Metal (100) Surfaces," *J. Phys. Chem. C*, vol. 120, no. 38, pp. 21617–21630, 2016, doi: 10.1021/acs.jpcc.6b07328.
[14] C. R. Stoldt, A. M. Cadilhe, C. J. Jenks, J. Wen, J. W. Evans, and P. A. Thiel, "Evolution of Far-From-Equilibrium Nanostructures Formed by Cluster-Step and Cluster-Cluster Coalescence in Metal Films," *Physical review*, Vol. 81 no. c, pp. 5–8, 1998. doi.org/10.1103/PhysRevLett.81.2950.
[15] P. A. Thiel, M. Shen, D. Liu, and J. W. Evans, "Coarsening of Two-Dimensional Nanoclusters on Metal Surfaces," *J. Phys. Chem. C* 113, 13, pp. 5047–5067, 2009.
[16] T. Kumagai *et al.*, "Water clusters on Cu (110): Chain versus cyclic structures, " *J. Chem. Phys.* 134, vol. 024703, no. 110, 134, 2011, doi: 10.1063/1.3525645.
[17] U. T. Ndongmouo and F. Hontinfinde, "Diffusion and growth on fcc(110) metal surfaces: A computational study," *Surf. Sci.*, vol. 571, no. 1–3, pp. 89–101, 2004, doi: 10.1016/j.susc.2004.08.010.
[18] F. Montalenti and R. Ferrando, "Dimers diffusion on (110) (1 × 2) metal surfaces," *Surface Science vol. 432, pp. 27–36, 1999.* doi.org/10.1016/S0039-6028(99)00496-3

[19] H. El Azrak *et al.*, "Investigation of fcc and hcp island nucleated during homoepitaxial growth of copper by molecular dynamics simulation," *Superlattices Microstruct.*, vol. 127, pp. 118–122, 2019, doi: 10.1016/j.spmi.2017.12.056.

[20] A. Hassani *et al.*, "Superlattices and Microstructures Statistical investigations of the fi lm-substrate interface during aluminum deposition on Ni (111) by molecular dynamics simulation," *Superlattices Microstruct.*, vol. 127, pp. 80–85, 2019, doi: 10.1016/j.spmi.2018.03.008.

[21] H. Y. Lin *et al.*, "Self-Organized Growth of Nanopucks on Pb Quantum Islands," *physical review letters,* vol. 136101, no. APRIL, pp. 2–5, 2005, doi: 10.1103/PhysRevLett.94.136101.

[22] T. Y. Fu, Y. J. Hwang, and T. T. Tsong, "Structure and diffusion of Pd clusters on the W(1 1 0) surface," *Appl. Surf. Sci.*, vol. 219, no. 1–2, pp. 143–148, 2003, doi: 10.1016/S0169.

[23] T. Y. Fu and T. T. Tsong, "Structure and diffusion mechanism of Ir and Rh tetramers on Ir (001) surfaces," *Surf. Sci.*, vol. 482–485, pp. 1249–1254, 2001, doi: 10.1016/S0039-6028(01)00906-2.

[24] D. Chen, W.Hu, J. Yang, H.Deng, L. Sun and F.Gao, "Diffusion of tungsten clusters on tungsten (110) surface." *Eur. Phys. J.* B , vol. 68, pp479–485, 2009, DOI: 10.1140/epjb/e2009-00100-1

[25] J.-S. Filhol, D. Simon, and P. Sautet, "Surface phase stability diagram for Pd deposits on Ni (110): A first-principles theoretical study," *Phy. Rev. B*, vol. 64, 085412, 2001, doi: 10.1103/PhysRevB.64.085412.

[26] T. Fu and T. T. Tsong, "Structure and diffusion of small Ir and Rh clusters on Ir (001) surfaces," *Surf. Sci.*, vol. 421, pp. 157–166, 1999. doi.org/10.1016/S0039-6028(98)00847-4

[27] H. Bulou and C. Massobrio, "Mechanisms of exchange diffusion on fcc(111) transition metal surfaces," *Phys. Rev. B - Condens. Matter Mater. Phys.*, vol. 72, no. 20, pp. 2–7, 2005, doi: 10.1103/PhysRevB.72.205427.

[28] M. Karimi, T. Tomkowski, G; Vidali, and O. Biham, "Diffusion of Cu on Cu surfaces," *Phy. Rev. B,* vol. 52, no. 7, 1995. doi.org/10.1103/PhysRevB.52.5364.

[29] E. Elkoraychy, K. Sbiaai, M. Mazroui, Y. Boughaleb, and R. Ferrando, "Numerical study of hetero-adsorption and diffusion on (100) and (110) surfaces of Cu, Ag and Au," *Surf. Sci.*, vol. 635, no. 100, pp. 64–69, 2015, doi: 10.1016/j.susc.2014.12.009;

[30] A. Karim, A. N. Al-rawi, A. Kara, and T. S. Rahman, "Diffusion of small two-dimensional Cu islands on Cu (111) studied with a kinetic Monte Carlo method," *Phy. Rev. B,* no. 111, pp. 1–11, 2006, doi: 10.1103/PhysRevB.73.165411.

[31] Z. Rehman, S. S. Hayat, "Thermal Diffusion Dynamic behavior of two-dimensional Ag-small clusters on Ag(1 1 1) surface," *Surface Review and Letters,* vol. 22, no. 5, pp. 1–11, 2015, doi: 10.1142/S0218625X15500675.

[32] F. Eddiai, M. Dardouri, A. Hassani, M. Badawi, K. Sbiaai, and A. Hassnaoui, " Structure, stability, and surface diffusion of clusters: Pt 4 /Cu (110) AND Au 4 /Ag (110) surface by molecular dynamics ," *Eur. Phys. J. Appl. Phys.*, vol. 91, no. 3, p. 31302, 2020, doi: 10.1051/epjap/2020200185.

[33] K. Sbiaai, Y. Boughaleb, M. Mazroui, A. Hajjaji, and A. Kara, "Energy barriers for diffusion on heterogeneous stepped metal surfaces: Ag/Cu(110)," *Thin Solid Films*, vol. 548, pp. 331–335, 2013, doi: 10.1016/j.tsf.2013.09.064

[34] K. Sbiaai, E. Elkoraychy, M. Mazroui, Y. Boughaleb, and I. Matrane, "Energetics and heterodiffusion of Cu on Ag(110) stepped surfaces," *Surf. Interface Anal.*, vol. 47, no. 9, pp. 855–862, 2015, doi: 10.1002/sia.5786.

[35] M. Imran, R. M. Arif, M. A. Javid, M. A. Sattar, S. A. Ahmad, and A. Rehman, "A study of surface diffusion of ternary (Cu - Ag - Zr) adatoms clusters for applications in thin film formation," *Surf. Interface Anal.*, vol.51. no. July 2018, pp. 1–9, 2019, doi: 10.1002/sia.6608.

[36] S. R. Acharya, S. I. Shah, and T. S. Rahman, "Diffusion of small Cu islands on the Ni(111) surface: A self-learning kinetic Monte Carlo study," *Surf. Sci.*, vol. 662, no. 111, pp. 42–58, 2017, doi: 10.1016/j.susc.2017.03.012.

[37] A. Kotri, E. El, M. Mazroui, and I. Achik, "Hetero-diffusion of small clusters on Ag (111) surface," *The European Physical Journal Plus* vol. 133, no. 7, pp. 270, 2018, doi: 10.1140/epjp/i2018-12088-8

[38] A. F. Voter, "Classically exact overlayer dynamics: Diffusion of rhodium clusters on Rh(100)," *Phys. Rev. B*, vol. 34, no. 10, pp. 6819, 1986. doi.org/10.1103/PhysRevB.34.6819.

[39] S. S. Hayat, Z. A. Shah, I. Technology, and R. Y. Khan, "A study of dynamical evolution of small two-dimensional Copper islands' diffusion on Ag(1 1 1) surface and observed surface effects," *Modern Physics Letters B* , vol. 31, no. 33., pp. 1–18, 2017, doi: 10.1142/S021798491750316X.

[40] F. Liu *et al.*, "Surface Science Atomic self-diffusion behaviors relevant to 2D homoepitaxy growth on stepped Pd (001) surface," *Surf. Sci.*, vol. 624, pp. 89–94, 2014, doi:10.1016 /j.susc.2014.01.011.

[41] R. Poteau, J. L. Heully, and F. Spiegelmann, "Structure, stability, and vibrational properties of small silver cluster," *Zeitschrift fur Phys. D-Atoms Mol. Clust.*, vol. 40, no. 1, pp. 479–482, 1997, doi: 10.1007/s004600050257.

[42] R. Wang and K. A. Fichthorn, "Kinetics of intermixing in Au/Ag(110) heteroepitaxy: A molecular-dynamics study," *Phys. Rev. B*, vol. 51, no. 3, pp. 1957–1960, 1995. doi.org/10.1103/PhysRevB.51.1957

[43] Michael I. Haftel, Mervine Rosen, Tameika Franklin, and Matthew Hettermann, "Molecular Dynamics Observations of Interdiffusion and Stranski-Krastanov Growth in the Early Film Deposition of Au on Ag(110)," *Physical Review Lett.*, vol. 72, no. 12, 1994.

[44] Michael I. Haftel and Mervine Rosen, "Molecular-dynamics description of early film deposition of Au on Ag(110)," *Phys. Rev. 8*, vol. 51, no. 7, 1995. doi.org/10.1103/PhysRevB.51.4426

[45] S. M. Foiles, M. I. Baskes, and M. S. Daw, "Embedded-atom-method functions for the fcc metals Cu, Ag, Au, Ni, Pd, Pt, and their alloys," *Phys. Rev. B*, vol. 33, no. 12, pp. 7983–7991, 1986, doi: 10.1103/PhysRevB.33.7983.

[46] C. L. Liu, J. M. Cohen, J. B. Adams, and A. F. Voter, "EAM study of surface self-diffusion of single adatoms of fcc metals Ni, Cu, Al, Ag, Au, Pd, and Pt," *Surf. Sci.*, vol. 253, no. 1–3, pp. 334–344 , 1991, doi: 10.1016/0039-6028(91)90604-Q.

[47] I. Baskes, "Molecular dynamics simulation of Co thin films growth on Cu(001) N.," *Surf. Sci.*, vol. 400, no. 12, pp. 54–62 , 1984. doi.org/10.1016/S0039-6028(97)00841-8

[48] A. Hassani, A. Makan, K. Sbiaai, A. Tabyaoui, and A. Hasnaoui, "Molecular dynamics study of growth and interface structure during aluminum deposition on Ni(1 0 0) substrate," *Appl. Surf. Sci.*, vol. 349, no. 100, pp. 785–791, 2015, doi: 10.1016/j.apsusc.2015.05.076.

[49] F. Eddiai, M. Dardouri, A. Hassani, H. El Azrak, K. Sbiaai, and A. Hassnaoui, "Structure Stability and Diffusion Mechanism of Au Tetramers on Ag (110) Surface," *Sens. Lett.*, vol. 16, no. 5, pp. 386–390, 2018, doi: 10.1166/sl.2018.3964

IOP Conf. Series: Materials Science and Engineering **948** (2020) 012012 doi:10.1088/1757-899X/948/1/012012

Techno-economic investigation of electricity and hydrogen production from wind energy in Casablanca, Morocco

Samir Touili [1], Ahmed Alami Merrouni [2], Youssef El Hassouani [1], Abdel-illah Amrani [1,*], Alae Azouzoute [2]

1 : ESIM, Département de Physique, Faculté des Sciences et Techniques, Université Moulay Ismail, Boutalamine BP 509, 52000 Errachidia, Morocco.

2: Research Institute for Solar Energy and New Energies (IRESEN), Green Energy Park, Benguerir, Morocco.

Corresponding author: a.amrani@umi.ac.ma

Abstract: The aim of this study is to investigate the technical and economic potential of electricity and hydrogen production in Casablanca, Morocco. For this reason, we simulated the performance of a 4.2 MW_p wind turbine if installed in Casablanca. The results show that the electricity and hydrogen production varies greatly through the year due to the high fluctuation in wind speed. The annual electricity and hydrogen production is 29.16 GWh and 555 Tons respectively. As for the levelized cost of production, the $LCOE$ was found to be 0.24 \$/kWh and the LCO_{H_2} were equal to 13.52 \$/Kg.

Keywords: Wind energy; Electricity production; Hydrogen production; Simulation; Economic analysis.

1. Introduction

Wind power technologies has known a significant development in the last decade. Indeed, between 2008 and 2017 several aspects of the technologies have been improved such as higher hub heights and larger areas swept by blades. This development was accompanied with a reduction of the wind turbines prices. The global installed wind capacity has increased as well, in fact from 2000 to 2016 the cumulative installed capacity increased by an annual rate of 15% to reach 467 GW by the end of 2016 [1].

In addition to electricity production, wind energy can also be used to generate hydrogen via the water electrolysis process, which consist of splitting the water molecule under the effect of electric current. In this context, the wind turbine is coupled with an electrolyzer to achieve that goal. The water electrolysis process has many advantages: it is a mature and well-established technology with a high energy efficiency, it produces hydrogen with high efficiency, and it has a has a perfect compatibility with renewable energy such as wind energy [2]. Therefore, the hydrogen produce in this case can be considered as "green and clean" and can be regarded as a sustainable substituent for fossil fuels.

Indeed, hydrogen can generate 3 time more energy than all other fuels, it can be used in most application that require fossil fuels and it release only water during its combustion [2]. Besides, a wide range of applications in the industrial sector uses hydrogen for their process such is the case for the fertilizer industry, food industry, petroleum industry, glass purification etc.[3].

Regarding the wind energy status in Morocco, the country has set an objective to produce 14% of its electricity need from wind energy by 2020. This ambitious goal was steam from the fact the Morocco is a fossil fuel deficient country (import more than 96% of its energy needs), but in the other hand, has a high wind energy potential especially in the coastline where the mean wind speed can reach up to 10 m/s [4–7]. Accordingly, various large scale wind farms are currently being constructed under a ten year project called the Moroccan Integrated Wind Energy Project (MIWEP) which aim to expand the wind energy capacity to 2000 MW by 2020 [6].

Many researchers has investigated the electricity and hydrogen production from wind energy in various locations. For instance, Mohsin et al. [8] conducted an assessment of wind energy potential for electricity and hydrogen production in Karachi region, Pakistan. They found a levelized cost of electricity in the range of 0.0844-0.0864 $/kWh and a supply cost of hydrogen between 5.30 $/Kg and 5.80 $/kg. Genç et al. [9] investigated the hydrogen production for 3 wind turbines with different hub height (50, 80 and 100 m) in Pınarbaşı-Kayseri, Turkey. The results shows that with the increase of the hub height, the hydrogen production increase as well while the cost of the production decrease. Also, the maximum hydrogen production and the minimum cost were found to be 14192 Kg/year and 8.5 $/Kg respectively, both in the case of the 100 m turbine. Siyal et al. [10] estimated the potential of hydrogen production from wind energy in Sweden. The authors found that a 25,580 ktons could be produce annually, which is equivalent to 860 TWh of energy. Rahmouni et al. [11] found that the potential of electricity and hydrogen production from wind energy in Algeria is in the range of 48-1074.8 GWh/km^2 and 842.62- 18447.6 Tons/km^2/year respectively.

In this study, we investigated the electricity and hydrogen production from wind energy in Casablanca. To this end, we simulated the performance of a 4.2 MW_p wind turbine under the climate of Casablanca. To achieve that goal first, we briefly describe the simulated system and the tool used, then the methodology followed for the estimation of the hydrogen production as well as the cost of production of electricity and hydrogen, and finally we present the results.

2. Methodology

In this part, we will first describe the general characteristics of the wind turbines than the technical data of the one selected for simulation, afterwards, we present the simulation software for the electricity production then the methodology to estimate the hydrogen production, and finally the metrics used for the economic analysis.

2.1 System description

The electricity production from wind energy is directly proportional to the wind speed, and in order to convert wind energy into electricity, wind turbines are used [12]. Depending on the axe of rotation, wind turbines can be divided into two categories: horizontal axis wind turbines (HAWTs) and vertical axis wind turbines (VAWTs). The HAWTs are the most used by the wind industry and almost entirely for the utility scale wind market since they offer higher efficiency and energy production than the VAWTs [1,12,13].

Typically, wind turbines are made of the following components [13]:

- Rotor: generally, its composed of three large blades similar to the airplane wing.
- Nacelle: connected to the rotor at the top of the wind turbine, and can rotate in accordance with the wind direction for a maximal harness of wind energy. In addition, it carries the principal technical parts like gearbox, and generator.
- Gearbox: increase the rotor speed to meet the generator requirement to produce electricity.
- Generator: used for the conversion of the mechanical energy from the rotor to electricity.

The main parameters that influence the electricity production from the wind turbine are height and the rotor diameter. Also, the wind turbine require a minimal wind speed to start generating electricity which is generally ranges between 3 and 5 m/s and for a wind speed around 25 m/s the wind turbine cut out, as for the maximum power production is reached for 11-12 m/s [1].

In this study, we simulated the performance of the Enercon E-126 EP4 under the climate of Casablanca, Morocco. The characteristics of the wind turbine are presented in table 1.

Table 1. Characteristics of the wind turbine (Enercon E-126 EP4)

Hub height (m)	135
Rated power (kW)	4200
Swept area (m^2)	12667
Cut-in speed Vcin (m/s)	3
Rated speed Vr (m/s)	14
Cut-out speed Vcout (m/s)	25

2.2 Simulation software

Simulation is a widely used method for the prediction of the performance of renewable energy systems. In this study, the Green Energy System Analysis Tool (Greenius) developed by the German Aerospace Center (DLR) is used to simulate the electricity production from the wind turbine. Greenius is a powerful tool for the simulation of renewable energy projects such as solar and wind that allow the entire definition of the energy system by the user, and provide a realistic simulation results in different time steps [14]. Accordingly, many studies has used Greenius to simulate various energy systems we cite for instance [15–19].

2.3 Hydrogen production

As mention above the electricity generated by the wind turbine can be used to drive the water electrolysis process, which is achieved by the mean of an electrolyzer. Currently, the Proton Exchange Membrane (PEM) is one of the most developed and used technology [20]. This due to PEM long life cycle, high efficiency and it compatibility with the fluctuation of electricity generation from intermittent sources such is the case for wind energy [2].

The hydrogen production is directly related to amount of electricity generated and it can be estimated with the following formula [2,11,19]:

$$M_{H_2} = \frac{E \times \eta_{ele}}{HHV_{H_2}} \quad \text{(Equation 1)}$$

M_{H_2} (Kg) is the amount of hydrogen produced; E (kWh) is the electricity production from the wind turbine; η_{ele} is the electrolyzer efficiency (75%); HHV_{H_2} is the hydrogen higher heating value (39.4 kWh/kg).

2.4 Economic analysis

In order to quantify the cost of electricity and hydrogen production, the levelized cost of production metric is used in this study. This metric is widely used for the economic assessment of renewable energy projects and it's computed from the technical and economic input of the energy system. In addition, it is very helpful when conduction a techno-economic investigation or for comparison of renewable energy systems between different locations.

The levelized cost of electricity $LCOE$ ($/kWh) is define as follow [2,21]:

$$LCOE = \frac{\sum_{i=0}^{N}\left(C_E(1+T)^{-i}\right)}{\sum_{i=0}^{N}\left(E_i(1+T)^{-i}\right)} \quad \text{(Equation 2)}$$

E_i (kWh/year) is the electricity production from the wind turbine in the year *i*. N and T are the project lifetime and the discount rate taken as 20 years and 6% respectively. C_E ($) represents the investment cost of the wind turbine and it's calculated by equation 3:

$$C_E = C_{wt} + C_{in} + C_{O\&M} \quad \text{(Equation 3)}$$

C_{wt} is the cost of the wind turbine considered as 1150 $/kW$_p$, C_{in} is the installation cost which represent 20% of the wind turbine cost, while $C_{O\&M}$ is the maintenance & operation cost and it represent 25% of the turbine cost [22–24].

Regarding the levelized cost of hydrogen LCO_{H_2} ($/Kg), its expressed by equation 4:

$$LCO_{H_2} = \frac{\sum_{i=0}^{N}\left((C_E + C_{elec})(1+T)^{-i}\right)}{\sum_{i=0}^{N}\left(M_{H_2,i}(1+T)^{-i}\right)} \quad \text{(Equation 4)}$$

$M_{H_2,i}$ (Kg/year) is the hydrogen production in the year *i*, C_{elec} ($) is the electrolyzer investment cost, which is given by equation 5:

$$C_{elec} = C_{el} + C_{O\&M} + C_{rep} \quad \text{(Equation 5)}$$

C_{el} represent the capital cost of the electrolyzer and its calculated by the following formula [25]:

$$C_{el} = C_{el,u} \times \left(\frac{M_{H_2} K_{el,th}}{8760\eta_{el}}\right) \quad \text{(Equation 6)}$$

$C_{el,u}$ is the unit price of the electrolyzer (368 $/kW$_e$), $K_{el,th}$ (52.5 kWh/Kg) is the theoretical energy required by the electrolyzer and η_{el} is the electrolyzer efficiency. The operation & maintenance ($C_{O\&M}$) and replacement (C_{rep}) cost are taken to be 2% and 25% respectively [2,25–27].

3. Results and discussion

3.1 Wind resources assessment

For wind energy based projects, the knowledge of the wind speed and direction frequencies are an important step toward evaluating the wind resource availability in a specific site. This due to the frequent changes in the wind direction caused by geographical effects and atmosphere circulation [6].

One of the most used methods for the assessment of wind resources is the wind rose diagram, because it provides both the wind direction and speed frequencies in a single polar diagram. The wind rose of Casablanca is presented in figure 1.

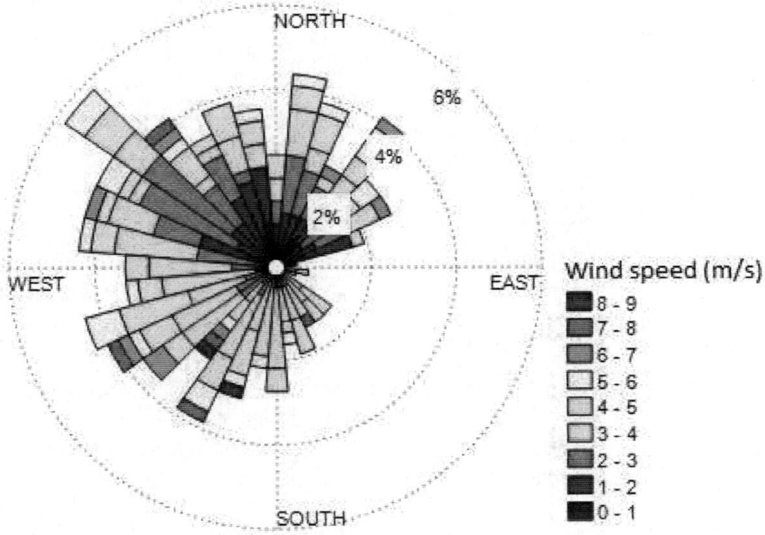

Figure 1. Casablanca wind rose.

As it can be seen from the wind rose, the most dominant wind direction is the west both west-north and west-south, also the north-east direction is observed with a relatively high frequency. As for the wind speed the most frequent values are in the range of 3-5 m/s, however for a better understanding of the wind speed variation we plotted the hourly wind speed frequency through the year at 10 m and at the hub height (135 m) in figure 2 and 3 respectively.

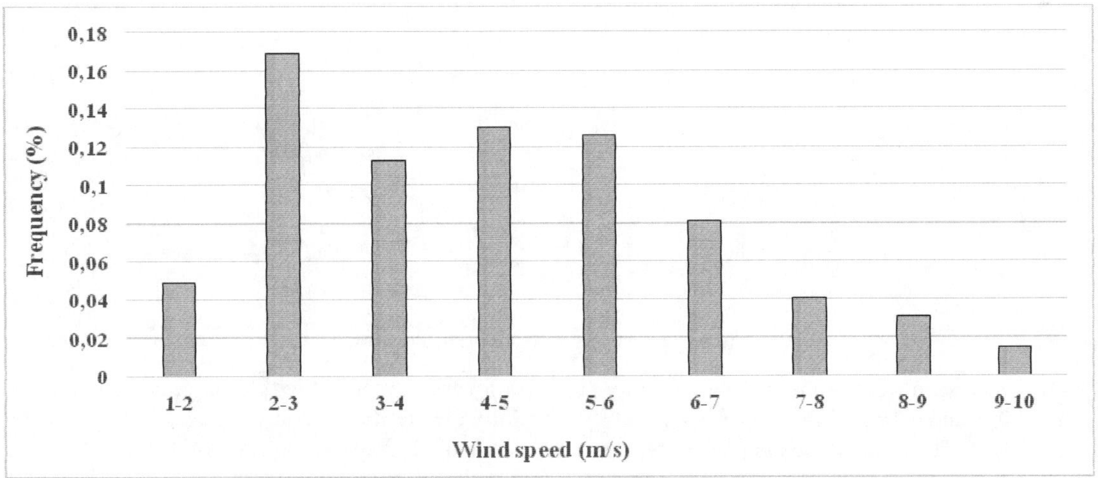

Figure 2. Wind speed frequency at 10 m.

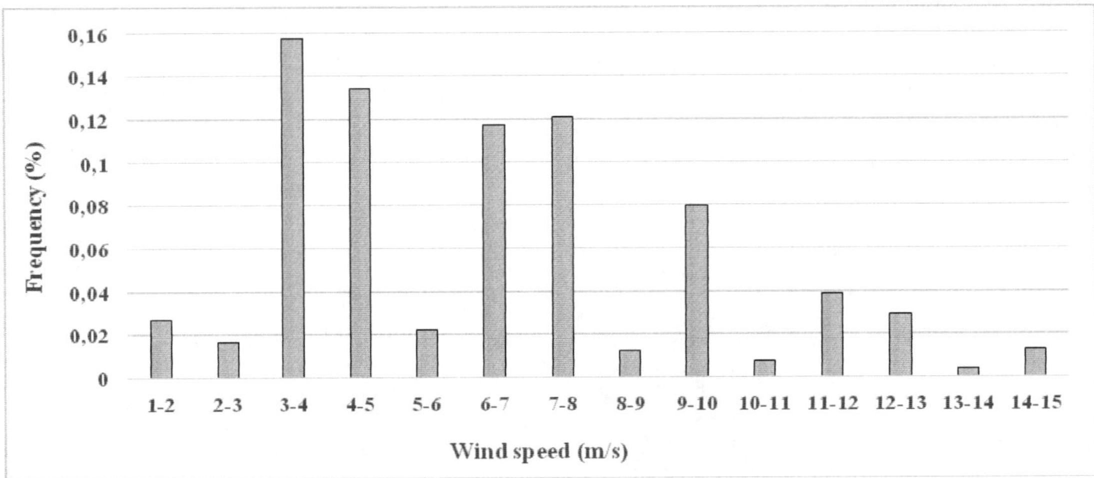

Figure 3. Wind speed frequency at hub height (134 m).

The first observation is that high wind velocity is more frequent at 134 m, which were expected since the wind speed increase with higher altitude. For the 10 m, the most frequent wind speed is between 2 m/s and 3 m/s, while for the hub height the most frequent is in the range of 3-4 m/s, however, wind speed that are in the range of 4-5 m/s and 6-8 m/s are frequent as well.

In what follow, we will present the simulation results for the electricity and hydrogen production in different time steps.

3.2 Electricity production

The daily electricity production from the simulated wind turbine is presented in figure 4.

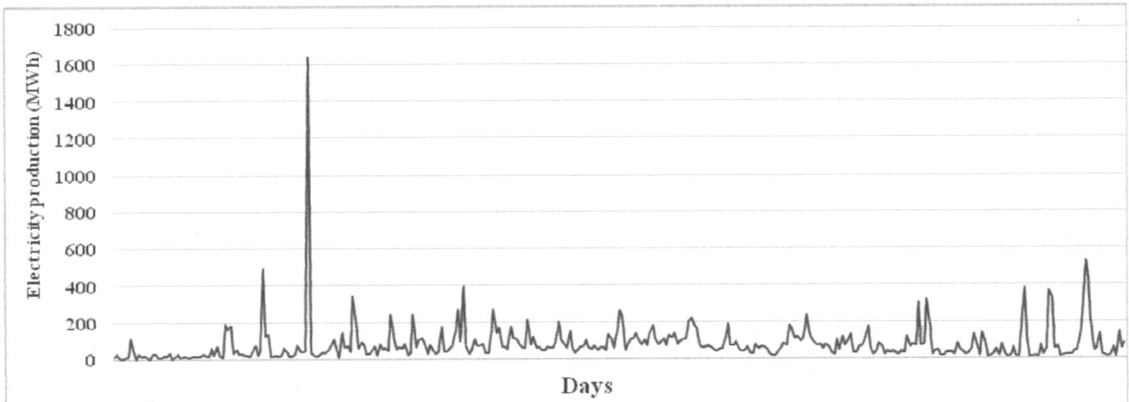

Figure 4. Daily electricity production.

As it can be seen from figure 4, the electricity production varies greatly through the year, which is due to the high fluctuation of the wind speed. Indeed, the daily electricity generation ranges between 0 and 1638 MWh. The value 0 means that, in that day the minimal wind speed required by the wind turbine to start operating (Cut-in speed) was not reached. As for the highest value, it was recorded on March 15 where the daily average wind speed was 7 m/s at 10 m, which correspond to 11 m/s at hub height. However, due the low frequency of high wind speed (see figures 2 and3), the electricity production does not exceed 100 MWh in most days; in fact, the daily average electricity production is 80 MWh.

Regarding the monthly electricity production, it's plotted in figure 5. The amounts of electricity generated also differ from one month to another. The highest amount is observed in July with 3814 MWh and the lowest in January with 570 MWh while the monthly average electricity production is

IOP Conf. Series: Materials Science and Engineering **948** (2020) 012012 doi:10.1088/1757-899X/948/1/012012

2430 MWh. As for the annual electricity production the wind turbine produces 29.16 GWh and the

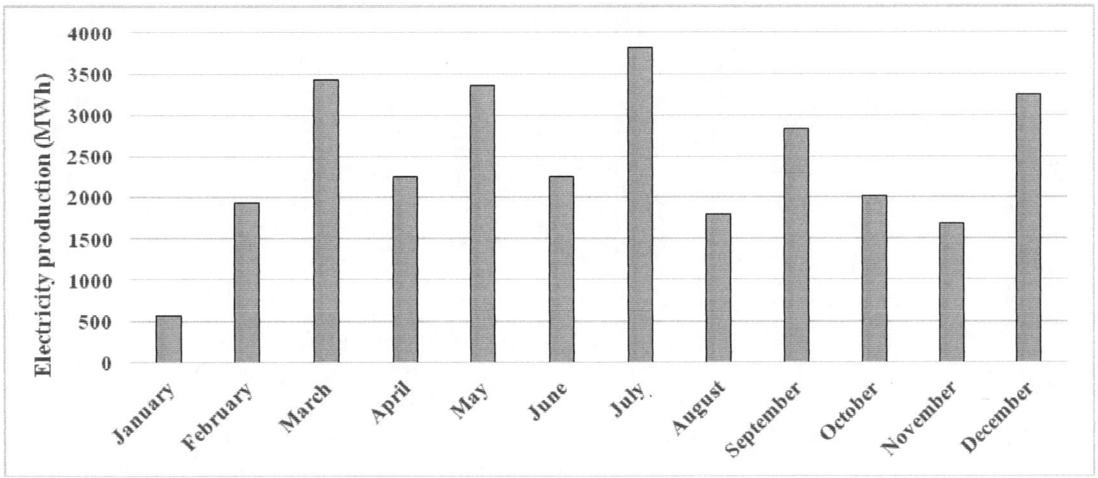

LCOE is 0.24 $/kWh.

Figure 5. Monthly electricity production.

3.3 Hydrogen production

The hydrogen production is directly proportional to the electricity production and therefore, it follows the same pattern as it can be notice in figures 6 and 7. The simulated wind turbine if coupled with a PEM electrolyzer can generate a daily amount of hydrogen in the range of 0 and 31.8 Tons with a daily average of 1.52 Tons. Also, the monthly hydrogen production between 10.52 Tons and 70.42 Tons, while the monthly average 46.25 Tons. Regarding the total hydrogen production, the simulated system can generate 555 Tons annually with an LCO_{H_2} equal to 13.52 $/Kg.

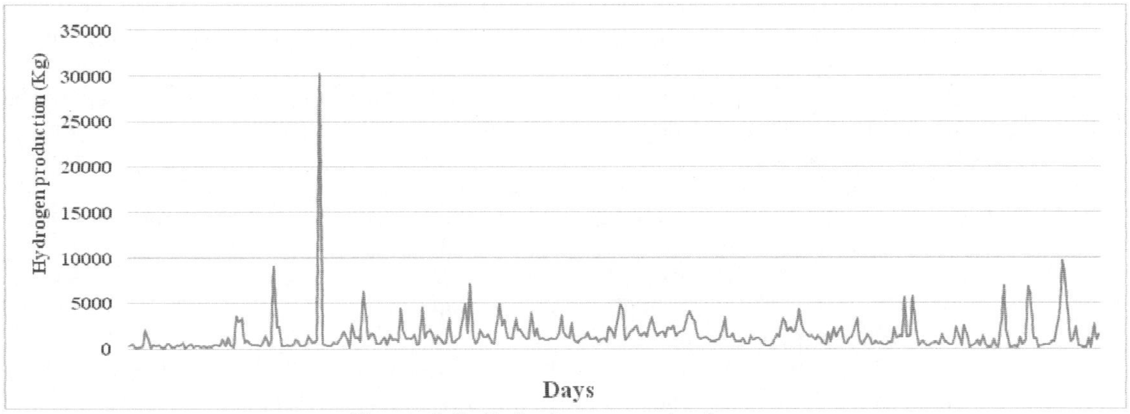

Figure 6. Daily hydrogen production.

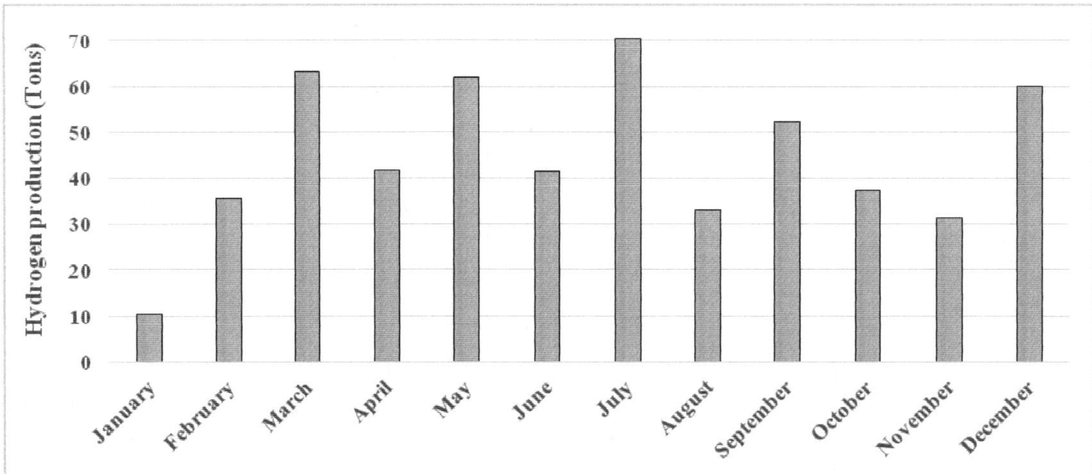

Figure 7. Monthly hydrogen production.

4. Conclusion

In this paper, a techno-economic simulation of electricity and hydrogen production from 4.2 MW_p wind turbine under the climate of Casablanca, Morocco. For this reason, we used Greenius simulation tool to simulate the electricity production, then the simulation outputs were used to estimate the hydrogen production from a wind turbine/PEM electrolyzer system.

The results show that, the most dominant wind direction is the west and the most frequent wind speed at 10 m in the range of 3-5 m /s. As for electricity production the wind turbine can generate 29.16 GWh/year with an $LCOE$ of 0.24 \$/kWh. In addition, if coupled with an electrolyzer the wind turbine can produce 555 Tons of hydrogen annually with an LCO_{H_2} equal to 13.52 \$/Kg.

References

[1] IRENA (2018), Renewable Power Generation Costs in 2017, International Renewable Energy Agency, Abu Dhabi

[2] Touili S, Alami Merrouni A, Azouzoute A, El Hassouani Y and Amrani A 2018 A technical and economical assessment of hydrogen production potential from solar energy in Morocco *Int. J. Hydrogen Energy* **43** 22777–96

[3] Abdalla A M, Hossain S, Nisfindy O B, Azad A T, Dawood M and Azad A K 2018 Hydrogen production, storage, transportation and key challenges with applications: A review *Energy Convers. Manag.* **165** 602–27

[4] Alami Merrouni A, Elwali Elalaoui F, Ghennioui A, Mezrhab A and Mezrhab A 2018 A GIS-AHP combination for the sites assessment of large-scale CSP plants with dry and wet cooling systems. Case study: Eastern Morocco *Sol. Energy* **166** 2–12

[5] Alami Merrouni A, Elwali Elalaoui F, Mezrhab A, Mezrhab A and Ghennioui A 2018 Large scale PV sites selection by combining GIS and Analytical Hierarchy Process. Case study: Eastern Morocco *Renew. Energy* **119** 863–73

[6] Allouhi A, Zamzoum O, Islam M R, Saidur R, Kousksou T, Jamil A and Derouich A 2017 Evaluation of wind energy potential in Morocco's coastal regions *Renew. Sustain. Energy Rev.* **72** 311–24

[7] Hajjaj C, Alami Merrouni A, Bouaichi A, Benhmida M, Sahnoun S, Ghennioui A and Zitouni H 2018 Evaluation, comparison and experimental validation of different PV power prediction models under semi-arid climate *Energy Convers. Manag.* **173** 476–88

[8] Mohsin M, Rasheed A K and Saidur R 2018 Economic viability and production capacity of wind generated renewable hydrogen *Int. J. Hydrogen Energy* **43** 2621–30

[9] Genç G, Çelik M and Serdar Genç M 2012 Cost analysis of wind-electrolyzer-fuel cell system for energy demand in Pnarbaş-Kayseri *Int. J. Hydrogen Energy* **37** 12158–66

[10] Siyal S H, Mentis D, Mörtberg U, Samo S R and Howells M 2015 A preliminary assessment of wind generated hydrogen production potential to reduce the gasoline fuel used in road transport sector of Sweden *Int. J. Hydrogen Energy* **40** 6501–11
[11] Rahmouni S, Negrou B, Settou N, Dominguez J and Gouareh A 2017 Prospects of hydrogen production potential from renewable resources in Algeria *Int. J. Hydrogen Energy* **42** 1383–95
[12] Kaplan Y A 2015 Overview of wind energy in the world and assessment of current wind energy policies in Turkey *Renew. Sustain. Energy Rev.* **43** 562–8
[13] Kumar Y, Ringenberg J, Depuru S S, Devabhaktuni V K, Lee J W, Nikolaidis E, Andersen B and Afjeh A 2016 Wind energy: Trends and enabling technologies *Renew. Sustain. Energy Rev.* **53** 209–24
[14] Balghouthi M, Trabelsi S E, Amara M Ben, Ali A B H and Guizani A 2016 Potential of concentrating solar power (CSP) technology in Tunisia and the possibility of interconnection with Europe *Renew. Sustain. Energy Rev.* **56** 1227–48
[15] Merrouni A A, Ouali H A L, Moussaoui M A and Mezrhab A 2016 Analysis and comparaison of different Heat Transfer Fluids for a 1MWe Parabolic Trough Collector *Proc. 2016 Int. Conf. Electr. Inf. Technol. ICEIT 2016* 510–5
[16] Ouali H A L, Merrouni A A, Moussaoui M A and Mezrhab A 2015 Electricity yield analysis of a 50 MW solar tower plant under Moroccan climate *Proc. 2015 Int. Conf. Electr. Inf. Technol. ICEIT 2015* 252–6
[17] Merrouni A A, Amrani A I and Mezrhab A 2017 Electricity production from large scale PV plants: Benchmarking the potential of Morocco against California, US *Energy Procedia* **119** 346–55
[18] Merrouni A A, Mezrhab A and Mezrhab A 2016 PV sites suitability analysis in the Eastern region of Morocco *Sustain. Energy Technol. Assessments* **18** 6–15
[19] Touili S, Alami Merrouni A, El Hassouani Y, Bennouna E G, Ghennioui A and Amrani A-I 2019 *A comparative study on hydrogen production from small-scale PV and CSP systems* vol 519
[20] Nikolaidis P and Poullikkas A 2017 A comparative overview of hydrogen production processes *Renew. Sustain. Energy Rev.* **67** 597–611
[21] Bahrami A, Okoye C O and Atikol U 2017 Technical and economic assessment of fixed, single and dual-axis tracking PV panels in low latitude countries *Renew. Energy* **113** 563–79
[22] Diaf S and Notton G 2013 Technical and economic analysis of large-scale wind energy conversion systems in Algeria *Renew. Sustain. Energy Rev.* **19** 37–51
[23] Douak M and Settou N 2015 Estimation of Hydrogen Production Using Wind Energy in Algeria *Energy Procedia* **74** 981–90
[24] Ayodele T R, Ogunjuyigbe A S O and Amusan T O 2016 Wind power utilization assessment and economic analysis of wind turbines across fifteen locations in the six geographical zones of Nigeria *J. Clean. Prod.* **129** 341–9
[25] Gouareh A, Settou N, Khalfi A, Recioui B, Negrou B, Rahmouni S and Dokkar B 2015 GIS-based analysis of hydrogen production from geothermal electricity using CO2as working fluid in Algeria *Int. J. Hydrogen Energy* **40** 15244–53
[26] Rahmouni S, Settou N, Negrou B and Gouareh A 2016 GIS-based method for future prospect of hydrogen demand in the Algerian road transport sector *Int. J. Hydrogen Energy* **41** 2128–43
[27] Dincer I, Ozgur Colpan C, Akif Ezan M and Kizilkan O 2015 Progress in clean energy, volume 2: Novel systems and applications *Prog. Clean Energy, Vol. 2 Nov. Syst. Appl.* **2** 1–1184

IOP Conf. Series: Materials Science and Engineering **948** (2020) 012013 doi:10.1088/1757-899X/948/1/012013

Investigation of a solid formed on the electrode surface of a damaged supercapacitor in order to determine the ageing process

A. Erradi [1,*], S. Touhtouh [*1], A. El Ballouti [1], A. Hajjaji [1]

1 Laboratory of Engineering Sciences for Energy, National School of Applied Sciences El Jadida, Morocco

* Email: touhtouh.s@ucd.ac.ma

Abstract: Supercapacitors are important energy storage devices with high power and long life. But they always age much faster than expected by some mechanisms. In addition, despite considerable efforts to understand these mechanisms, there is still a lack of a clear picture of the parasitic reactions that occur in operating devices. Therefore, in this work, based on an experimental study of a damaged supercapacitor, we analyzed a white powder occurring on the surface of the electrodes due to aging. The identification of the elements of this powder is carried out by X-ray diffraction, ^{13}C and ^{1}H NMR spectroscopy and thermogravimetric analysis. The results obtained allow us to determine the causes of ageing and thus to understand their mechanisms.

KEYWORDS: Supercapacitors; Mechanisms; Electrode-Electrolyte; Ageing; Cases.

1. Introduction

In many application areas, among the strongest and most practical technologies for the conversion and storage of electrochemical energy are batteries, fuel cells and electrochemical supercapacitors (ES)[1]. In recent years, ES or supercapacitors have attracted considerable attention, mainly due to their high power density, long life cycle and bridging function for the power/energy difference between dielectric capacitors (which have high output power) and fuel cells (which have high energy storage)[2]. Due to these properties, they are of major importance in sustaining the voltage of a system during increased loads for a large number of applications, from portable equipment to hybrid electric vehicles, passenger cars, buses and locomotives[3]. Starting and accelerating the engine are examples of the exploitation of high specific power. Due to the almost symmetrical charge and discharge behaviour, supercapacitors are suitable for regenerative braking, which means that they can recover energy that would otherwise be lost during deceleration and braking[4].

However, after long periods of operation, supercapacitors based on activated carbon and organic electrolytes can exhibit a decrease in capacity and an increase in resistance, increase in electrode mass, local separation of the metal collector coating layer, as well as macroscopic phenomena such as gas evolution[5], [6]. In addition, several aspects of supercapacitor performance are strongly influenced by operating temperature, including power, energy capacity, reliability, lifetime and cost[7]. Therefore, a much better understanding of the mechanisms of ageing is necessary, in order to define strategies to circumvent these drawbacks. The surface functionality and porosity of activated carbons should certainly be optimized in order to increase long-term electrochemical performance in organic milieu[8].

IOP Conf. Series: Materials Science and Engineering **948** (2020) 012013 doi:10.1088/1757-899X/948/1/012013

To date, only a few reviews have focused on studying the aging mechanisms of activated carbon-based SC in an acetonitrile electrolyte[8]–[11]. Among them, Zhu et al. have studied the ageing mechanisms of supercapacitor materials in more detail[8]. Kotz and co-workers detected CO_2, propene and H_2 in the gas space above a double layer capacitor in carbon electrodes and a solution of tetraethylammonium tetrafluoroborate in propylene carbonate (PC)[12]. Azaïs et al. also observed that activated carbons should have small amounts of surface functionality, appropriate pores and absence of moisture to ensure stable long-term performance[10]. In addition, Azaïs et al. suggested an understanding of the aging mechanism in 2007 [10]. Using two distinct carbon mechanisms, they proposed that ageing of SC is closely related to the decomposition of the organic electrolyte at the surfaces of the activated carbon. In addition, Zhu et al and Bittner et al have specifically noted the chemical and electrochemical ageing of activated carbon in electrodes by various characterization methods[8], [13]. They identified discoloration of carbon materials, particularly the sealing or depletion of micropores, as the main reason for the deterioration of the performance of AC-based SCs. At the same time, they also identified other components of SC and published many illuminating results and related analytical methods. They found that the separator had turned yellow and that the aluminum foil was corroded. Their work was remarkable and contributed significantly to the aging mechanism of SC. Furthermore, other researchers have reached similar conclusions [14]–[17].

However, despite all the above-mentioned studies, the aging mechanisms of supercapacitors have so far been very unclear. Therefore, in this work, based on an experimental study of a damaged supercapacitor, we have tried to analyze a white powder occurs on the surface of the electrodes. The identification of the elements of this powder is carried out by X-ray diffraction (XRD), ^{13}C and ^{1}H NMR spectroscopy and thermogravimetric analysis (TGA). The results obtained allow us to determine the causes of ageing and thus to understand their mechanisms.

2. Materials and method

The supercapacitors used in this work were supplied by Maxwell Technologies, Switzerland, and had a nominal capacity of 350F (BCAP0350) (figure 2). The structure of the supercapacitor cell, examined in this paper, is composed of two activated carbon electrodes and a separator impregnated with an organic electrolyte. The electrodes consist of a metal collector, covered on both sides with an active material, which has a large specific surface area necessary for the double layer. The two electrodes are separated by a porous membrane (separator), which prevents electronic conduction between the electrodes but allows ionic conduction between them[1]. The electrolyte was 1M tetraethylammonium tetrafluoroborate (mol/L1) (ET_4NBF_4) in acetonitrile (CH_3CN); and the separator was a porous polypropylene (PP) membrane. Since this technological specification was the standard specification for commercial supercapacitors.

Fig 1. Maxwell Technologies, Switzerland, and had a nominal capacity of 350F (BCAP0350)

A damaged supercapacitor is used in this study.after disassembling the supercapacitor according to the protocol illustrated in the figure above (figure 2); we recovered the product on the surface of the electrodes in solid form (white powder). Then, a series of analysis is carried out on the recovered powder (X-ray diffraction, and thermogravimetric analysis).

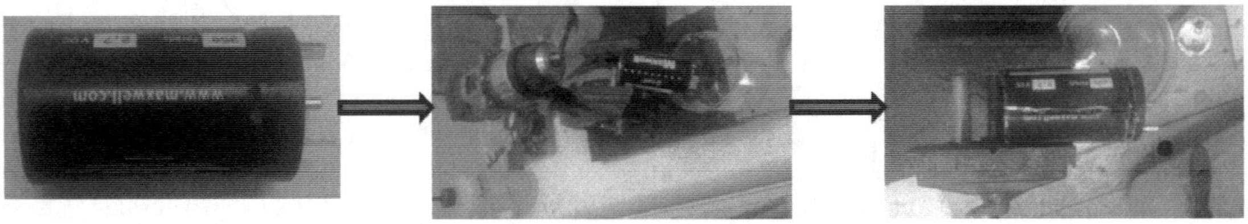

Fig. 2 Protocol for disassembling supercapacitor

3. Results and discussion

3.1 Powder XRD

X-ray diffraction (XRD) pattern ($10^\circ - 60^\circ$) was obtained using PANalytical X'Pert Pro diffractometer with CuKα monochromatized radiation (λ = 1.5418 Å). To study our powder in more detail, we applied X-ray powder diffraction (XRD) (to check for impurities, identify the phases and verify their crystallinity). The results of the X-ray diffraction show many peaks at different angles (2Theta), which represents its high crystallinity.Using phase identification software, we were able to find the majority of the peaks (Peaks 1; 2; 3; 4; 5 (figure 3)). The results obtained show that our powder probably consists of acetamide, unsaturated aromatic organic acids, fluoroacetic acid derivatives, oximes and macromolecular compounds.

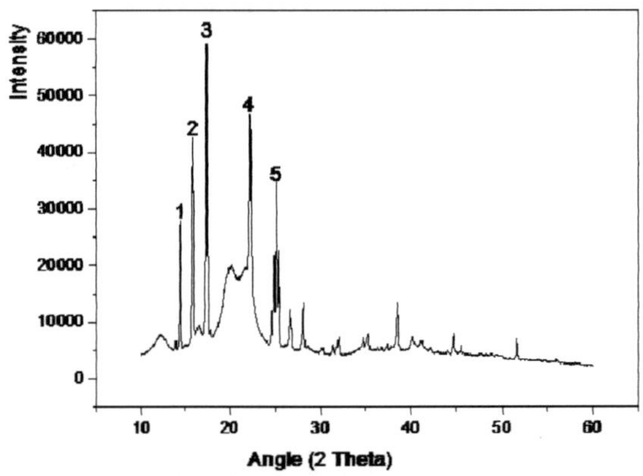

Fig 3. X-ray diffractogram (DRX) of the formed powder on the surface of electrodes

3.2 Solid-state NMR spectroscopy

Nuclear Magnetic Resonance Spectroscopy (NMR) is an approach that has shown promise for the study of materials. NMR has the advantage of being selective with respect to the elements, allowing independent observation of different ionic species. In this study the powder is placed in a 4 mm ZrO_2 rotor, making sure that the rotor is always correctly filled. The volume is adjusted with the calibrated tools used to fill the MAS-NMR rotors. The spectra were recorded with a 500 MHz Bruker Avance III (Karlsruhe, Germany) apparatus. During signal acquisition, spinal ^1H decoupling was applied. The MAS frequency was 12.5 kHz. The results obtained are represented on ^{13}C; 1H spectra (Figures 3 and 4). The figures above show the ^{13}C NMR and ^1H NMR spectra. In the first ^{13}C NMR spectrum (fig 4), we can see three indexed signals, one signal at d=-0.4ppm, the literature allows to attribute this chemical shift to the acetonitrile trace[18]. Another signal is observed at d= 46 ppm corresponding to the acetylene carbon CH_2=CH_2, and the last signal at d=123.7 ppm (light peak) corresponds to the carbon of an amide carbonyl group, then the spectrum obtained in ^{1}H NMR (Fig 5) is composed of two signals, which are attributed to protons of the acetonitrile molecular group as detailed in the article by Borchart et al[18]. These results confirm the hypotheses suggested by the XRD analysis, in particular the presence of amide and acetylene.

On the basis of these conclusions, it is clear that the aging process of supercapacitors certainly takes place through chemical reactions. These reactions are considered to take place within the framework of well-defined thermodynamic and kinetic factors. However, the environment and the operating conditions of these components may favour certain reactions. We can therefore propose that our powder, which consists mainly of amide, acetylene and other compounds, is obtained by hydrolysis and thermal decomposition of acetonitrile (electrolyte). It is well noted that the water molecule that reacts with acetonitrile during hydrolysis is obtained from the trace amounts of water that are resistant during carbon activation[15].

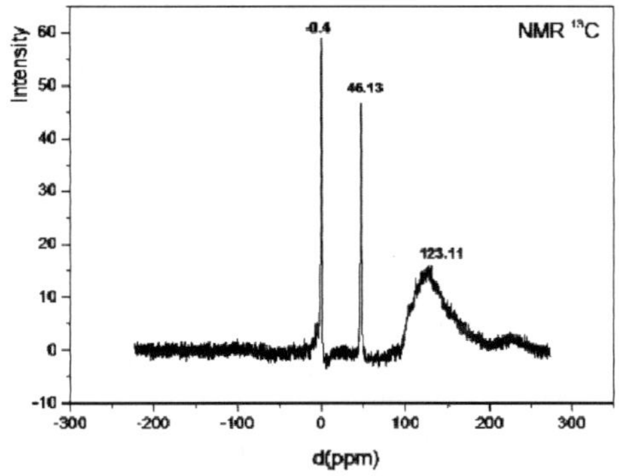

Fig 4. MAS ^{13}C NMR of the powder formed on the surface of electrodes

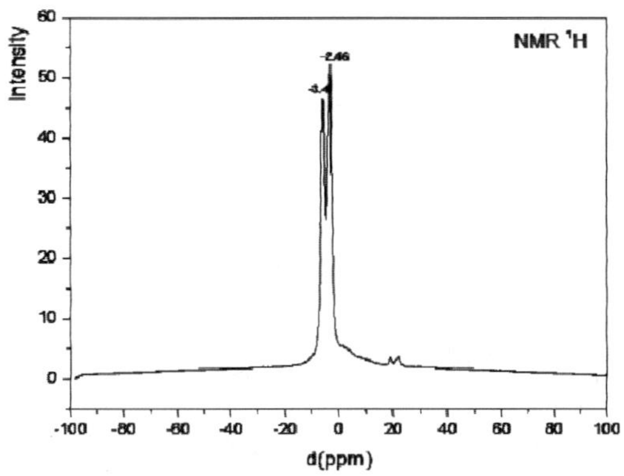

Fig 5. MAS ^{1}H NMR of the powder formed on the surface of electrodes

3.3 Thermogravimetric analysis (TGA)

In order to confirm the previously proposed hypotheses, we conducted a thermogravimetric analysis study. Mass loss measurements are performed on our sample (formed powder) in a temperature range of 20 to 800°C using a

SETSYS (SETARAM, Caluire, France) analyzer (heating rate of 5 °C·min^{-1} under 40 mL·min^{-1} flow of air)[19]. The thermogravimetric curve shows that the mass loss of the sample was progressive and occurred at three temperature levels. There is a significant peak at 90°C, which is due to the loss of moisture and the evaporation of acetylene. Then, until the temperature reached 350 °C, a mass loss was observed, which was due to the decomposition of the acetamide[15]. The third step at 540 °C corresponds to the decomposition of traces of the electrolyte, mainly the alkylammonium ion, followed by the more stable tetrafluoroborate[14]. These results definitely confirm our previous explanations.

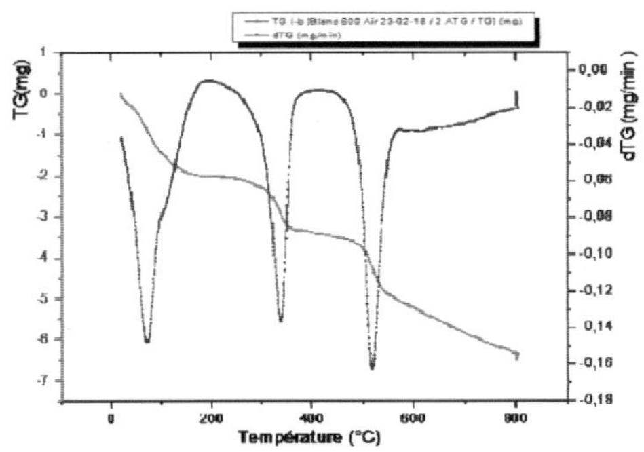

Fig 6. TGA curve of the powder formed on the electrodes

4. Conclusion

Aging of supercapacitors based on carbonaceous materials in the acetonitrile electrolyte is very slow, and even under conditions of forced aging (increase in temperature and electrochemical potential).After the life cycles of supercapacitors expire, they are damaged, therefore a crystalline mass is produced on the electrode surface. The study of this crystalline mass in the present work is carried out by X-ray diffraction (XRD), ^{13}C and ^{1}H NMR spectroscopy and Thermogravimetric Analysis (TGA). Thanks to the results obtained, we have been able to collect the first information on the mechanisms of thermal decomposition of acetonitrile, as well as on their hydrolysis reaction. Our experiments also gave clear indications on the formation of amides, acetylene and other compounds. This finding should lead to new considerations in the design of supercapacitors, such as the addition of additives that are capable of blocking or inhibiting chemical reactions, and in the case of temperature gradients, placing the electrodes in a colder environment.

REFERENCES

[1] H. Gualous *et al.*, "Supercapacitor ageing at constant temperature and constant voltage and thermal shock," *Microelectronics Reliability*, vol. 50, no. 9, pp. 1783–1788, Sep. 2010, doi: 10.1016/j.microrel.2010.07.144.

[2] H. Gualous, D. Hissel, S. Bontour, F. Harel, and J. M. Kauffmann, "Power management of an embedded fuel cell - supercapacitor APU," in *2005 European Conference on Power Electronics and Applications*, Sep. 2005, p. 8 pp.-P.8, doi: 10.1109/EPE.2005.219618.

[3] R. Kötz and M. Carlen, "Principles and applications of electrochemical capacitors," *Electrochimica acta*, vol. 45, no. 15–16, pp. 2483–2498, 2000.

[4] S. Hmam, J.-C. Olivier, S. Bourguet, and L. Loron, "Techno-economic optimization of a supercapacitor-based energy storage unit chain: Application on the first quick charge plug-in ferry," *Applied Energy*, vol. 153, pp. 3–14, 2015.

[5] Y. Ayoubi, M. Elsied, A. Oukaour, H. Chaoui, Y. Slamani, and H. Gualous, "Four-phase interleaved DC/DC boost converter interfaces for super-capacitors in electric vehicle application based onadvanced sliding mode control design," *Electric Power Systems Research*, vol. 134, pp. 186–196, 2016.

IOP Conf. Series: Materials Science and Engineering **948** (2020) 012013 doi:10.1088/1757-899X/948/1/012013

[6] H. Gualous, H. Louahlia-Gualous, R. Gallay, and A. Miraoui, "Supercapacitor thermal modeling and characterization in transient state for industrial applications," *IEEE Transactions on industry applications*, vol. 45, no. 3, pp. 1035–1044, 2009.

[7] H. Gualous, D. Bouquain, A. Berthon, and J. M. Kauffmann, "Experimental study of supercapacitor serial resistance and capacitance variations with temperature," *Journal of power sources*, vol. 123, no. 1, pp. 86–93, 2003.

[8] M. Zhu *et al.*, "Chemical and electrochemical ageing of carbon materials used in supercapacitor electrodes," *Carbon*, vol. 46, no. 14, pp. 1829–1840, 2008.

[9] A. M. Bittner *et al.*, "Ageing of electrochemical double layer capacitors," *Journal of Power Sources*, vol. 203, pp. 262–273, 2012.

[10] P. Azaïs *et al.*, "Causes of supercapacitors ageing in organic electrolyte," *Journal of power sources*, vol. 171, no. 2, pp. 1046–1053, 2007.

[11] P. Kurzweil and M. Chwistek, "Electrochemical stability of organic electrolytes in supercapacitors: Spectroscopy and gas analysis of decomposition products," *Journal of Power Sources*, vol. 176, no. 2, pp. 555–567, 2008.

[12] R. Kötz, M. Hahn, and R. Gallay, "Temperature behavior and impedance fundamentals of supercapacitors," *Journal of Power Sources*, vol. 154, no. 2, pp. 550–555, 2006.

[13] A. M. Bittner *et al.*, "Ageing of electrochemical double layer capacitors," *Journal of Power Sources*, vol. 203, pp. 262–273, 2012.

[14] Y. Liu, B. Soucaze-Guillous, P.-L. Taberna, and P. Simon, "Understanding of carbon-based supercapacitors ageing mechanisms by electrochemical and analytical methods," *Journal of Power Sources*, vol. 366, pp. 123–130, 2017.

[15] Y. Huang *et al.*, "Experimental and correlative analyses of the ageing mechanism of activated carbon based supercapacitor," *Electrochimica Acta*, vol. 228, pp. 214–225, 2017.

[16] P. Sharma and T. S. Bhatti, "A review on electrochemical double-layer capacitors," *Energy conversion and management*, vol. 51, no. 12, pp. 2901–2912, 2010.

[17] P. W. Ruch, D. Cericola, A. Foelske, R. Kötz, and A. Wokaun, "A comparison of the aging of electrochemical double layer capacitors with acetonitrile and propylene carbonate-based electrolytes at elevated voltages," *Electrochimica Acta*, vol. 55, no. 7, pp. 2352–2357, 2010.

[18] L. Borchardt, M. Oschatz, S. Paasch, S. Kaskel, and E. Brunner, "Interaction of electrolyte molecules with carbon materials of well-defined porosity: characterization by solid-state NMR spectroscopy," *Physical Chemistry Chemical Physics*, vol. 15, no. 36, pp. 15177–15184, 2013.

[19] S. Komaty *et al.*, "A Facile Route toward the Increase of Oxygen Content in Nanosized Zeolite by Insertion of Cerium and Fluorinated Compounds," *Molecules*, vol. 23, no. 2, p. 37, 2018.

IOP Conf. Series: Materials Science and Engineering **948** (2020) 012014 doi:10.1088/1757-899X/948/1/012014

Theoretical study and Simulation method for optimizing the performance of advanced Energy Harvesting techniques

Y. Elhmamsy, C. Ennawaoui, A. Hajjaji*, Y. Boughaleb

Laboratory of Engineering Sciences for Energy (LabSIPE)
National School of Applied Sciences EL Jadida, Chouaib Doukkali University, Morocco.
*Corresponding authors: hajjaji.a@ucd.ac.ma

Abstract: This paper investigates and compares the efficiencies of four different interfaces for vibration based energy harvesting systems. In some cases, the increase in power was found to be in the order of ten times the output power compared to the standard approach to energy harvested by piezoelectric materials. The authors updated the review of synchronous switching techniques used in the conversion of ambient mechanical energy to useful electrical energy using piezoelectric materials is given. The basic concepts involved in the standard energy harvesting approach and synchronous switch harvesting techniques are presented. A comparative analysis of these techniques, highlighting the strengths and limitations of each approach in terms of power conversion efficiency, load independence, implementation complexity and adaptability for system applications are discussed.

Keyword: Energy harvesting systems, standard approach, piezoelectric materials, synchronous switching techniques.

1. Introduction

Energy harvesting [1, 2] has become important in the last few years due to the amount of methods of harvesting energy and the different fields where they already implemented. The energy harvested as small as it is important to power the sensors [3-5] and make them independent [6]. Converting mechanical energy to electrical energy using smart materials has been the choice for many energy harvesting applications such as autonomous devices [7-10].

The most basic energy harvesting topology to increase the harvested power is the standard technique, which directly connects the piezoelectric element to the load R [1]. A regulated output voltage Vc is obtained in Standard technique DC mode (STD DC) by a full-wave diode bridge. The capacitor Cr is used to filter the output voltage Vc. Power output is related to the load impedance R. Therefore, impedance matching is used to optimize the output power. Gi et al. proposed a DC-DC converter to optimize the power output of a piezoelectric harvesting device [11]. Since a piezoelectric energy system is capacitive in most cases, perfect impedance matching is impossible with a resistive load.

In addition, the inductance required for a perfect match is usually unrealistic. To solve this problem, Guyomar et al. [12] proposed an innovative non-linear circuit named as "Synchronized Switch Harvesting on Inductor" (SSHI). This technique used to control vibration by using a piezoelectric damping element and inducing a phase shift between voltage and strain by a simple vibration-synchronized switching process. It consists of inverting the voltage each time it reaches an extremum. Inversion is achieved by a brief oscillation triggered by an electronic switch . It can also be considered as a periodic change in the stiffness of the system due to a continuous change in boundary conditions. The vibration energy is converted into electrical energy stored on the piezoelectric element and dissipated as heat with each switching operation.

The so-called p-SSHI technique modifies the Standard technique by adding in parallel with the piezoelectric element a branch composing an inductor L in series with a switch SW, The switching operation will only take place when the movement of the harvester or the voltage of the piezoelectric element reaches its extreme. Once the voltage resulting from this switching process is in phase with the deformation speed. Harvested electrical power can be increased by more than 400% under low coupling conditions [13].

A similar topology design is the s-SSHI [14], which places the inductor in series with a piezoelectric element. A switch is also connected with the inductor in series following the same control strategy as previous SSHI. This indicates that most of the time the circuit is open circuit and the piezoelectric voltage is determined by the displacement. Once the reverse action occurs, the charge is driven with the piezoelectric voltage.

Therefore, Lallart et al [15] proposed a Double Synchronized Switch Harvesting (DSSH) and Enhanced Synchronized Switch Harvesting (ESSH), These techniques behave equivalently, the main difference between DSSH and ESSH being in the technological solution to emulate the output load.

In DSSH techniques, energy is transferred to the first storage capacitor Cint as series-SSHI does. Then the second switch SW2 is closed to transfer the energy from the intermediate capacitor Cint to the inductance load. In all this process, the charge is totally independent of the energy extraction by the piezoelectric element [16].

The SECE and DSSH techniques exhibit better harvesting ability with load-independent behavior. The harvesting ability of DSSH is determined by γC, the intermediate inversion factor mainly determined by Cint and L2 [17].

This paper is organized as follows. Section II introduces the goal and the compositions of the proposed energy harvesting topologies. Section III provides the theoretical model and simulation study on the energy harvesting circuits. Then, the simulations are conducted to evaluate the comparison between the topologies performance. The same section investigate the energy harvesting ability of this type of circuits. Finally, conclusions from this study are presented in discussions.

2. Energy harvesting Circuit Topologies

2.1 Mechanical and electrical modeling

In previous parts, we talked about the general functionality of piezoelectric energy harvesters. These types of harvesters are only useful and liable for today's high demanded industry if only the output power is increased while keeping their lengths and volumes as small as possible. However, to increase the output power, other ways must be sought. The most promising approach is to design an efficient interface charging circuits, storage buffers and hot structure configuration. In this part, we will provide a review on different approaches and see how the output power can be enhanced efficiently by changing electrical components of the interface circuits. As it was mentioned before, cantilever beams are the most used configurations in piezoelectric energy harvesters. Their popularity is because they enable relatively high stress levels on the piezoelectric material while reducing the dimensions of the devices [18]. Therefore, in this part all the optimization methods are based on configuration of cantilever beams. Because the piezoelectric power generator is excited at a sinusoidal force, where the first natural frequency causes the maximum power generator with small displacement and the movements are linear, it can be modeled mechanically with a rigid mass (M), spring (K), damper (C) and piezoelectric material. The following figure shows the spring mass model.

IOP Conf. Series: Materials Science and Engineering **948** (2020) 012014 doi:10.1088/1757-899X/948/1/012014

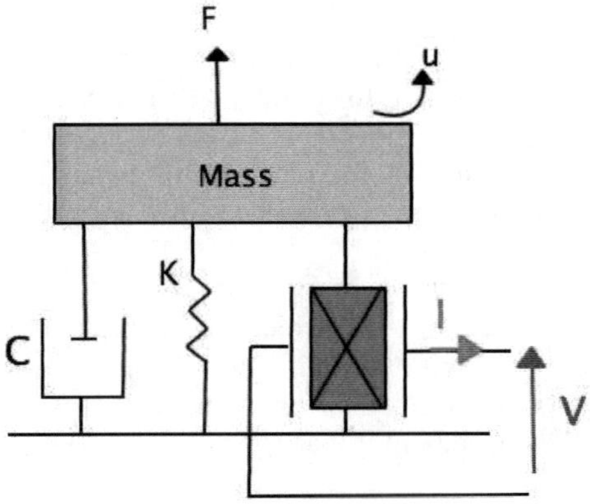

Fig. 1. Mechanical representation of the piezoelectric system

The piezoelectric element is modeled by equations (1), F is the force applied to the piezoelectric element. These equations are derived from the constitutive relations [19-21]. They involve the geometrical dimensions of the piezoelectric element.

$$\begin{cases} M\ddot{u} + C\dot{u} + Ku = F - \alpha V \\ I = \alpha\dot{u} - C_0\dot{V} \end{cases} \quad (1)$$

With:

- K stiffness of the piezoelectric element when short-circuited.
- α (N/V) is the factor of the piezoelectric element and V is the output voltage.
- C (F) is the piezoelectric capacitance.

Multiplying this equation by the speed ($\dot{u}$) and then integrating the expression will yield:

$$M\ddot{u} = F - Ku - C\dot{u} - \alpha V \quad (2)$$

$$\int F\dot{u}\,dt = \frac{1}{2}M\dot{u}^2 + \frac{1}{2}Ku^2 + \int C\dot{u}^2\,dt + \int \alpha V\dot{u}\,dt \quad (3)$$

The following table 1 represents the type of energy for each term

Type of Energy	Equation
Provided	$\int F\dot{u}\,dt$
Kinetic	$\frac{1}{2}M\dot{u}^2$
Potential	$\frac{1}{2}Ku^2$
Dissipated	$\int C\dot{u}^2\,dt$
Transferred	$\int \alpha V\dot{u}\,dt$

Table 1. Definition of the piezoelectric energies from equations

From the constitutive equations of piezoelectricity, the current flowing out of the piezoelectric element, I, can be expressed as:

$$I = \alpha\dot{u} - C_p\dot{V} \quad (4)$$

The energy transferred can then be rewritten as:

IOP Conf. Series: Materials Science and Engineering **948** (2020) 012014 doi:10.1088/1757-899X/948/1/012014

$$\int \alpha V \dot{u}\, dt = \frac{1}{2} C_p V^2 + \int VI\, dt \quad (5)$$

Wherein the energy transferred is the sum of the electrostatic energy stored on the piezoelectric element and the energy delivered to the connected load [22-24]. The second term of the above equation is the collected energy that needs to be improved for better efficiency and output power.

$$E = \int VI\, dt \quad (6)$$

$$P = \frac{1}{t} \int VI\, dt \quad (7)$$

The following figure shows the electrical representation of the piezoelectric material. There is also a load connected parallel to the capacitance, which shows the loaded piezoelectric output voltage in the frequency domain as a function of R.

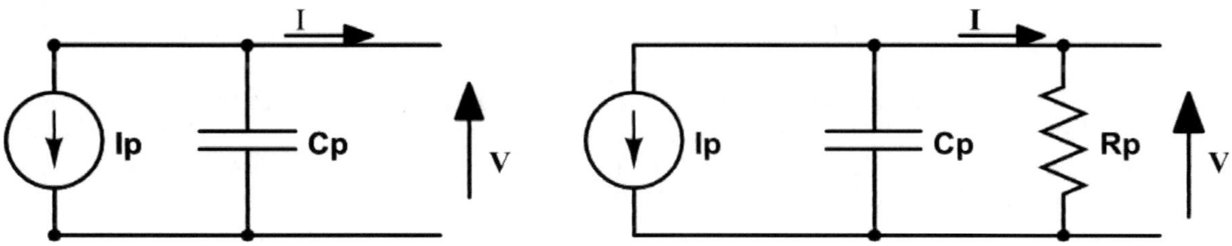

Fig. 2. Uncoupled equivalent circuit of the piezoelectric harvester without and with the the load resistance

The following formulas show the current and voltage where $\tilde{V}$ and u show the maximum voltage and the maximum displacement.

$$\tilde{V} = \frac{\alpha R}{1 + j\omega r C_p} j\omega \tilde{u} \quad (8)$$

The average power delivered to the resistive load of the piezoelectric material (P_R) is:

$$P_R = \frac{\alpha^2 R^2 \omega^2 \tilde{u}^2}{2(1 + (j\omega R C_p)^2)} \quad (9)$$

2.2 Standard circuit description

The configuration of energy harvesting systems are as following. However, the power generated by the piezoelectric devices cannot be used directly as it is generated. Thus, some other electric interfaces have to be used to make the output power compatible with other devices. To maintain this compatibility a basic charging circuit is connected the piezoelectric to convert the AC voltage to DC or in other word a rectifier is connected and then the generated electrical energy goes to a storage buffer to be stored for the later on uses. The most commonly used circuit is called classic interface. The classic interface consist of a diode rectifier, filter capacitor and load P_R at the terminal. The following figures show the simple configuration of cantilever beam system.

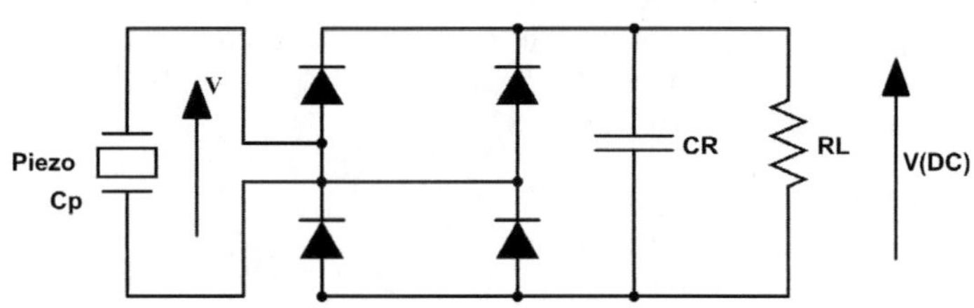

Fig. 3. Standard full bridge rectifier

To calculate the output power, we can assume the mechanical displacement of u is supposed to be a sinusoidal wave and the open circuit V on the piezoelectric is also sinusoidal. In this method, the diode rectifier is on the blocked mode thus the piezoelectric element is on an open circuit when the absolute value of V is lower than VQR across the capacitor. When the value of V is more than the VQR a current flows through the capacitor and then to the load. The generated power is the energy produced in a period T which can be calculated by integrating the product of the voltage and the current under one period. The power is equivalent to:

$$P = \frac{4\alpha^2 U^2 \omega^2 R_L}{(2R_L C_P \omega + \pi)} \tag{10}$$

However, the power can also be measured experimentally when the circuit is operating. By simply using the power formula, we have:

$$P = \frac{V_{DC}^2}{R_L} \tag{11}$$

The disadvantage of this circuit is that the current flowing to the load is discontinuous and the speed is not in phase with the piezoelectric voltage, so the conventional interface circuit will not work effectively. As mentioned, this circuit is only a standard circuit that will not increase power, so the following circuits and the topology network have been suggested to increase power.

2.3 Synchronized Switch Harvesting on Inductor in parallel (parallel-SSHI)

The schematic diagram of Synchronized switching harvesting on inductor in parallel (parallel-SSHI) piezoelectric energy harvesting device with full-bridge rectifier to a simple resistor load. This is called SSHI technique. Assuming the structure is excited at resonant frequency and from the governing equation, the piezoelectric energy harvesting device can be modeled as a current source parallel with a clamped capacitor and the equivalent circuit of entire system is shown as Figure x. In this technique, a bi-directional switch and an inductor L are added in parallel with the piezoelectric patch. The switch is conducted at each maximum and minimum of the displacement or at the zero crossing of the vibration velocity, in order to reverse the voltage across the piezoelectric element and put it in phase with velocity.

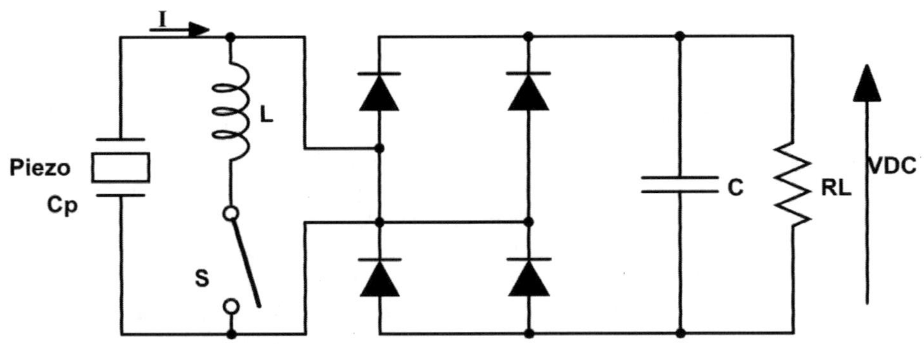

Fig. 4. Piezoelectric energy harvesting system with parallel SSHI technique

The result is that the energy stored in the structural clamped capacitor (C_0) is extracted by the LC resonance and achieve to a minimum value, and thus the piezoelectric voltage can be increased [17]. The harvested energy of the system with the SSHI technique is similar to that using the standard interface under the strongly coupled condition [18].

$$P_{p-SSHI} = \frac{4R\alpha^2}{(RC_p\omega(1-\gamma)+\pi)^2}\omega^2 u^2 \tag{12}$$

3. Synchronized Switch Harvesting on Inductor in Series (Series-SSHI)

The schematic diagram of synchronized switching harvesting on inductor in series (Series-SSHI) piezoelectric energy harvesting device with full-bridge rectifier to a simple resistor load.

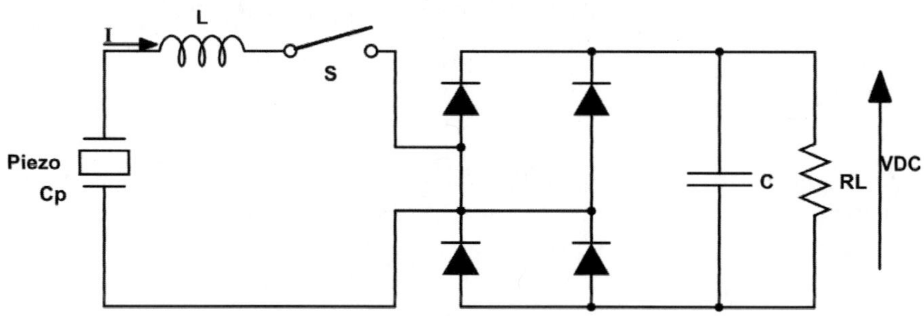

Fig. 5. Piezoelectric energy harvesting system with series SSHI technique

The piezoelectric energy harvesting device can be modeled as a current source parallel with a clamped capacitor and the equivalent circuit of entire system.

$$P_{S-SSHI} = \frac{4R\alpha^2(1+\gamma)^2}{(RC_p\omega(1+\gamma)+\pi(1-\gamma))^2}\omega^2 u^2 \tag{13}$$

4. DSSH technique

The DSSH technique optimizes the electromechanical conversion. It also gives a fixed harvested energy regardless of the load. This technique uses a capacitor as an intermediate energy storage stage. The intermediate stage stores energy and transmits the additional energy to the resonant inductor and finally to the load via the smoothing capacitor. This helps the phase alignment of voltage and current. In addition, we get a fixed output and the dead zone present in previous models is removed.

Next, the operation of the DSSH circuit is explained. For the majority of the time, the piezoelectric element is kept in open circuit form. When the voltage on the piezoelectric element reaches the maximum point, the switch S1 is turned on. The time period for which this switch is closed is very small but enough for the capacitor to get charged to its maximum point. At this point, energy is transferred to the capacitor from the piezo element.

When maximum energy transfer is complete, the switch S1 is turned off and simultaneously the switch S2 is turned on. At this point, energy transfers from the capacitor to the inductor. When total energy is transferred to the inductor, the switch S2 opens and directly the diode conducts due to its natural nature.

When the diode conducts, energy is transferred to the smoothing capacitor and finally to the load. As all these processes get completed, the circuit starts operating for the initial operation.

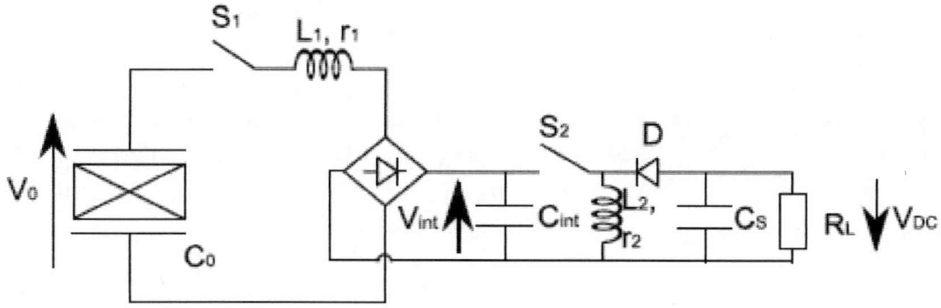

Fig. 6. Piezoelectric energy harvesting DSSH technique

$$E_{DSSH\_ind} = \frac{1}{2}\gamma_c C_{int}\left(\frac{C_0}{C_{int}+C_0}(1+\gamma_0)\, V_0\right)^2 \quad (13)$$

With γ C, the overall efficiency of the converter, given by

$$\gamma_c = e^{-\xi_{int}\pi} \quad (14)$$

With ω_{int} and ξ_{int}, the natural angular frequency and the damping coefficient of the system, respectively, defined as

$$\omega_{int} = \sqrt{\frac{1}{L_2 C_{int}}} \quad (15)$$

$$\xi_{int} = \frac{1}{2} r_2 \sqrt{\frac{C_{int}}{L_2}} \quad (16)$$

Finally, it can be seen that the harvested energy does not depend on the rectified voltage VDC, and is therefore independent of the value of the load resistance RL.

5. Modelization and Simulation of energy harvesting techniques

5.1Energy harvesting techniques modelization

The deformation of a piezoelectric membrane induces a polarization of the piezoelectric material and a charge transfer proportional to the stress:

$$q = \alpha . u \quad (17)$$

Where q is the displaced load (C), α the coefficient of deformation (C / m) and u the deformation (m).

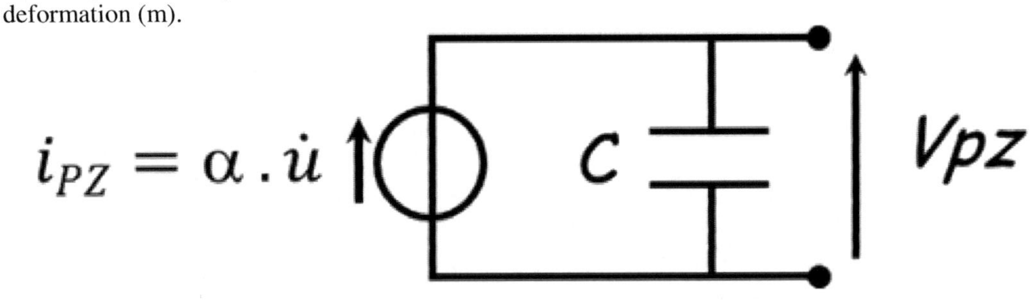

Fig. 7. Electrical resonance model

Conventionally, in a first approach, the beam and piezoelectric membrane device is modeled in the near of resonance by a current source i_{PZ} in parallel with a capacitor C (fig. 7).

5.2 Simulation of energy harvesting techniques

This model is then used in PSIM by adding the recovery device. We only simulate a rectifier delivering on a constant direct current source IS. In order to understand the operation and to calculate the harvested power in steady state, this part is important for the comparison and understanding of the system because the currents are difficult to measure in practice.
With the gain of a circuit X $G_{circuit\,X} = \frac{circuit\,X}{standard}$ and □= $\frac{P_{out}}{P_{int}}$

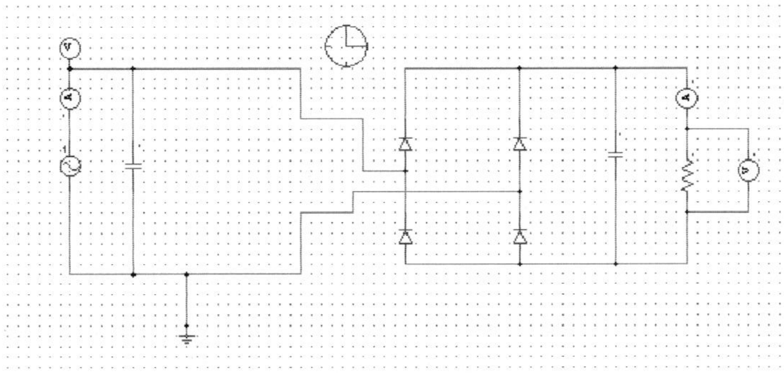

Fig. 8a. Standard Circuit compositions

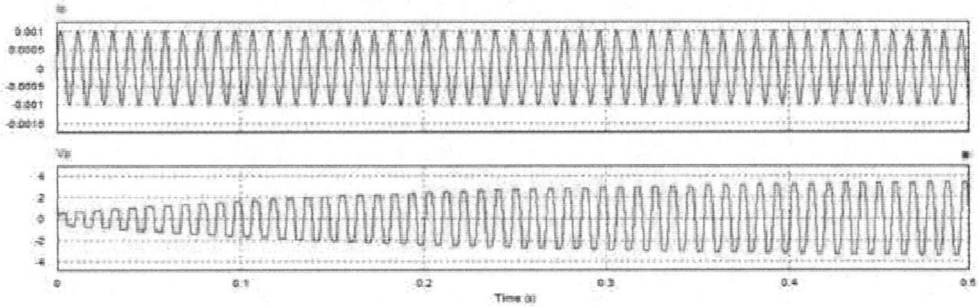

Fig. 8b. Piezoelectric Input voltage and current

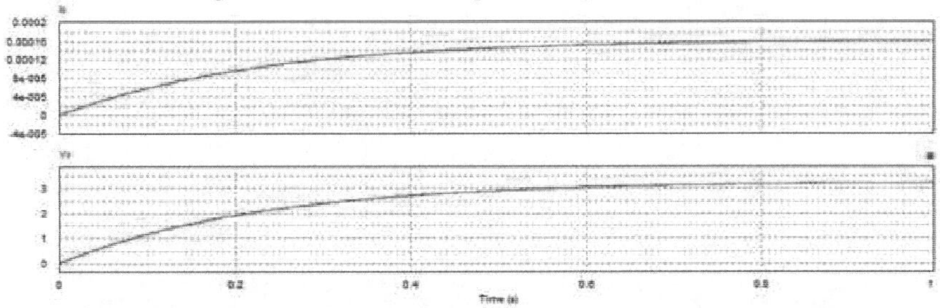

Fig. 8c. Piezoelectric input power and Standard output power

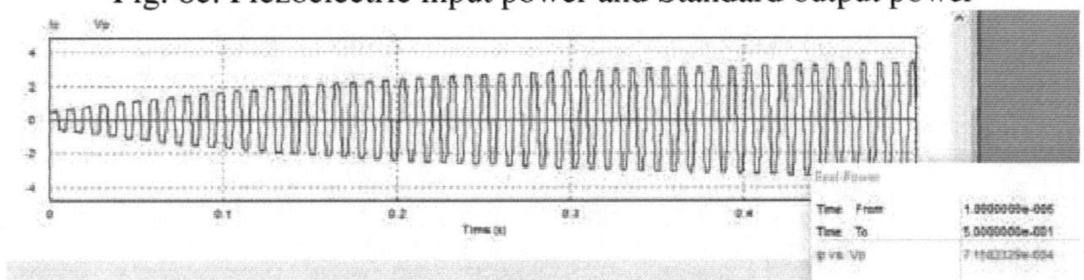

Fig. 8d. Piezoelectric input power

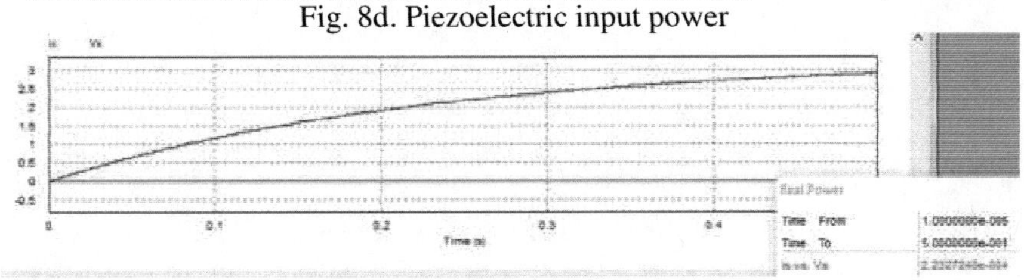

Fig. 8e. Standard output Power

The Standard technique performance

$$P_{int} = 71,5mW \tag{18}$$

$$P_{out} = 22,3mW \tag{19}$$

$$stantard = \frac{P_{out}}{P_{int}} = 0{,}3 \tag{20}$$

$$P_{dis} = P_{int} - P_{out} = 71{,}5mW - 22{,}3mW = 49{,}2mW \tag{21}$$

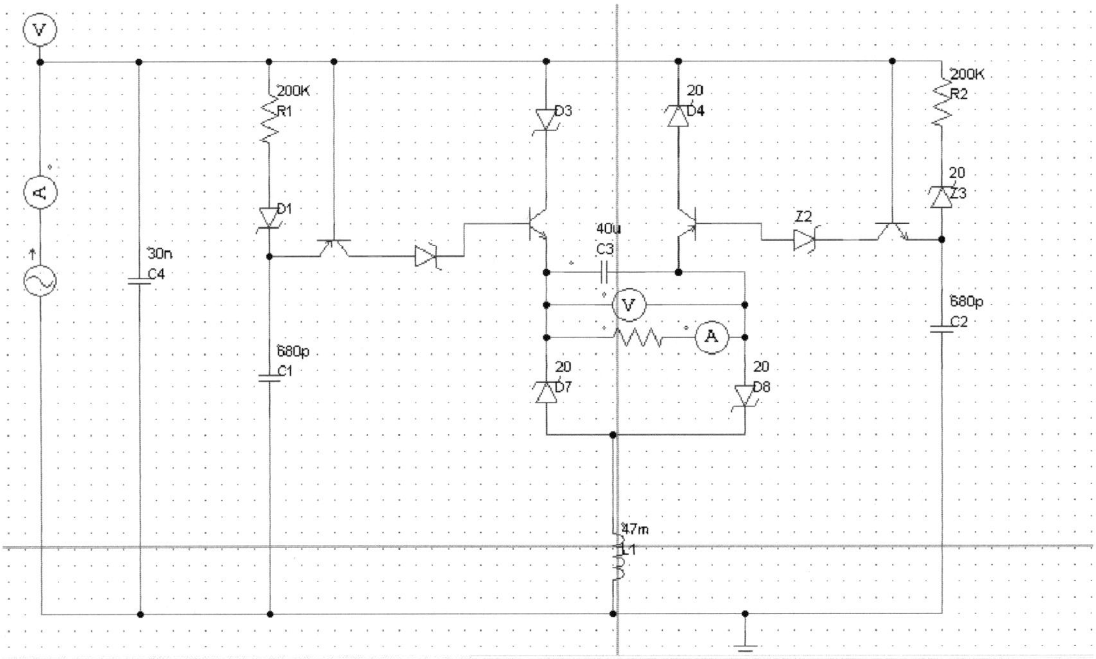

Fig. 9a. SHHL-s Circuit compositions

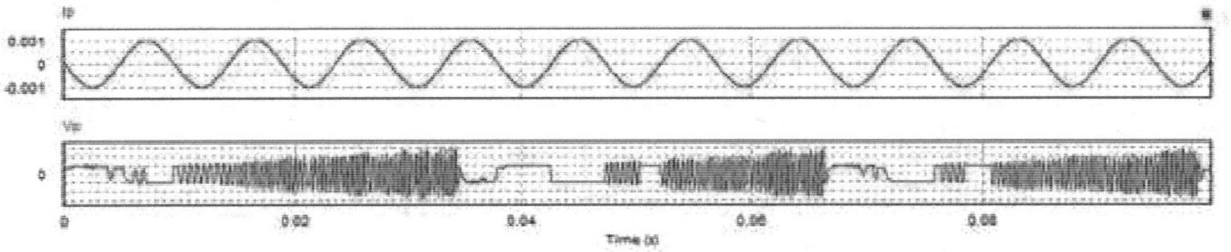

Fig. 10b. Piezoelectric Input voltage and current

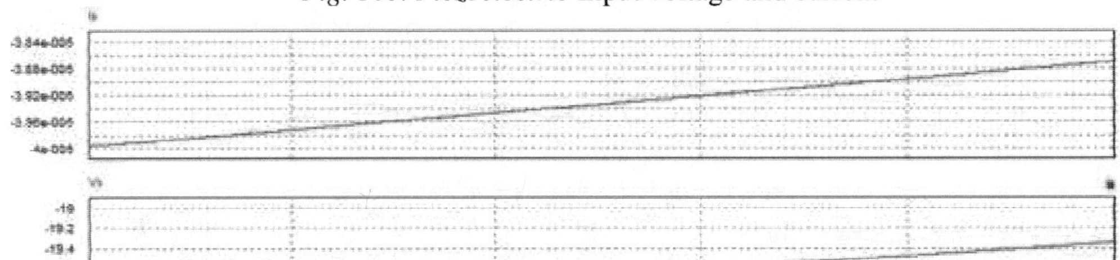

Fig. 11c. SHHL-s Output voltage and current

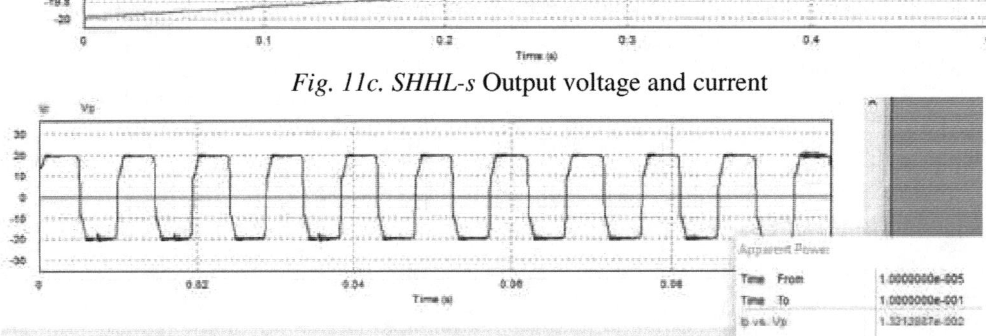

Fig. 12d. Piezoelectric Input power

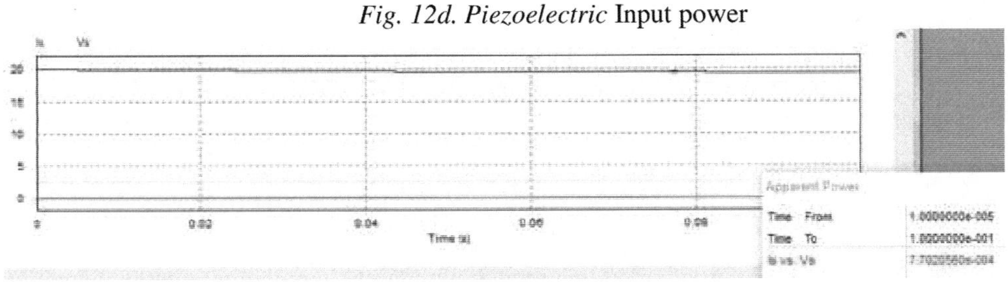

Fig. 13e. SHHL-s Power output

The SHHL-s performance

$$G_{SSHL-s} = \frac{SSHL-s}{stantard} \quad (22)$$

$$SHHL-s = 6{,}5 \; stantard \quad (23)$$

$$SHHL-s = 650\% \; stantard \quad (24)$$

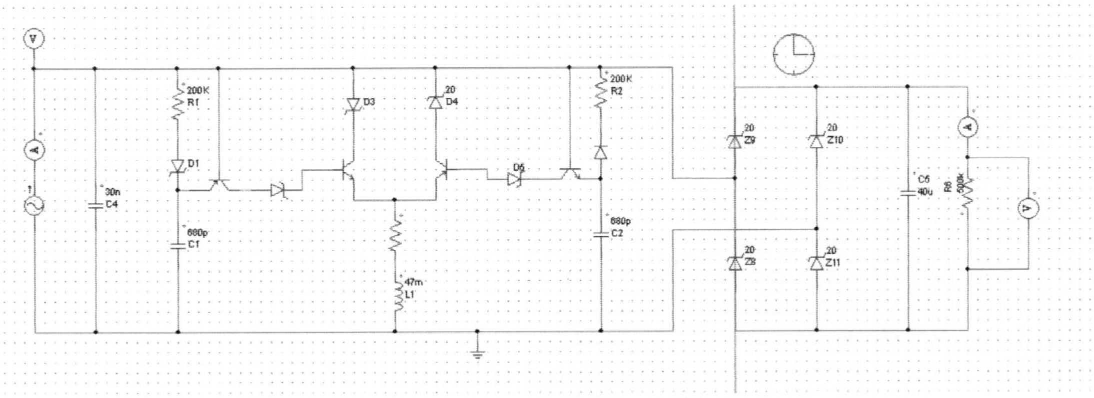

Fig. 14 a. SHHL-p Circuit compositions

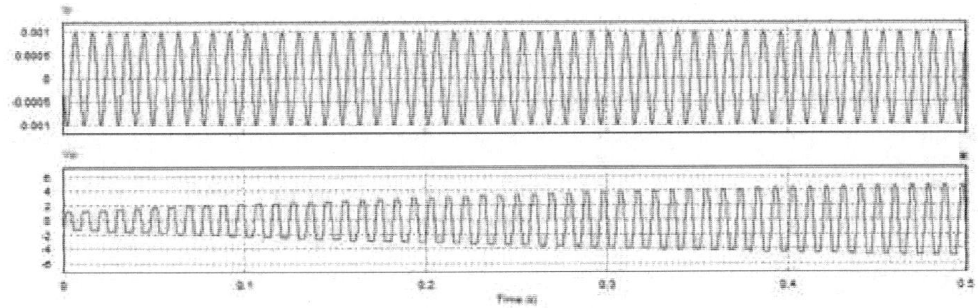

Fig. 15 b. Piezoelectric Input voltage and current

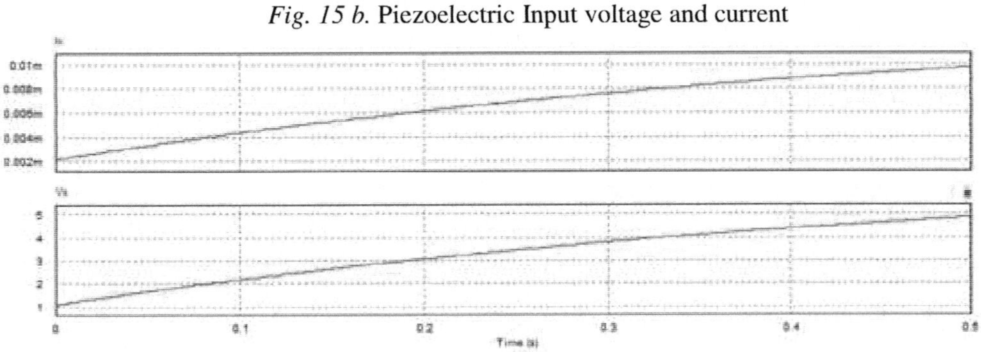

Fig. 16 c. SHHL-p Output voltage and current

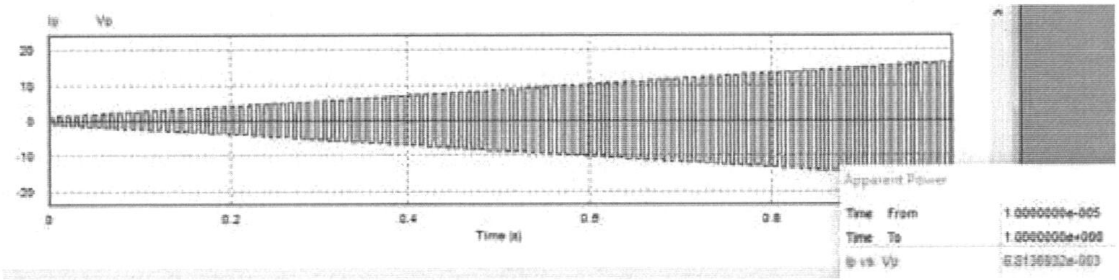

Fig. 17 d. Piezoelectric Input power

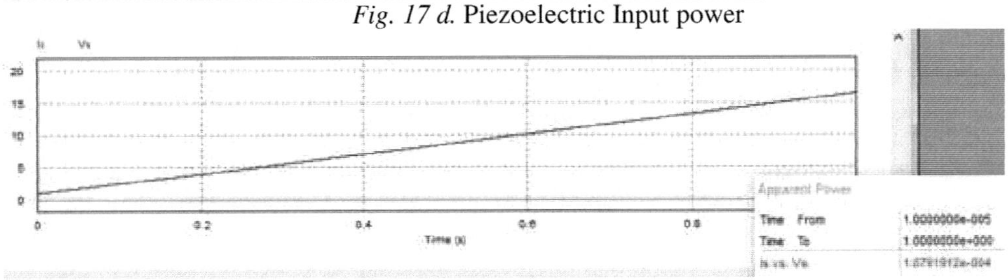

Fig. 18 e. SHHL-p Power output

The SSHL-p performance

$$G_{SSHL-p} = \frac{SSHL\text{-}p}{stantardp} \quad (25)$$

$$_{SHHL-p} = 7.4_{stantard} \quad (26)$$

$$_{SHHL-p} = 740\%\ _{stantard} \quad (27)$$

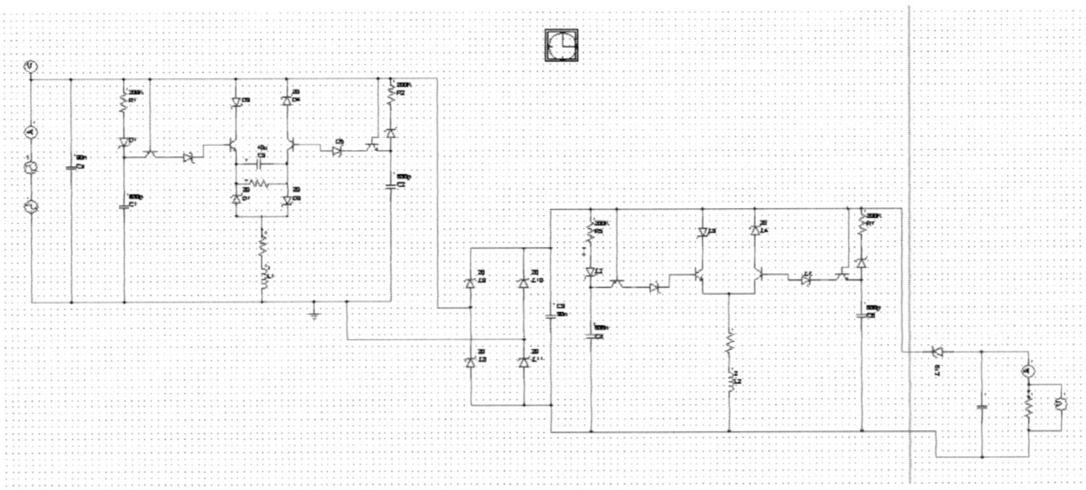

Fig. 19a. DSSH Circuit compositions

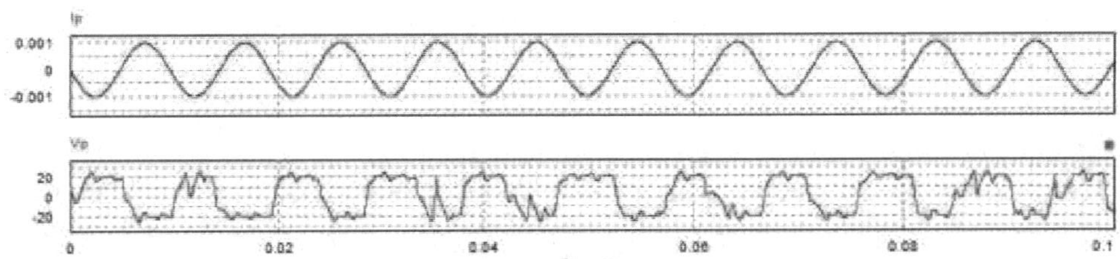

Fig.11b. *Piezoelectric* Input voltage and current

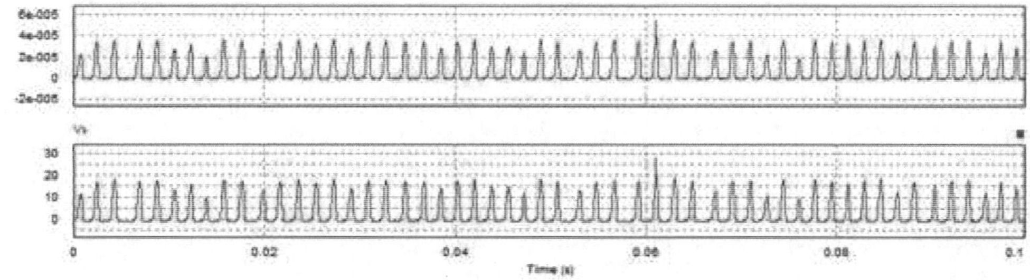

Fig.11c. *DSSH* Output voltage and current

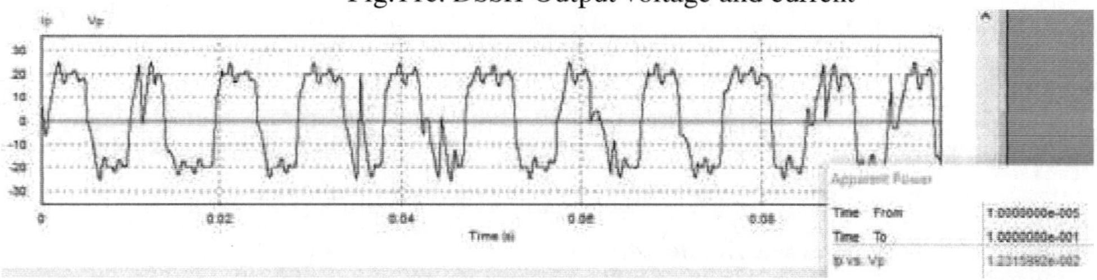

Fig.11d. Piezoelectric Input power

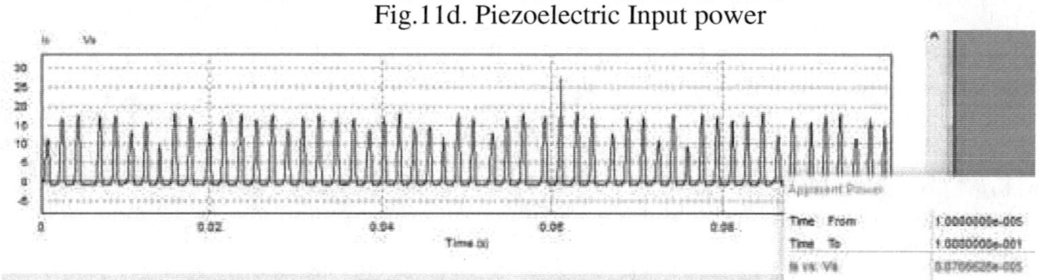

Fig.11e. *DSSH* Power output

The DSSH performance:

$$G_{DSSH} = \frac{DSSH}{S_{stantard}} \tag{28}$$

$$_{DSSH} = 6\ _{stantard} \tag{29}$$

$$_{DSSH} = 600\ \%\ _{stantard} \tag{30}$$

6. Discussions

The Figure 11 illustrates the harvested power as a function of the optimal load and the comparison with non-linear techniques and standard circuit,

All these topologies of non-linear circuits show an improvement in harvesting power compared to the STD technique. The gain is around 800% for p-SSHI, 700% for s-SSHI, 600% for DSSH and 400% for SECE. SSHI parallel and SSHI series depend on the frequency of variation, energy harvesting cannot be optimal over a large frequency band, which contradicts the fact that the system can oscillate over a wide frequency range. The DSSH technique does not present optimal resistance; it allows an efficient recovery whatever the frequency of variation.

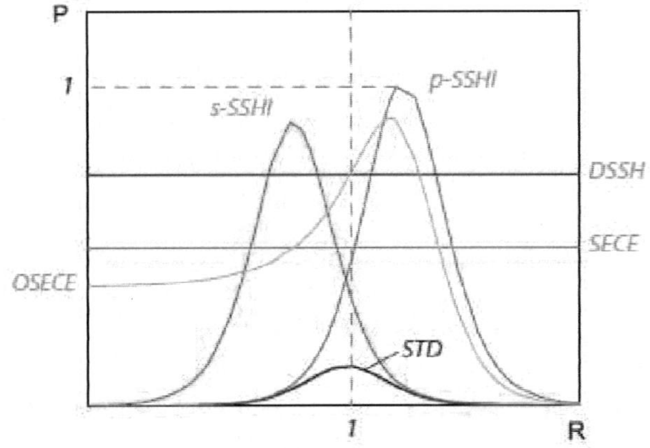

Fig. 20.The illustration the power harvested as a function of the optimal load and the comparison with non-linear techniques and standard circuit

7. Comparison of energy harvesting techniques

A comparison between the parameters is needed for the final result. Table 2 gives an overview of the balance between the energy harvesting techniques.

Technique	**Interface Standard**	**SHHI-s**	**SHHI-p**	**DSSH & ESSH**
Normalized power under constant vibration amplitude	Low	Good	Good	Medium
How is the energy harvested under low electromechanical coupling regime?	Poor	Good	Good	Good
How is load independence?	Poor	Poor	Poor	Very Good
How is low voltage harvesting ?	Poor	Good	Poor	Good
Implementation complexity ?	Any	Low	Low	Medium

Table 2. *Summarizes the performances of these techniques in terms of harvested power, advantages and drawbacks.*

8. Conclusion

This paper investigates and compares the efficiencies of four different interfaces for vibration based energy harvesting systems. This paper illustrates the power versus the load impedance for a given piezoelectric energy harvester excited in forced displacement mode. Both SSHI techniques have significant power harvesting ability. To harvest a given level power, p-SSHI and s-SSHI have a wider load range compared with STD technique. The SECE and DSSH techniques exhibit better harvesting ability with a load-independent behavior. The harvesting ability of DSSH is determined by γC, the intermediate inversion factor mainly determined by Cint and L2, but with load sensitive SSHI performance for high load impedance operation. All of these nonlinear circuit topologies show improvement harvested power compared to STD technique. The gain is around 800% percent for p-SSHI, 700% for s-SSHI, and 600% for DSSH . Many problems arise when deploying these non-linear energy harvesting topologies. They rely on signal detection, switch operation, inductors, complex circuitry, and optimal load. Even the load-independent DSSH and ESSH have optimal load region due to the oscillation and losses in circuitry. On one hand, the tradeoff between feasibility and high harvested power should be considered. On the other hand, it is worthy to find an energy harvesting circuit with simple topology and strong power harvesting capability.

References

[1] Ennawaoui, C., Lifi, H., Hajjaji, A., Elballouti, A., Laasri, S., & Azim, A. (2019). Mathematical modeling of mass spring's system: Hybrid speed bumps application for mechanical energy harvesting. *Engineering Solid Mechanics*, *7*(1), 47-58.

[2] Zhang, Y., Bowen, C. R., Ghosh, S. K., Mandal, D., Khanbareh, H., Arafa, M., & Wan, C. (2019). Ferroelectret materials and devices for energy harvesting applications. *Nano Energy*, *57*, 118-140.

[3] Hajjaji, A., Guyomar, D., Touhtouh, S., Pruvost, S., Boughaleb, Y., Rguiti, M., ... & Benkhouja, K. (2010). Nonlinearity and scaling behavior in a soft lead zirconate titanate piezoceramic. *Journal of Applied Physics*, *108*(6), 064103.

[4] Yang, L., Chi, S., Dong, S., Yuan, F., Wang, Z., Lei, J., ... & Wang, J. (2020). Preparation and characterization of a novel piezoelectric nanogenerator based on soluble and meltable copolyimide for harvesting mechanical energy. *Nano Energy*, *67*, 104220.

[5] Wilson, S. A., Jourdain, R. P., Zhang, Q., Dorey, R. A., Bowen, C. R., Willander, M., ... & Johansson, C. (2007). New materials for micro-scale sensors and actuators: An engineering review. *Materials Science and Engineering: R: Reports*, *56*(1-6), 1-129.

[6] Hajjaji, A., Benayad, A., Sebald, G., Pruvost, S., Guiffard, B., Qiu, J., & Guyomar, D. (2008). Synthesis and characterization of 0.65 Pb (Mg1/3Nb2/3) O3–0.35 PbTiO3 fibers with Pt core. *Materials Research Bulletin*, *43*(3), 493-501.

[7] Zhukov, S., von Seggern, H., Zhang, X., Xue, Y., Ben Dali, O., Pondrom, P., ... & Kupnik, M. (2020). Microenergy Harvesters Based on Fluorinated Ethylene Propylene Piezotubes. *Advanced Engineering Materials*, 1901399.

[8] Hajjaji, A., Guyomar, D., Pruvost, S., Touhtouh, S., Yuse, K., & Boughaleb, Y. (2010). Temperature/electric field scaling in Ferroelectrics. *Physica B: Condensed Matter*, *405*(13), 2757-2761.

[9] Wegener, M., & Bauer, S. (2005). Microstorms in cellular polymers: A route to soft piezoelectric transducer materials with engineered macroscopic dipoles. *ChemPhysChem*, *6*(6), 1014-1025.

[10] Ennawaoui, C., Lifi, H., Hajjaji, A., Samuel, C., Rguiti, M., Touhtouh, S., ... & Courtois, C. (2019). Dielectric and mechanical optimization properties of porous poly (ethylene-co-vinyl acetate) copolymer films for pseudo-piezoelectric effect. Polymer Engineering & Science, 59(7), 1455-1461.

[11] Gi, H., Park, J., Yoon, Y., Jung, S., Kim, S. J., & Lee, Y. (2020). A Soft-Charging-Based SC DC-DC Boost Converter With Conversion-Ratio-Insensitive High Efficiency for Energy Harvesting in Miniature Sensor Systems. *IEEE Transactions on Circuits and Systems I: Regular Papers*.

[12] Guyomar, D., & Lallart, M. (2011). Recent progress in piezoelectric conversion and energy harvesting using nonlinear electronic interfaces and issues in small scale implementation. Micromachines, 2(2), 274-294.

[13] Garbuio, L., Lallart, M., Guyomar, D., Richard, C., & Audigier, D. (2009). Mechanical energy harvester with ultralow threshold rectification based on SSHI nonlinear technique. IEEE Transactions on Industrial Electronics, 56(4), 1048-1056.

[14] Nechibvute, A., Chawanda, A., Luhanga, P., & Akande, A. R. (2012). Piezoelectric energy harvesting using synchronized switching techniques. International Journal of Engineering and Technology, 2(6), 936-46.

[15] Lallart, M., Garbuio, L., Petit, L., Richard, C., & Guyomar, D. (2008). Double synchronized switch harvesting (DSSH): A new energy harvesting scheme for efficient energy extraction. *IEEE transactions on ultrasonics, ferroelectrics, and frequency control*, *55*(10), 2119-2130.

[16] Lallart, M., & Guyomar, D. (2008). An optimized self-powered switching circuit for non-linear energy harvesting with low voltage output. *Smart Materials and Structures*, *17*(3), 035030.

[17] Lallart, M., Anton, S. R., & Inman, D. J. (2010). Frequency self-tuning scheme for broadband vibration energy harvesting. *Journal of Intelligent Material Systems and Structures*, *21*(9), 897-906.
[18] Erturk, A., & Inman, D. J. (2008). Issues in mathematical modeling of piezoelectric energy harvesters. *Smart Materials and Structures*, *17*(6), 065016.
[19] Ennawaoui, C., Hajjaji, A., Azim, A., & Boughaleb, Y. (2016). Theoretical modeling of power harvested by piezo-cellular polymers. Molecular Crystals and Liquid Crystals, 628(1), 49-54.
[20] Meddad, M., Eddiai, A., Cherif, A., Hajjaji, A., & Boughaleb, Y. (2014). Model of piezoelectric self powered supply for wearable devices. *Superlattices and Microstructures*, *71*, 105-116.
[21] Belhora, F., Cottinet, P. J., Hajjaji, A., Guyomar, D., Mazroui, M. H., Lebrun, L., & Boughaleb, Y. (2013). Mechano-electrical conversion for harvesting energy with hybridization of electrostrictive polymers and electrets. *Sensors and Actuators A: Physical*, *201*, 58-65.
[22] Lefeuvre, E., Audigier, D., Richard, C., & Guyomar, D. (2007). Buck-boost converter for sensorless power optimization of piezoelectric energy harvester. *IEEE Transactions on Power Electronics*, *22*(5), 2018-2025.
[23] Meddad, M., Eddiai, A., Guyomar, D., Belkhiat, S., Cherif, A., Yuse, K., & Hajjaji, A. (2012). An adaptive prototype design to maximize power harvesting using electrostrictive polymers. *Journal of Applied Physics*, *112*(5), 054109.
[24] Lifi, H., Ennawaoui, C., Hajjaji, A., Touhtouh, S., Laasri, S., Yessari, M., & Benjelloun, M. (2019). Sensors and energy harvesters based on (1−x) PMN-xPT piezoelectric ceramics. *The European Physical Journal Applied Physics*, *88*(1), 10901.

Permeate flux in air gap membrane distillation for seawater desalination

S. ROCHD, L. SALAMA, K.EL GHAZAOUY, S. MIZANI, H. ZERRADI, M. EL MARSI, R.MOULTIF, A. DEZAIRI*,
Laboratory of condensed matter, faculty of sciences Ben M`sick (URAC.10). University Hassan II of Casablanca, Morocco
* aouatif.dezairi@univh2c.ma ; a.dezairi@gmail.com

Abstract: Membrane distillation (MD) is a thermally driven process since only vapor molecules are transported through porous hydrophobic membranes. This process leads to the separation into different desired phases. Air gap membrane distillation (AGMD) is considered one of the most used configurations of the membrane distillation. In this work, the impacts of NaCl concentration, porosity, and membrane pore size on the production of the flux have been investigated. The results have been carried using polynomial approximations through MATLAB. These latter are in agreement with the experiments.

Keywords: Desalination, Membrane, Distillation, Air gap, NaCl concentration, Diffusion, Pore size, porosity.

1. Introduction:

Freshwater is becoming an increasingly scarce global resource, and its availability is expected to influence the long-term wealth and well-being of nations [1]. Furthermore, the common use of unhealthy water in developing countries is estimated to be causing 80-90% of all diseases and 30% of all deaths [1]. As a definition, MD is a thermally driven process, in which only vapor molecules are transported through porous hydrophobic membranes. The liquid feed to be treated by MD must be maintained in direct contact with one side of the membrane without penetrating its dry pores unless a trans-membrane pressure higher than the membrane liquid entry pressure is applied. The hydrophobic nature of the membrane prevents liquid solutions from entering its pores due to the surface tension forces. As a result, liquid/vapor interfaces are formed at the entrances of the membrane pores. Various MD modes differing in the technology applied to establish the driving force could be used [2-3].
A variety of methods may be employed in MD, such as direct contact membrane distillation (DCMD) in which the membrane is in direct contact with liquid phases in both sides [4-6]. Then, the air gap membrane distillation (AGMD) in which an air layer is interposed between the membrane and the condensation surface [7-8]. At that time, a vacuum membrane distillation (VMD) where a vacuum is applied to increase or establish the vapor pressure difference between the membrane sides and the condensation takes place in an external condenser [9-10]. At that point, sweeping gas membrane distillation (SGMD) in which a stripping gas is used on the cold side to sweep the permeate away, with Condensation in a separate device[11-15].
At this juncture, it is worthwhile to underline that the membrane technique is considered promising since it takes place at temperatures range (30 to 90) °C and could use the solar energy [16]. A survey of the state-of-the-art of membrane distillation (MD) and it is various and detailed applications was presented by Alklaibi and Lior [17]. Not far, Ding et al presented a model for predicting the rate of mass transfer in a membrane distillation unit to direct contact (DCMD) [18]. Other researchers, including Meindersma [19], Guijt [20], Payo [15], and Chouikh [21,22] worked on AGMD, but no one of them used solar as the energy source. Likewise, many authors have studied three different types of mass transfer modes through the membrane and possible combinations of this type [23, 24].

Also, Morteza Asghari and Rochd studied the effect of the thickness of the membrane in three different types of mass transfer [24-25]. ~~An~~ experience in the effect of feed saline concentration on permeate flux was studied by Dahiru [26]. At the sometimes, Lucy Mar Camacho [28] have studied the effect of pore size in the production of flux. Furthermore, many other authors produced a simulation with aspen plus in the effect of porosity [29-34]. Thus, the objective of this work is to study the effect of NaCl concentration, porosity, and membrane pore size on the production of the flux. The present article is organized as follows: In Section 2, we present the mechanism of mass transfer. In Section 3, the results and discussion. In section 4, the conclusion.

2. Mechanism of mass transfer:

The numerical solution methodology employed in this work is described in detail in some of our recent studies [24]. In order to study the mass transfer in this method, the models cited in our previous work have been used, which are Knudsen diffusion, viscous diffusion, and molecular diffusion. Thus, to get rid of the problem of the infinity of solutions, the polynomial approximation has been used. At this point, the three types of mechanisms of mass transfer are presented to show the importance and the equations governing each kind of mechanism.

. Knudsen diffusion: is based on the collisions between molecules and the-wall [24], this type of distribution is important in the systems with high temperature and pressure [31].

$$J_K = K_K(P_{hm} - P_{mg}) \quad (1)$$

With: $$K_k = \frac{2}{3}\frac{\varepsilon}{\tau}\frac{r_P}{\delta_m}\sqrt{\frac{8M_v}{\pi R T_m}} \quad (2)$$

. Molecular diffusion: is based on the collision between the molecules [24], this type of distribution is important in the systems of intermediate temperature and pressure [31].

$$J_{M,S} = K_{M,S}(P_{hm} - P_{mg}) \quad (3)$$

With: $$K_{M,S} = \frac{\varepsilon p D_{v/a} M_v}{\tau \delta_m R T_m P_a} \quad (4)$$

. Viscous diffusion: is based on both types of collisions [24], and this type of distribution is also useful in the systems of low temperature and pressure [31].

$$J_p = K_p(P_{hm} - P_{mg}) \quad (5)$$

With: $$K_p = \frac{1}{8\gamma_g}\frac{r_p^2\varepsilon}{\tau}\frac{P_m M_v}{\delta_m R T_m} \quad (6)$$

To throw more light into the effect of the reduction of vapor pressure caused by the dissolved species, Raoul's law [26] proposed the following equation

$$P_{hm} = (1 - C_s)p_v \quad (7)$$

With:

$$C_s = \frac{\frac{x}{M_{NaCl}} V_{Solution}}{\frac{x}{M_{NaCl}} V_{Solution} + n_{water}} \quad (8)$$

Where: x is the concentration of NaCl, the number of moles of water n_{water} is 55.6 moles for 1 liter, M_{NaCl}=58.44mol/l., and C_s is the mole fraction of solute or salinity. The difference in the partial pressure of the saturated vapor of both sides of the membrane may be calculated from the law of Antoine [15] using the following equation:

$$p_v = \exp(23.328 - \frac{3841}{T - 45}) \quad (9)$$

The average temperature of the membrane is given by the following equation [22]

$$T_m = \frac{T_{mh} + T_{mg}}{2} \quad (10)$$

Furthermore, Qtaishat et al [27] proposed the expression of the amount of steam/air: P Dv/a ($Pa.m^2.s^{-1}$) depending on the temperature [25]

$$PD_{v/a} = 1.985.10^{-5} T^{2.072} \quad (11)$$

With: P is the total pressure, and $D_{v/a} = \frac{k_v}{\rho_v C_P}$ is a coefficient of vapor diffusion in the air

$$D_{v/a} = \frac{k_v}{\rho_v C_P} \quad (12)$$

3-Results and Discussion:

Based on the present numerical approximations, reliable simple predictive graphics have been presented permitting the estimation of the flux as a function of NaCl concentration for different temperatures. The modelisation using numerical approximations in the figures1, 2 and 3 were carried with the following input data: $K_m=0.05 Wm^{-1}K^{-1}$; $\delta_m=4.10^{-4}m4$; $\varepsilon=0.75$; $\tau=1.5$; r_p =0.11µm. When the temperature is fixed, and the molar fraction x is swept between [0: 30]g/l the following results are observed:

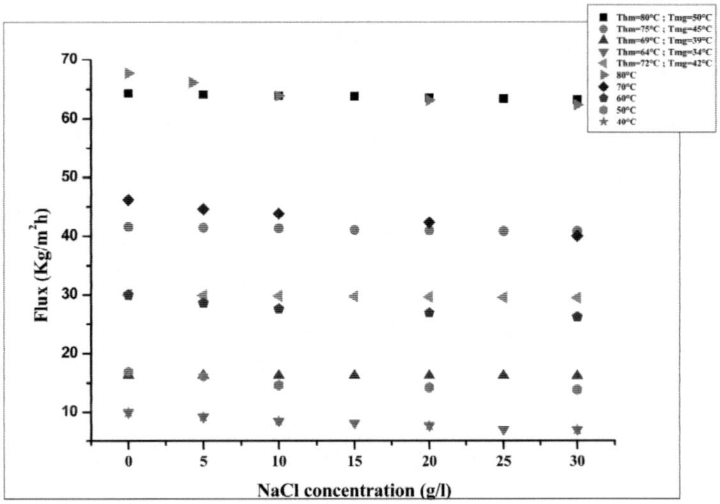

Figure 1: The variation of the flux depending on the NaCl concentration using the molecular mechanism at different temperatures and comparing the results obtained with the experience of Dahiru. [26].

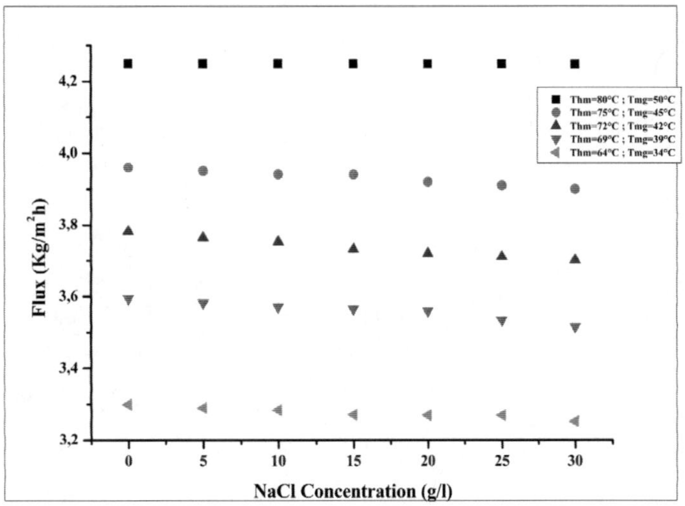

IOP Conf. Series: Materials Science and Engineering **948** (2020) 012015 doi:10.1088/1757-899X/948/1/012015

```
ERROR: ioerror
OFFENDING COMMAND: imagemask

STACK:
```

IOP Conf. Series: Materials Science and Engineering **948** (2020) 012017 doi:10.1088/1757-899X/948/1/012017

Extraction of the natural indigo carmine pigment from the Isatis plant, characterization and dyeing of wool

Y Raji [1,2,3], I Mechnou [2], W Yassine [2], Z Kadri [2], K Oumghar [1], O Cherkaoui [1], S Zyade [2,*]

[1] REMTEX Laboratory, Higher School of Textile and Clothing Industries (ESITH), Casablanca, Morocco

[2] LIPE laboratory, High Technology School, ESTC .Casablanca, Morocco

[3] Author to whom any correspondence should be addressed

*souad.zyade@univh2c.ma

Abstract. Natural indigo is hardly used anymore, replaced by synthetic indigo. Considered to be more powerful and more intense. The aim of this study is to optimize the method of extracting indigo carmine from the Isatis plant located in the Timhdit region of Morocco, during the fermentation period of the extracted liquid (pH = 10) and by adding 10% sugar it was possible to increase the yield by 0.5% indigo carmine blue powder to 0.9%. The main effect of applying centrifugation at 7000 rpm instead of conventional filtration is the reduction of the time from 30 min to 5 min as well as the reduction of the loss of traces of indigo in the filter papers by easily recovering the blue powder in the form of sediment in the centrifuge tubes. The extracted pigment is characterized by the infrared which confirmed the structure of the indigo carmine obtained, by visible UV spectroscopy, the maximum absorption of which is approximately 598 nm. This blue pigment was applied to the wool fibres by optimizing the dyeing procedure (T = 50 ° C and pH=10) with a 1/100 bath ratio and a 35 min soaking time. Dyed wool sample has reached a rigid fixation confirmed by the test for fastness washing 105-C06 and to friction 105 -X12.

1. Introduction

Indigo pigment was the first extracted dye used in natural dyeing. The chemical synthesis of indigo by Adolph Von Baeyer in 1883 extended the use of indigo as the preferred blue in dyeing [1]. Being a non-polluting dye extracted from a natural source that gives a deep and solid coloration, it is interesting to think about its reintegration. However, the best source of blue dyes resistant to discolouration is indigo plants, dispersed in many different botanical families present in very diverse natural environments, the main ones belonging to Indigofera, Isatistinctorial and Polygonum (Table 1) [1].

According to our work, indigo is extracted as pigment from the leaves of Isatis located in the Timhdit region of the Middle Atlas Mountains of Morocco. The extraction of indigo is done by macerating the leaves in hot water at 50°C for 2 hours, then oxidation, fermentation and finally filtration [2-4]. According to previous studies, during maceration, it is advisable to stay in a temperature range between 40 and 50 ° C. This condition corresponds to the maximum plateau of activity of the enzyme responsible for the kinetics of the reaction that transforms the indican present in the leaves into indoxyl [3] and maceration at 48 ° C over a period ranging from 1 h 30 to 2 hours would extract enough indigo while limiting the diffusion of impurities and as a last step following the technique of oxidation with compressed air, indigo is formed from two molecules of indoxyl during the fermentation period (Figure 1). By this method, we were able to extract 0.5% indigo pigment. Based on these experiments already in use, this next work aims to optimize the fermentation step by adding sugar to the fermented baths to improve the yield of the indigo extract. Subsequently, the centrifugation technique facilitated the recovery of the indigo without any loss in place of the filtration used in previous experiments. Indigo is a vat dye, known for its water insoluble properties. However, under the action of a reducing agent, sodium dithionite ($Na_2S_2O_4$) and in an alkaline medium, indigo gives its water-soluble leuco derivative and once out of the bath, by oxidation in air or with an oxidizing agent, the initial water-insoluble dye is regenerated in the fiber [3, 4], the chemistry of the indigo dye involves the redox reaction [8].In order to control the different factors for dyeing wool fibres with natural indigo, these factors need to be studied. In this work, we are interested in these main factors: the temperature of the dye bath, the pH of the solution, the dyeing time and the effect of the salt on the fixation of this pigment.

Table1. Sources of major natural blue colorants from plant species

Botonicalname	**Common name**	**Origin**
Indigofera	Indigo, Nila	Tropical countries in Asia
Isatis tinctoriaL	Woad	Europe.
Polygonmtinctorium)	Polygonum,orJapanese indigo	Northern/Central Italy, Germany, Japan, China, Korea

2. Material and methods

2.1 Extraction method

The leaves of Isatis were collected from cultivated plants in early May. The leaves (1kg) were washed and cut into small pieces of 1cm^2. The dye was then extracted using 4 liters of hot water (50 ° C) for 90 minutes, the sheets were removed and the sample was quickly cooled to 30 ° C, then 40 ml of 25% ammonia was added to reach a pH of 9, the extraction liquor was then aerated for 1 h using a compressor to allow the complete oxidation of the indigo precursors, up to the appearance of the color blue. The liquid rested for 24 h, different concentrations of sugar (5%,10%,15% and 20%) were added to the water of the fermented baths (see figure 3) then centrifuged at a speed of 7000 rpm for 5 min, and finally dried in the oven for 12 h at a temperature of 105 ° C. Once the pigment has been collected, it is stored in tubes in a controlled laboratory environment (20 ° C, 65% relative humidity).

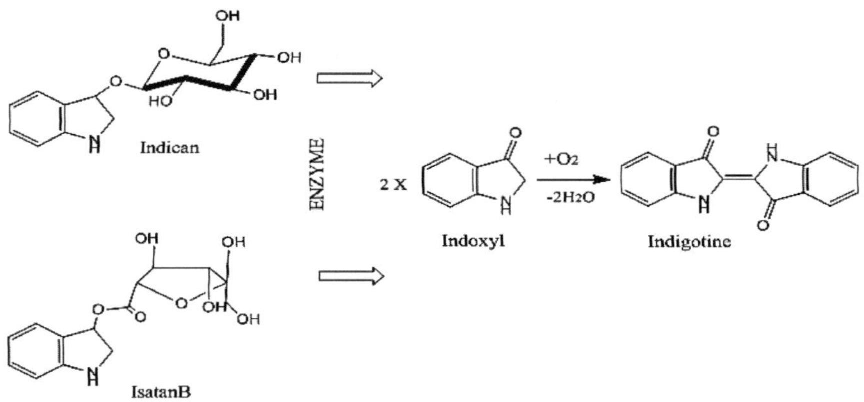

Figure 1. Formation of indigo from its precursors Indican and Isatan B

2.2 Infrared

The IR spectrum of the indigo carmine extracted from the Isatis plant was recorded using the Nicolet IS10 FTIR-ATR spectrophotometer by recording the scans in transmission mode (%) in the range from 4000 to 1000 cm-1. FTIR spectroscopy was used to determine and confirm the functions present in the extracted indigo molecule.

2.3 Visible UV spectroscopy

The spectrophotometer used in the laboratory is of the Thermo Evolution 300 brand, Variable bandwidth from 0.2 to 4nm, Dual-beam technology Autonomous or controllable by PC, Photodiodes detector or photomultiplier.

Using the absorption spectrum of the coloring solution A = f (λ), we started by identifying the peak of its maximum absorbance; the wavelength for which the absorbance of the solution is maximum $\lambda_{max.}$ (See figure 5).

2.4 Dyeing conditions

The wool fibres are dyed on a Datacolor "AHIBA NUANCE" programmable laboratory dyeing machine. The procedures for dyeing the wool fibres followed by rinsing, oxidation, neutralization and drying [5] are described below in the diagram (figure 2), we have tested two samples of dyed wool fibres under one condition (ph=10,T=50°C), the first in the absence of NaCl and the second in the presence of 5% Nacl.

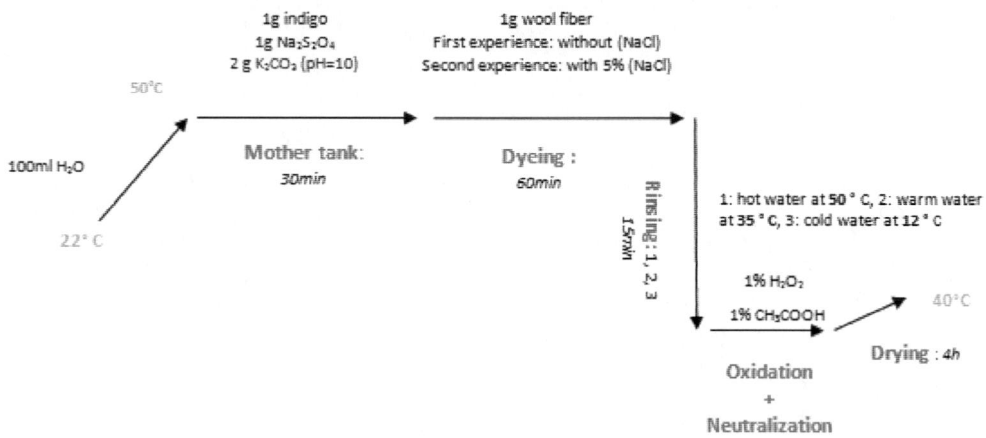

Figure 2. Wool dyeing process by indigo

2.5 Color fastness

The quality controls of the wool samples dyed with natural indigo were carried out according to international standards. For the washing and friction tests, the degrees of solidity are indicated by a gray scale numbered from 1 to 5 showing disgorgement and degradation: "1" representing the lowest values in quality and "5" the highest .These standards meet strict and well-defined criteria: color fastness to washing: ISO 105-C06, and stain resistance to friction: ISO 105-X12. [6].

2.5.1 Washing test

1 g of dyed wool fiber is brought into contact with a specified control fabric and washed under appropriate conditions (concentration of detergent 5 g / l and temperature 60 ° C in a container for 1 hour, once the sample has been recovered, it is rinsed and dried. After the 12 wash cycle, the degradation of the coloration of the dyed sample and its disgorgement on the control fabric are evaluated using gray levels (see table 2).

2.5.2 Friction test

the friction tests consists in subjecting a 100 x 25 mm test piece to friction (from front to back) under constant pressure with a dry or wet standardized felt for 1 hour. Degradation and disgorgement are then evaluated using gray scales. This test makes it possible to check whether, by repeated rubbing, the color of the dyed wool can on the one hand deteriorate and on the other hand if it can degrade (transfer) on the control supports. (Table 2).

3. Results and discussion

3.1 Optimization of the extraction

Although of natural origin, indigo is not found as such in plants but in the form of precursors, the two main ones being Indican and Isatan B [3]. The maximum yield of indigo depends on the complete release of the Indican precursor from the plant biomass during fermentation (figure 1). We evaluated the role of sugar on the yield of indigo dye during biomass fermentation. The yield of indigo with 10% sugar during 24-hour fermentation is 0.9%, these parameters were adequate to increase the yield by 0.5% (figure 3). This result can be explained by the fact that during the fermentation period the bacteria feed on the sugar and better allow to predigest and transform all the precursors of the plant, the sugar then influences the fermentation kinetics of the plant. After fermentation, as indigo is insoluble in water, the centrifugation technique was adequate for the separation of the blue pigment from the water, the centrifuge with a rotation speed of 7000 rpm for 5 minutes separated the mixture well according to their difference in density, easily recovering the indigo blue as sediment after emptying the supernatant of clear water at the top of the tubes. It was found that this method had a higher efficiency compared to the conventional filtration technique which caused a large loss of pigment in the filter papers.

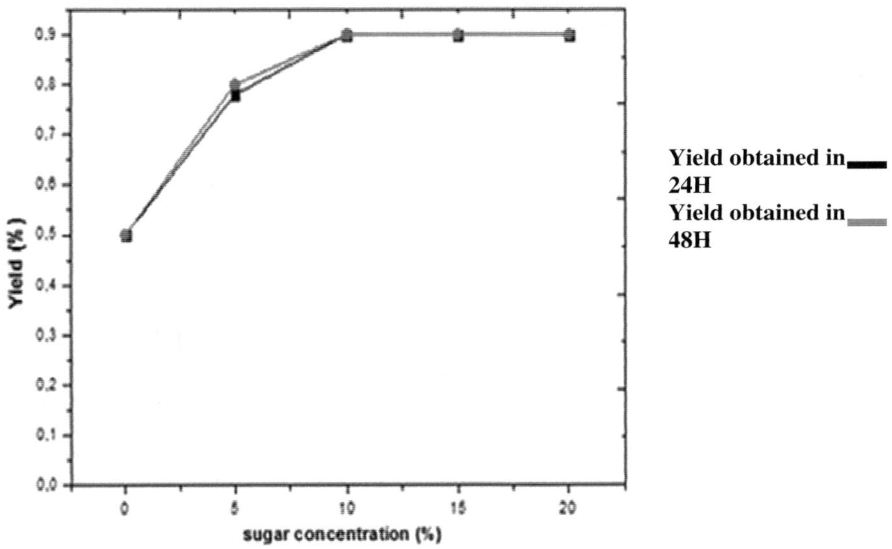

Figure 3. Indigo yield obtained as a function of sugar concentration in fermented baths

3.2 Characterization

3.2.1 Infrared IR

The infrared spectrum of the extracted indigo dye is shown in the figure 4, the values of the IR vibration bands found in this case are in agreement with previous studies carried out on the indigo molecule [7]. The latter comprises three main groups whose bands characteristic of the different vibration modes are easily identifiable. 3600-3150 cm-1 very wide peak bands of high intensity indicate the presence of elongation vibrations of the associated O-H bonds. 3280 cm-1 very thin peak bands indicate elongation vibrations of N-H bonds noted with high intensity. 3950-2800 cm-1st bands of several very thin peaks indicate elongation vibrations of Csp^2-H bonds noted with variable intensities. 1700 cm-1 several modes of C = C bond elongations, they are found here at Medium intensity .1650 cm-1st band of a single low intensity peak, corresponds to elongations of the links C = O . 1490 cm-1 bands of very fine average intensity, corresponds to the presence of deformations in the plane of the links N-H. 1480 cm-1 bands of very fine average intensity, of high intensity, corresponds to the presence of elongation vibrations S = O. 1400-1200 cm-1 bands of very fine average intensity, corresponds to the presence of elongation vibrations C-N . 1050 cm-1 bands of very fine average intensity, corresponds to the presence of elongation vibrations S-O. It is concluded that the product

IOP Conf. Series: Materials Science and Engineering **948** (2020) 012017 doi:10.1088/1757-899X/948/1/012017

obtained is exactly indigo carmine (or sodium sulfoindigotate) $C_{16}H_8N_2Na_2O_8S_2$, its structure is almost identical to that of indigo $C_{16}H_{10}N_2O_2$, only two sulphate groups, in positions 5 and 5' differentiate them [7].

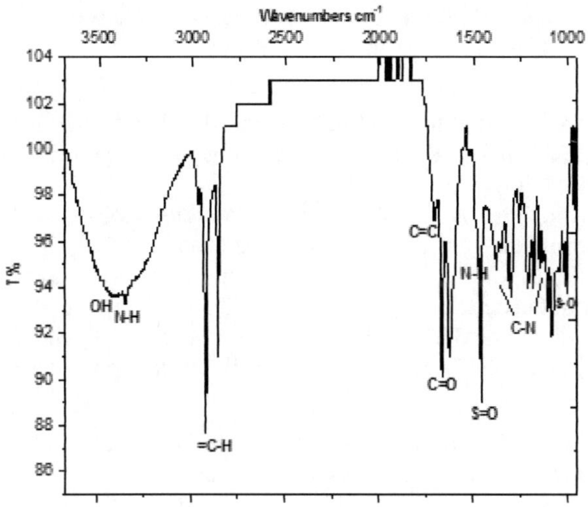

Figure 4. Infrared spectrum of indigo carmine extracted from the plant Isatis

3.2.2 UV-visible spectrophotometer

The measurement of the absorbance of an indigo carmine solution, as a function of wavelength leads to the spectrum below (figure 5); we place ourselves at the wavelength λmax corresponding to the maximum absorption in order to obtain the greatest precision for the dosage. The maximum absorption is around 598 nm which explains that indigo is blue and that the indigo molecule absorbs mainly in the orange-yellow, this result was confirmed by the literature and previous studies. [7].The 600 nm band, present in a highly diluted medium, is attributed to the indigo carmine monomer. [7]

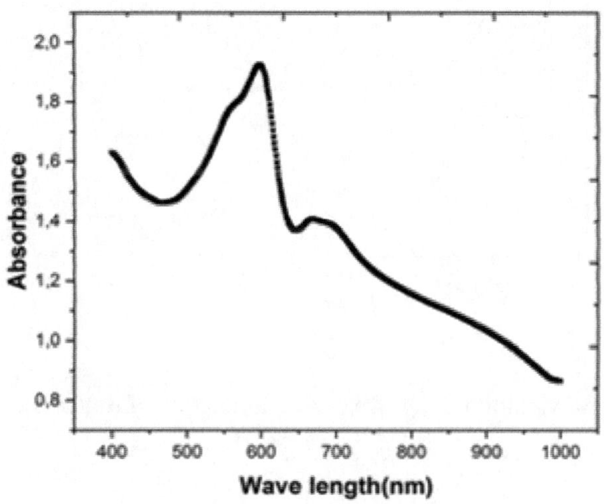

Figure 5: Absorption spectra for carmine indigo

3.3 Influence of dye parameters

3.3.1 Temperature

For this dye, we avoided working at high temperatures to avoid oxidation of the reduced dye and the temperature of 50 ° C is sufficient to saturate the wool fibers with dye provided that the soaking time of 35min is respected.

3.3.2 pH

The indigo carmine dye being very slightly soluble in water, the variation in pH slightly improves the solubility of the dye and the existence of several acid functions in the dye molecule results in a change in color of the solution as a function of pH

- In the range of pH from 0 to 10, the solution is blue and insoluble.
- In the range of pH from 9 to 12.6, the solution is green and soluble.
- pH greater than 12.6, the solution becomes yellow but the color is unstable and discoloration occurs quickly even in the dark.

pH to be chosen for dyeing the wool fiber must therefore be between 9 and 12.6 or the predominant leuco-indigo carmine molecule is soluble in water and capable of incorporating wool fiber [8].

3.3.3 Contact time

The contact time is the time needed to dye the wool fiber, depending on the optimum conditions (pH = 10, T = 50 °C). Samples of the dye bath were taken every 5 minutes and the absorbance measurement was carried out at □ max = 598 nm, in order to follow the kinetics of the dye, it can be noted that the duration of 35 min of dyeing allowed the maximum absorbance of the dye on the wool fiber to be reached (Figure 6), it can be concluded that this time is sufficient to saturate the fiber with indigo blue dye and beyond 35 min the dye bath is saturated, in fact, the dye adsorbs on the easily accessible sites on the fiber surface. Moreover, the diffusion of the dye towards the less accessible sites is favored by a regular agitation until saturation.

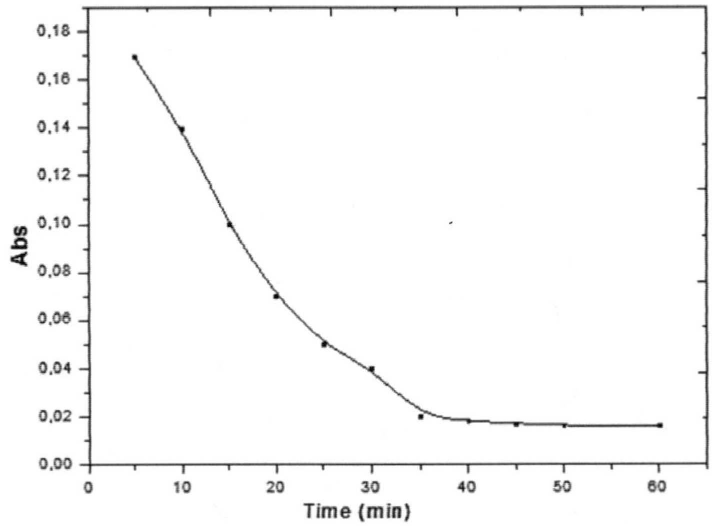

Figure 6. Monitoring the variation of absorbance as a function of time in the optimal dye bath (pH=10, T=50°C)

IOP Conf. Series: Materials Science and Engineering **948** (2020) 012017 doi:10.1088/1757-899X/948/1/012017

3.3.4 Dye concentration (light, medium, dark)

1 g of pastel dye was used to dye 1 g of wool with a bath ratio Rb: 1/100 in order to obtain a dark shade for 40 min at 50 ° C. The amount of dye used corresponds to 100% relative to with wool fibers. However, to obtain other shades with this pigment, only 40% is required for the medium shade, and 20% of the dye is required for a light shade. (Figure 7).

Figure 7. The shade of the wool fiber color (light, medium and dark) as a function of the concentration of the blue pigment.

3.3.5 Influence of salt NaCl on dye fixation

By using grey scales in the wash test, it was found that the indigo dye disintegrated on the cotton and wool multifiber giving a violet color for the first sample dyed in the absence of NaCl in the dye bath, which explains the poor fixation of the dye on the surface dyed fiber (table 2, sample 1'). However, the wool fabric dyed under these same conditions (T=50°C ,pH=10) by adding 5% NaCl to the dye baths (sample 2) did not bleed in different control samples (acetate, cotton, nylon and wool) keeping their initial colors, the dye that was not eliminated in the washing stage depends on the dyeing process in the presence of salt. In fact, the NaCl acting as a fixative converts the dye into an insoluble molecule within the fiber and offers increased resistance properties in the dyeing process compared to the other test on fabric dyed in the absence of NaCl; the salt has the function of catalyzing the reaction between the dye and the salt on the fiber to permanently fix the color. The rubbing test then confirmed that the salt fixed the indigo color very well on the surface of the fiber (sample 2), which not leave traces of blue in the control sampled while being rub-resistant.

Table 2. Evaluation of indigo dye fixation by washing and rubbing tests

Test	**Disgorgement(sample 1)**	**evaluation**	**Disgorgement(sample 2)**	**evaluation**
Washing test *105-C06*	Acetate Cotton Nylon Wool 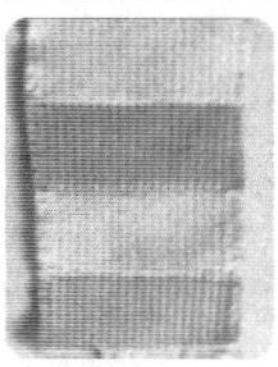	**4** **2** **5** **3**	Acetate Cotton Nylon Wool 	**5** **4** **5** **5**
Friction test *105 -X12*	dry friction 	**2/3**	dry friction 	**3**
	wet friction 	**3**	wet friction 	**3/4**

IOP Conf. Series: Materials Science and Engineering **948** (2020) 012017 doi:10.1088/1757-899X/948/1/012017

4. Conclusion

In summary, the yield of indigo is increased by 0.4% by extracting the leaves of Isatis by adding 10% sugar during fermentation, as indigo is insoluble in the extraction water. The centrifugation method had also reduced the separation time of the mixture: organic matter (indigo) and extraction water compared to conventional filtration, which caused a large loss of traces of indigo in the filter papers. Once the pigment was obtained, the factors influencing its dyeing by the wool fiber were temperature and Ph in an alkaline medium, the rinsing strength of the dyed wool fibers was considerably improved on the surface of the dyed fibers in the presence of NaCl. Natural and traditional dyeing can fundamentally change the current situation of severe wastewater pollution for denim production, we can also say that this dyeing technology will greatly expand the imagination of cowboy product designers for environmental protection and ecology.

Reference

[1] Angelini L G, Tozzi S, and Nassi o Di Nasso N 2004 Environmental factors affecting productivity, indican content, and indigo yield in Polygonum tinctorium Ait., a subtropical crop grown under temperate conditions *Journal of agricultural and food chemistry* **52(25)** 7541-7547

[2] Comlekcioglu N, Efe L, and Karaman S 2015 Extraction of indigo from some Isatis species and dyeing standardization using low-technology methods *Brazilian Archives of Biology and Technology* **58(1)** 96-102

[3] Vuorema A, John P, Keskitalo M, Mahon M F, Kulandainathan M A and Marken F 2009 Anthraquinone catalysis in the glucose-driven reduction of indigo to leuco-indigo *Physical Chemistry Chemical Physics* **11(11)** 1816-1824

[4] Park S, Ryu J Y, Seo J, and Hur H G 2012 Isolation and characterization of alkaliphilic and thermotolerant bacteria that reduce insoluble indigo to soluble leuco-indigo from indigo dye vat *Journal of the Korean Society for Applied Biological Chemistry* **55(1)** 83-88

[5] Luo Y, Pei L, and Wang J 2020 Sustainable indigo dyeing and improvement of rubbing fastness of dyed cotton fiber using different fixing agents for obtaining eco-friendly cowboy products *Journal of Cleaner Production* **251** 119728

[6] Mongkholrattanasit R, Kryštůfek J and Wiener J 2010 Dyeing and fastness properties of natural dyes extracted from eucalyptus leaves using padding techniques Fibers and Polymers 11(3) 346-350

[7] Komorsky-Lovrič Š, Mirčeski V, and Scholz F 1999 Voltammetry of organic microparticles *Microchimica Acta* **132(1)** 67-77

[8] Venil C K, Zakaria Z A and Ahmad W A 2013 Bacterial pigments and their applications *Process Biochemistry* **48(7)** 1065-1079

Structural properties and spontaneous polarization behaviour in Cu-doped LiTaO3

M.Tahiri[1, 2*], A. Jennane[2], N. Masaif[3] and E.M. Lotfi[4]

[1]Laboratory of LBGIM, Physics and chemistry department, University Hassan 2, ENS- Casablanca, Casablanca, Morocco.
[2]Laboratory of RM&I, Physics department, Faculty of Sciences and Technics, University Hassan 1, 26000 Settat, Morocco.
[3]Physics department Faculty of Sciences Kenitra, University Ibn Toufail, Kenitra, Morocco.
[4]Department of Energy and Environment, University Mohammed V, ENSET, B.P: 6207, Rabat Institutes, Morocco

* m.tahiri@usms.ma

Abstract. In this work, we have study theoretically the effect of cooling modes on the changes of spontaneous polarisation (Ps) of non-stoichiometric solid solutions of Cu-doped $LiTaO_3$ in correlation with the defects structure. The changes of spontaneous polarization in rapid and slow cooling modes with the rate of % Cu^{2+} and with the Curie temperature, allows us to analyse the defect structure of solid solutions synthesized in ternary system Li_2O-Ta_2O_5-$(CuO)_2$ in vicinity of $LiTaO_3$. The theoretical results obtained by application of our approach combined with the proposed vacancy models; show that the cooling modes have an effect on the spontaneous polarization, consequently on the defects structure of Cu-doped $LiTaO_3$. In other hand, the proposed vacancy models investigated by our theoretical approach are able to describe the defect structure and the substitution mechanism in each solid solutions of Cu-doped lithium tantalate.

KEYWORDS: Defect Structure; Vacancy Model; Cu; $LiTaO_3$, Spontaneous polarization; Substitution mechanism.

1. Introduction

Generally, a crystal having a spontaneous polarization is composed of positive and negative ions in a certain temperature range. These ions are at in their equilibrium position for which the free energy of a crystal is minimal and no coincidence of the positive and the negative charges center. Knowing that its dielectric properties strongly depend on the crystalline structure of the material, the axis of spontaneous polarization is in general, one of the crystallographic axes of the material. Spontaneous polarization in ferroelectric materials varies with temperature and it is also sensitive to the electric field created in the material. Abrahams et al [1] have been explained the origin of ferroelectricity in $LiTaO_3$ as a results of deplacement of Ta^{5+} and Li^+ inside their octaedron sites. In paraelectric phase, the spontaneous polarization disappears because the ions Ta^{5+} are in center of the oxygen octahedral and the ions Li^+ are in common faces of octahedron along the c-axis (see figure 1.a). In ferroelectric phase, the spontaneous polarization appears when the ions Ta^{5+} and Li^+ are in motion along the c-axis (see figure 1.b). Lithium tantalate ($LiTaO_3$) is one of the most used ferroelectric materials and has been the subject of extensive studies, due to their properties in electro-optics, electro- acoustic and nonlinear optics. Their physical properties have influenced by defects (intrinsic or extrinsic) structures [2–7].

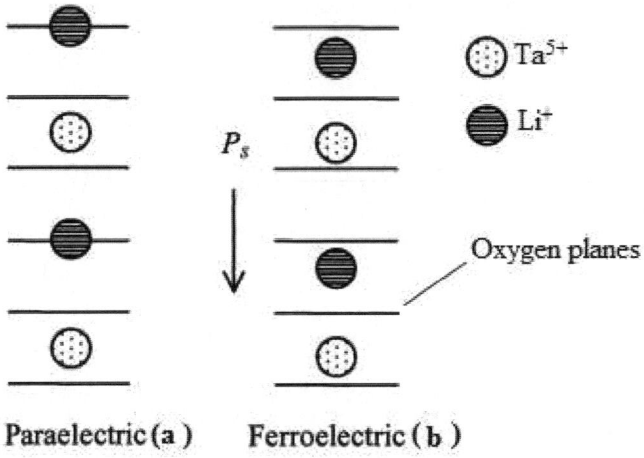

Figure 1. (a) Ferroelectric and (b) paraelectric of the $LiTaO_3$ structure.
The arrow indicates the direction of the spontaneous polarization in the ferroelectric phase.

Lithium tantalate undergo only one structural phase transition corresponding to a ferroelectric - paraelectric phase transition at temperature Tc equal to 938K [8]. His ferroelectric structure belongs to the R3c space group and the atomic arrangement consists of oxygen octaedra sharing faces along the polar trigonal axis. In this work, based on the ferroelectric phase transition theory developed by Safaryan [9], we are study the effect of slow and rapid cooling modes on spontaneous polarization in correlation with defects structure of solid solutions synthesized in the vicinity of $LiTaO_3$ in the ternary system Li_2O-Ta_2O_5-$(CuO)_2$.

2. Theoretical approach

The structure of the ferroelectric material $LiTaO_3$ as well as that of $LiNbO_3$ belongs to space group R3c and can be, considered as a superstructure of the $\alpha - Al_2O_3$ corundum structure, with Li^+ and Ta^{5+} cations along the c-axis [10]. The formula of spontaneous polarization proposed by Safaryan [9] on ferroelectric phase transition in the crystal $LiNbO_3$ and investigated by Tahiri et al [6] in $LiTaO_3$ and $LiNbO_3$ doped with the nickel is tested also for Cu- $LiTaO_3$, it can apparently, write:

$$P_0 = \frac{Ne}{V}(q_2 R_{20} + q_1(\frac{a}{2} - R_{10}) \quad [6, 9] \qquad (1)$$

Where N is the number of Li or Ta ions in the elementary cell of the crystal, V is the volume of the elementary cell, a is the distance between two oxygen planes, R_{10} and R_{20} are respectively the distances between Ta, Li and O and q_1 et q_2 are respectively the charges of Ta and Li.
In this work, while the content of Cu occupy the sites of Li or Ta with the variable rates, we are only three plans and the above formula is applicable successfully in this case of Cu doped $LiTaO_3$. Than we can write

$$P_s^* = \frac{Ne}{V^*}(q_2^* R_{20} + q_1^*(\frac{a}{2} - R_{10}) \quad [6, 9] \qquad (2)$$

Where q_1^* and q_2^* are respectively the charges of Ta and Li ions in nonstoichiometric case. V^* is the volume of elementary cell which changes with the changes of cell parameters a_H and c_H by doping of Cu.
The proposed and simplified vacancy models of solid solutions described the defects structure of Cu doped $LiTaO_3$ in the cases of slow and rapid cooling modes are in Table 1.

Table 1: Analytical formulae of lines B and D in cases of slow and rapid cooling modes

Lines	Analysed models	Proposed and simplified models

Line B	$Li_{1-x}Ta_{1-x}Cu_{3x}O_3$	$[Li_{1-x}Cu_{3x/2}][Ta_{1-x}Cu_{3x/2}]O_3$
Line D	$Li_{1-x}Ta_{1-x/3}Cu_{4x/3}O_3$	$[Li_{1-x}Cu_{2x/3}][Ta_{1-x/3}Cu_{2x/3}]O_3$

Our approach combined with the above vacancy models allows us the calculation of spontaneous polarization as a function of % Cu. For each vacancy model, we substitute the charges with her expressions in non-stoichiometric case, for example:

Line B : $$q_1^* = (1-x)q_1 + \frac{3x}{2}q_{Cu} \text{ et } q_2^* = (1-x)q_2 + \frac{3x}{2}q_{Cu}$$

Line D : $$q_1^* = (1-\frac{x}{3})q_1 + \frac{2x}{3}q_{Cu} \text{ et } q_2^* = (1-x)q_2 + \frac{2x}{3}q_{Cu}$$

The injection of those quantities in the formula of spontaneous polarization gives the analytical formulae situated in table 2.

Table 2: Analytical formulae of lines B and D in cases of slow and rapid cooling modes, where: V_{Sc} and V_{Rc} are respectively the volume of unit cell of B and D in slow and rapid cooling modes.

Lines	Proposed and simplified models	Analytical formulae
Line B	$[Li_{1-x}Cu_{3x/2}][Ta_{1-x}Cu_{3x/2}]O_3$	$P_s(B) = \frac{1.52+0.78x}{V_{Rc}(B)}$
		$P_s(B) = \frac{1.52+0.78x}{V_{Sc}(B)}$
Line D	$[Li_{1-x}Cu_{2x/3}][Ta_{1-x/3}Cu_{2x/3}]O_3$	$P_s(D) = \frac{1.52+0.13x}{V_{Rc}(D)}$
		$P_s(D) = \frac{1.52+0.13x}{V_{Sc}(D)}$

3. Results and discussion

Our approach, which combines a ferroelectric phase transition theory and the above vacancy models allow us to calculate and to analyse the behaviour of the spontaneous polarization of Cu doped $LiTaO_3$. The analytical formulae corresponding of each vacancy model in two cases of cooling modes are in table 2. Generally, in the two cases of slow and rapid cooling modes, the theoretical results obtained by our approach show that the spontaneous polarization depends strongly of the quantity of dopant for the two lines B and D with some difference.

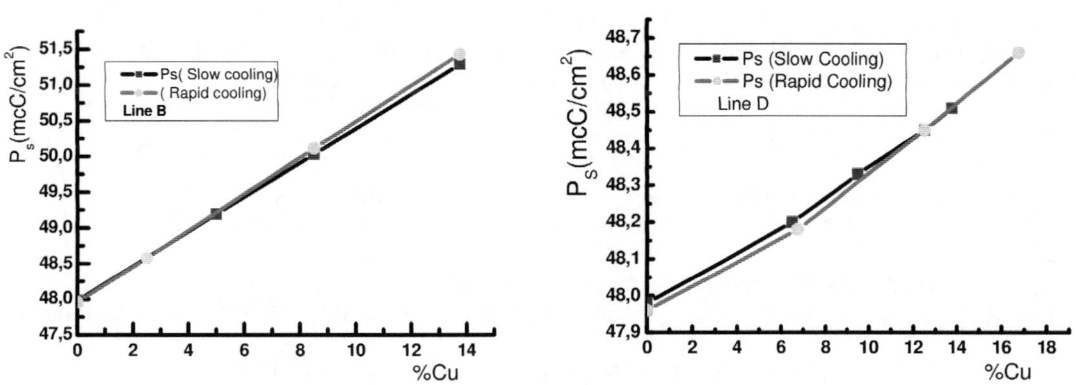

Figure 2. Variation of spontaneous polarization for Cu-doped $LiTaO_3$ as a function of % Cu in rapid and slow cooling modes

The figure 2 represents the spontaneous polarization of lines B and D in cases of slow and rapid cooling modes, we see that the spontaneous polarization increases with the rate of Cu, this result is similar to that of $LiNbO_3$ and $LiTaO_3$ doped by the nickel [6]. This phenomenon may be explained by the substitution of the vacant sites of Li and Ta with the variable rate of Cu. In this work we choose to study the case where 5O percentage of Cu occupy the site of Li and other rate occupy the sites of Ta. According to this results we assume that the spontaneous polarization increases when Cu occupy totally the site of Li, because this latest become big than the opening of triangle oxygen and is not able to pass through it. Consequently, the Li occupy both sites above and below the oxygen triangle with equal rates and the Ta ions occupies the center of its octahedron of oxygen, this result is similar to that concluded by Safaryan when, He analysed $LiNbO_3$ based on T_C.
We have confirmed our proposition by study of the spontaneous polarization as function of the Curie temperature in two cases of cooling modes; the results are gathered in figure 3. It shows that the spontaneous polarization increases with decreasing of Curie temperature.

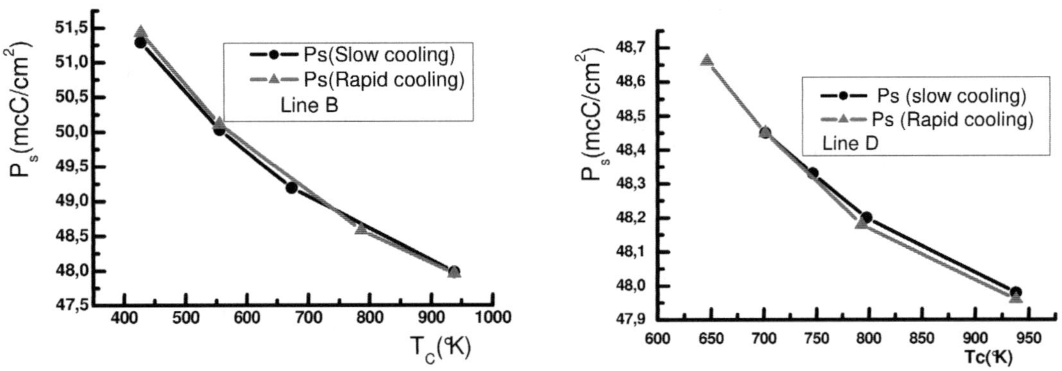

Figure 3. Variation of spontaneous polarization for Cu-doped $LiTaO_3$ as a function of T_C in rapid and slow cooling modes

This result is logical because generally, at low temperatures and sometimes at temperatures below of Tc, $LiTaO_3$ is ferroelectric material.

4. Conclusion

In this paper, we have study and analyze in the cases of rapid and slow cooling modes the defect structure of non-stoichiometric solid solutions of Cu doped $LiTaO_3$, in correlation with the changes of the spontaneous polarization at different rates of Cu content and as a function of T_C. The results shows the proportionality between the increasing of Ps with increasing of Cu content in Li site and allows us to predict and to propose the suitable vacancy models that describing this defect in the structure of Cu doped $LiTaO_3$. In other hand, this study shows that the Ps increases when T_C decreases.

References

[1] Abrahams S C and Bernstein J L 1967 *J. Phys. Chem. Solids* **28** 1685
[2] Dash B N, Dash P, Rath H, Mullick P, Biswal R, Kulriya P K, and Mishra N C 2010 *Indian J. Phys.* **84** 1315
[3] Sanyal P, Tarat S and P Majumdar 2010 *Indian J. Phys.* **84**, 1379
[4] Modi K B, Pathak T K, Vasoya N H, Lakhani V K, Baldha G J and Jha P K 2011 *Indian J. Phys.* **85** 411
[5] Tahiri M, Jennane A, Masaif N and. Lotfi E. M 2019 *Ferroelectrics* **540** 222-226, DOI: 10.1080/00150193.2019.1611098
[6] Tahiri M, Masaif N and. Jennane A 2015 Ferroelectrics, 481 189–195 DOI: 10.1080/00150193.2015.1051931

IOP Conf. Series: Materials Science and Engineering **948** (2020) 012018 doi:10.1088/1757-899X/948/1/012018

[7] Tahiri M, Jennane A, Masaif N, Lemdek E.M and Lotfi E M 2018 *Ukr. J. Phys*. **63** 347
[8] Razbirin B. S 1964 *Sov. Phys. Solid State* **6** 254
[9] Safaryan F. P 1999 Phys. *Lett. A* **255** 191
[10] Abrahams S C and Bernstein L J 1967 *Phys. Chem. Solids* **28** 1685

IOP Conf. Series: Materials Science and Engineering **948** (2020) 012019 doi:10.1088/1757-899X/948/1/012019

Structural and optical characterization of sprayed Mg and Ni co-doped CdS thin films for photovoltaic applications

T Garmim[1], L Soussi[2], A Louardi[1,*], M Monkade[1], M Khaidar[1], A Zradba[1], H Erguig[2], E Bouabdali[3], R Mghaioiuini[1,4], and A Elmlouky[1].

[1] Department of Physics, Laboratory of Condensed Matter Physics, Faculty of Science, Chouaib Doukkali University, El Jadida, Morocco
[2] Engineering Laboratory of Electrical Systems and Telecommunications, National School of Applied Sciences of Kenitra (ENSAK), Morocco.
[3] Laboratory of Engineering Sciences for Energy (LabSIPE), National School of Applied Sciences, Chouaib Doukkali University, El Jadida, Morocco.
[4] Department of Chemistry, Laboratory of Physics Chemistry of Applied Materials, Faculty of Sciences Ben M'sik, Hassan II University, Casablanca, Morocco

*Corresponding author: Email: louardi.a@ucd.ac.ma

Abstract. In the present work, Mg and Ni co-doped CdS thin films are grown on glass substrates at a temperature of 400 °C through spray pyrolysis. The influence of Mg-doping on structural, morphological and optical properties of CdS:Ni thin films are examined. Mg level is changed from 0 % to 7 % for CdS: Ni samples just as Ni concentration is fixed 2 % for all CdS thin films. It is observed from the X-ray diffraction spectra that all the samples exhibit hexagonal structure of CdS thin films. The scanning electron microscopy (SEM) images showed a dense surface structure composed of crystallites whose average size increase with increasing the Mg doping. The optical transmission curves demonstrate that CdS:Ni thin films exhibit a best optical transparency in the visible range for 1 % Mg content compared to other specimens. The energy of the optical band gap tends to decrease from 2.46 to 2.40 eV with increasing Mg concentration in CdS: Ni.

Keywords: CdS, (Ni-Mg) co-doping, Structural properties, Optical properties, spray pyrolysis.

Abbreviations

ε_s: structural strain.

σ_{st}: compressive stress (GPa).

T_C: texture coefficient of the hkl planes.

D_{hkl}: Average crystallites size (nm).

β_{hkl}: (FWHM) width half-height.

θ_{hkl}: diffraction angle

α : absorption coefficient

h: Planck constant $6.62607004 \times 10^{-34}$ m^2 kg / s

Eg: optical band gap (eV)

T: Optical transmission

1. Introduction

Cadmium sulphide CdS among the II-IV semiconductors with CdTe and CdSe, are materials of great interest in many practical applications, such as solar cells, optoelectronic devices, optical detectors, and so on. CdS characterized by good transmission, and also a direct gap with a forbidden bung width of the order of 2.42 eV, and n-type conductivity. The CdS uses, in the solar cells, as optical buffer layer for the heterojunctions, CdS / ZnS [1], CdTe / CdS [2], CdTe / CdS / ZnS [3], CdS / Cu (In, Ga) Se_2 [4].

The CdS has been developed by forming thin layers by several chemical methods such as chemical deposition bath [5], spray [6], dip caoting [7], spin coating [8] CVD [9], Some of these are pulsed laser deposition (PLD)[10],SILLAR [11] and metal organic chemical vapour deposition (MOCVD) [12]. The liquid phase spray technique is suitable for thin film deposition on large surfaces, it is also characterized by low cost and easy production process, and also provides high quality films and good adhesion to substrates.

The CdS doped by several elements to improve these physico-chemical properties, the literature shows several works on the doping of CdS by a single element, such as the elements of group III Al [13], In [14], Ga [15] and B [16] to improve the structural and electrical properties of CdS , Furthermore, by means of doping II-VI semiconductors with alkali metals like Na [17], K [18], and transition metals as Cu [19], Ni [20], Zn [21], Ag [22], Mn [23], Fe[24]. The co-doping of CdS with two elements of different natures such as a transition element Ni and an element Mg is not widely studied. Ni^{2+} and Mg^{2+} have standard ion rays of 0.069 nm and 0.072 nm, respectively, slightly lower than that of Cd^{2+} (0.097 nm). This information makes it possible to predicting that Mg and Ni ions can substitute for Cd^{2+} ions in the crystalline structure of CdS may improve the opto-electrical properties of CdS thin films.

The work of Mg and Ni co-doping in CdS has not been reported in the literature until recently. Hence, in this work, the investigation of the influence of Mg doping on structural, morphological and optical proprieties of 2 at% Ni doped CdS thin films prepared by spray pyrolysis has been carried out for first time. The technique of spray pyrolysis is used in this work successfully, this technique is quite simple and the required configuration is less time consuming and flexible for process modifications. In addition, using this technique, it is possible to produce large-area films without the need for a high vacuum and the films produced can be controlled step by step.

2. Experimental procedures

1.1 Preparation of thin films by spray pyrolysis

The nickel-magnesium co-doped CdS thin layers are prepared from a solution containing the basic precursors namely cadmium chloride hexahydrate ($CdCl_2$, $6H_2O$), as a source of cadmium, and thiourea ($CS(NH_2)_2$), as a source of sulfur. To this initial solution we add a solution of nickel chloride hexahydrate ($NiCl_2$, $6H_2O$) of a fixed concentration, such that the ratio [Ni] / [Cd] equal to 2%. Add a solution of magnesium sulfate $MgSO_4$ of different concentration by varying the ratio [Mg] / [Cd] between 0 and 7%. The solutions obtained are stirred vigorously with a magnetic stirrer for 20 minutes and then sprayed into fine droplets by means of a perfume atomiser on the preheated glass substrates at 400°C.

1.2 Characterizations

IOP Conf. Series: Materials Science and Engineering **948** (2020) 012019 doi:10.1088/1757-899X/948/1/012019

The as deposited films show good adhesion to the substrate and exhibit a yellow color. The crystal structure and phase transition for undoped and Mg and Ni co-doped CdS thin films were determined with a diffractometer 'XPERT-PRO' model using CuKα radiation with 2θ ranging from 10° to 70°. The surface morphologies of the deposited films were observed using scanning electron microscopy (SEM). Optical measurements were carried out with a SHIMADZU 3101 PC UV-VIS-NIR spectrophotometer in the wavelength range of 300-2500 nm.

3. Results and Discussion

2.1 Structural properties

The results of the X-ray diffraction on the thin layers of cadmium sulphide give very clear peaks, Figure 1 shows the spectra of the X-ray diffraction of the thin layers of CdS co-doped with nickel and magnesium. These spectra show several diffraction lines indicating that the resulting CdS layers are polycrystalline. The lines of the CdS sample doped with 2% Ni are located at 2θ = 25.00°, 26.70°, 28.37°, 36.88°, 43.94°, 48.06°, and 52.06° (Figure 1. (a)) showing that CdS exhibits polycrystalline behavior with a hexagonal crystalline structure [25], due to the diffraction planes (100), (002), (101), (102), (110), (103), (112), with the most intense peak (002) which correspond to the preferred orientation [26]. After the doping of $Cd_{0.98}Ni_{0.02}S$ with 1% Mg, the peak (002) remains more intense, which implies that the preferential orientation is retained, with a remarkable increase in the peak (101). But, when $Cd_{0.98}Ni_{0.02}S$ doped with 3% Mg the peak (101) becomes more intense, which shows that the preferential orientation is changed towards the plane [101] with the decrease of the peaks (102) and (103), and the birth of a new peak located at 2θ = 53.11°. After the doping of the CdS at 5% and 7% there is a decrease in the peak (101) which can be attributed to the low degree of crystallinity. It should be noted that no peak related to Ni and Mg metal elements and their compounds including NiS, MgS, MgO, NiO, etc. can be observed. The values of the inter-reticular distances d_{hkl} corresponding to the (100), (002), (101), (102), (110) and (103) planes are compared with those reported in reference [25]. A mapping of the calculated d_{hkl} values to the standard values confirms the hexagonal crystal structure (wurtzite) of CdS for all deposited layers. It is well known that CdS has two crystal structures; hexagonal (wurtzite) and cubic (blende). The latter is considered the most stable phase between 50 and 900°C. In addition, the hexagonal structure of CdS thin films is preferable for the applications of this material in solar cells because of the parameters of the CdS and $CuInSe_2$ cells (absorber material in the solar cell) which are very close.The difference is 0.7% while that between cubic CdS and $CuInSe_2$ is 1.2% [27].

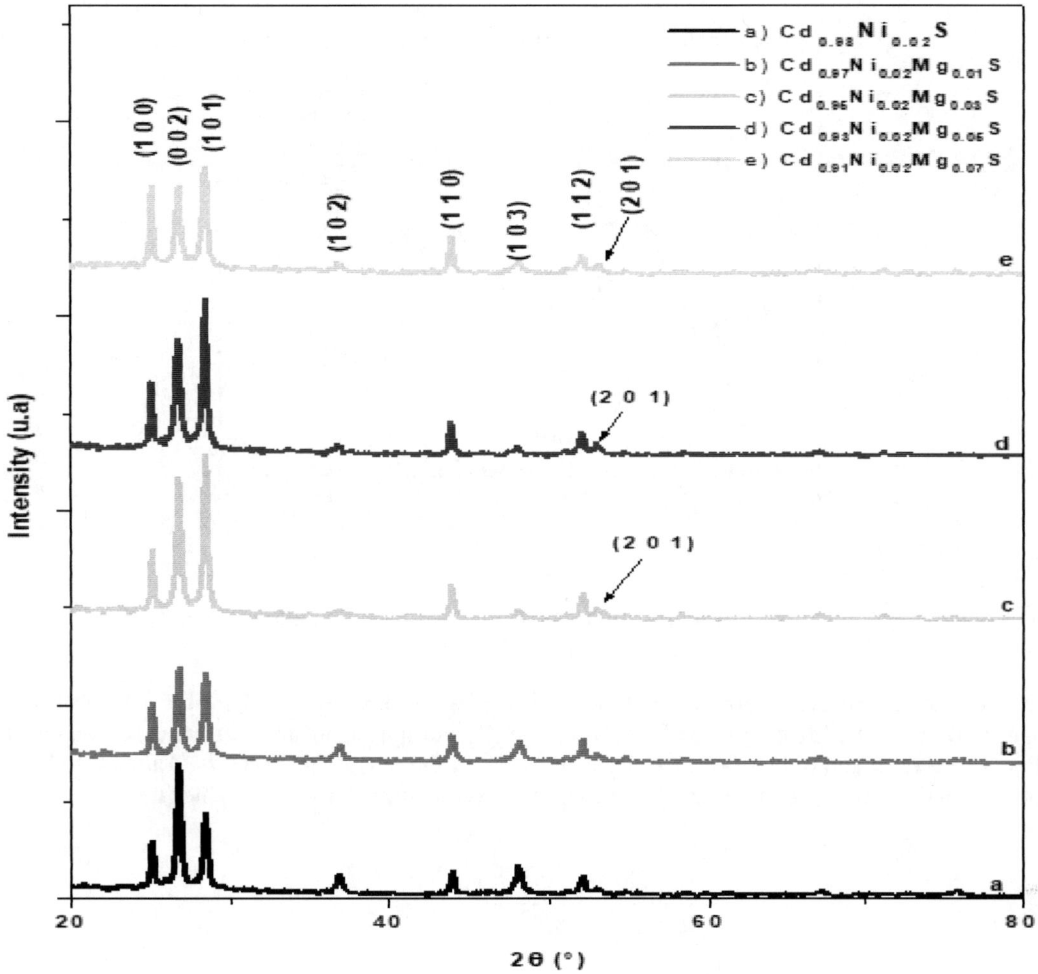

Figure 1 . XRD data of a) $Cd_{0.98}Ni_{0.02}S$, b) $Cd_{0.97}Ni_{0.02}Mg_{0.01}S$, c) $Cd_{0.95}Ni_{0.02}Mg_{0.03}S$, d) $Cd_{0.93}Ni_{0.02}Mg_{0.05}S$ and e) $Cd_{0.91}Ni_{0.02}Mg_{0.07}S$ thin films.

Figure 2 shows a displacement Δ ($2\theta_{101}$) of the position of the diffraction line (101) of CdS co-doped with Ni and Mg towards the large Bragg angles. It goes from 28.37° for a layer of $Cd_{0.098}Ni_{0.02}S$ to 28.42° for $Cd_{0.95}Ni_{0.02}Mg_{0.03}S$. This peak displacement results from the structural stress associated to the co-doping with Ni and Mg, calculated by the relation [28].

$$\varepsilon_s = -\Delta(\theta_{101})\text{cotg}(\theta_{101}) \quad (1)$$

This structural stress is caused by a compressive stress σ_{st} which can be estimated by:

$$\sigma_{st} = (\varepsilon_{st})B \quad (2)$$

With B a ''bulk modulus'' constant of CdS estimated by the value of 69.44 GPa [28].

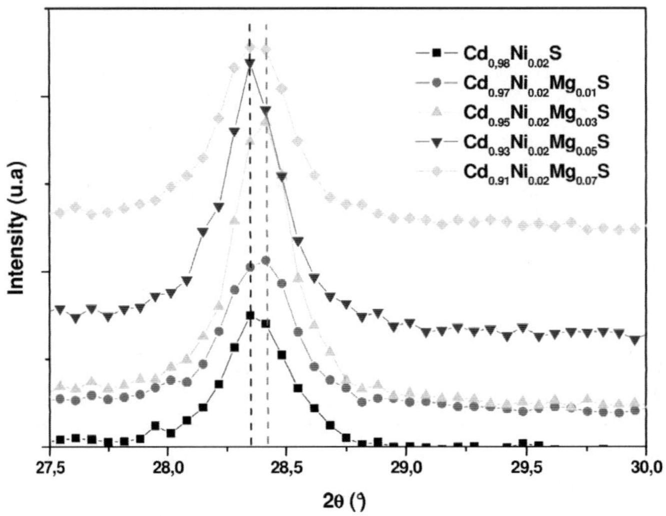

Figure 2. *Variation of the (101) peak position of $Cd_{0.98}Ni_{0.02}S$, $Cd_{0.97}Ni_{0.02}Mg_{0.01}S$, $Cd_{0.95}Ni_{0.02}Mg_{0.03}S$, $Cd_{0.93}Ni_{0.02}Mg_{0.05}S$, and $Cd_{0.91}Ni_{0.02}Mg_{0.07}S$ thin films*

Figure 3 also shows that the compressive stress varies between 0.08 to 0.14 GPa for 1 % and 3 % Mg doped CdS:Ni. This small decrease in the value of the constraint is not able to create a change in the crystalline structure, it can only produce a slight decrease in the lattices parameters (Table 1). The lattice parameters, a and c, are calculated from the peaks position using the formula of hexagonal system (Eq. (3)).

$$\frac{1}{d_{hkl}^2} = \frac{4(h^2+k^2+hk)}{3a^2} + \frac{l^2}{c^2} \quad (3)$$

Table 1: Lattice parametersof$Cd_{0.98}Ni_{0.02}S$, $Cd_{0.97}Ni_{0.02}Mg_{0.01}S$, $Cd_{0.95}Ni_{0.02}Mg_{0.03}S$, $Cd_{0.93}Ni_{0.02}Mg_{0.05}S$, and $Cd_{0.91}Ni_{0.02}Mg_{0.07}S$ thin films.

Lattices parametrs / Sample	a=b (□)		c(□)	
	Calculated	ASTM	Calculated	ASTM
$Cd_{0.98}Ni_{0.02}S$	4.112	4.120	6,678	6.682
$Cd_{0.97}Ni_{0.02}Mg_{0.01}S$	4.111	4.120	6.674	6.682
$Cd_{0.95}Ni_{0.02}Mg_{0.03}S$	4.109	4.120	6.674	6.682
$Cd_{0.93}Ni_{0.02}Mg_{0.05}S$	4.115	4.120	6.684	6.682
$Cd_{0.91}Ni_{0.02}Mg_{0.07}S$	4.114	4.120	6.680	6.682

However, with increasing doping up to 5 % and 7 % with Mg, the 'a' and 'c' 'values increase, which can be explained by the fact that Mg^{2+} ions take up the interstitial sites in the CdS structure. The interstitial site occupancy by Mg^{2+} ions causes an enlargement in the volume of the crystal lattice of CdS, causing a decrease in the quality of the crystal.

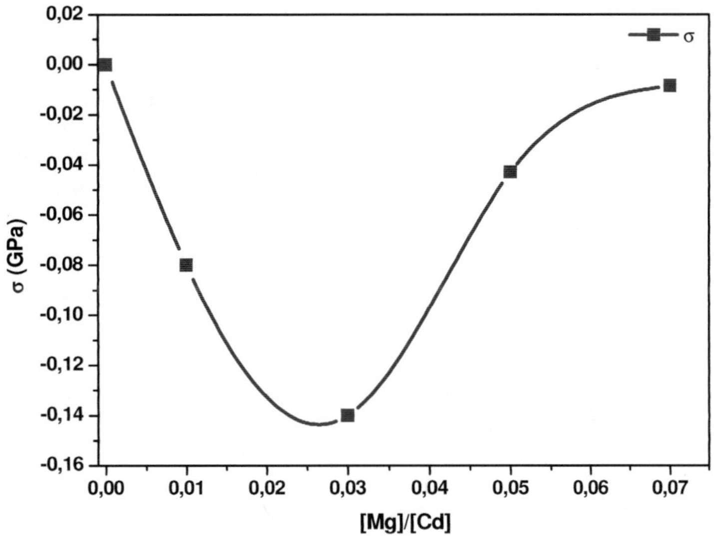

Figure 3. Structural strain and stress of Mg and Ni co-doped CdS.

The preferred orientations of Mg and Ni co-doped CdS layers, giving an indication of the degree of crystallization of the lattice planes (hkl) were evaluated by the coefficient of texture (T_C) of the two main planes (002) and (101), this coefficient is calculated from the data of the X-ray diffraction spectra, using the formula [29].

$$T_C(hkl) = \frac{I(hkl)/I_0(hkl)}{\frac{1}{N}\sum I(hkl)/I_0(hkl)} \quad (4)$$

Where T_C (hkl) is the texture coefficient of the planes (hkl), I is the measured intensity, I_0 is the standard intensity of CdS, and N is the number of reflection peaks observed on the spectrum.

Figure 4 represents the evolution of the crystallization coefficients of the two main planes (002) and (101) as a function of the percentage of doping of Ni and Mg in the starting solution. It can be observed that the coefficient of texture is maximum for both planes (002) and (101) with a doping level equal to 3% of Mg (2% Ni is fixed). This reflects the high crystallization degree of these two crystallographic planes.

Figure 5 shows the variation of the half-height width (FWHM) as a function of the Mg doping concentration of the diffraction line (101). This figure shows that the width at half height has a minimum value for the two doping rate values at Mg 0% and 3%. The fineness of the X-ray diffraction line is the consequence of an improvement of crystallinity. This will result in the formation of crystallites with a large size in the layers $Cd_{0.98}Ni_{0.02}S$ and $Cd_{0.95}Ni_{0.02}Mg_{0.03}S$.

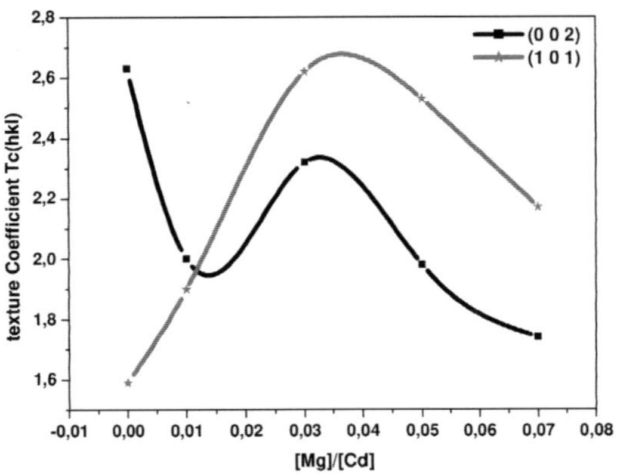

Figure 4. T_C (101) and T_C (002) plotted against the [Mg]/[Cd] ratio (Ni fixed at 2 at%).

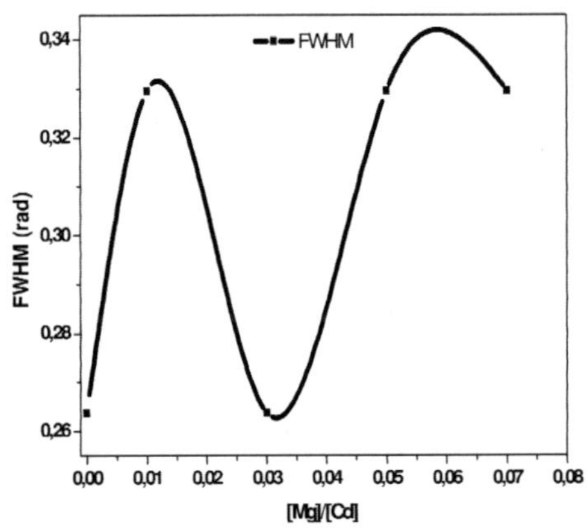

Figure 5. Variation of the width at half height according to the rate of doping

The average size of Mg and Ni co-doped CdS crystallites (D_{hkl}) is determined from the X-ray diffraction patterns using Scherrer's formula [30].

$$D_{hkl} = 0.9\frac{\lambda}{\beta_{hkl}\cos\theta_{hkl}} \quad (5)$$

With λ the wavelength of the X-ray (λ = 1.5406 Å), β_{hkl} the width at half-height and θ_{hkl} the diffraction angle of Bragg.

Figure 6 shows the variation of the crystallite size as a function of the [Mg] / [Cd] ratio in the solution, with the [Ni] / [Cd] ratio set at 2%. We notice that its average value takes a maximum value for both the samples $Cd_{0.98}Ni_{0.02}S$ and $Cd_{0.095}Ni_{0.02}Mg_{0.03}S$.

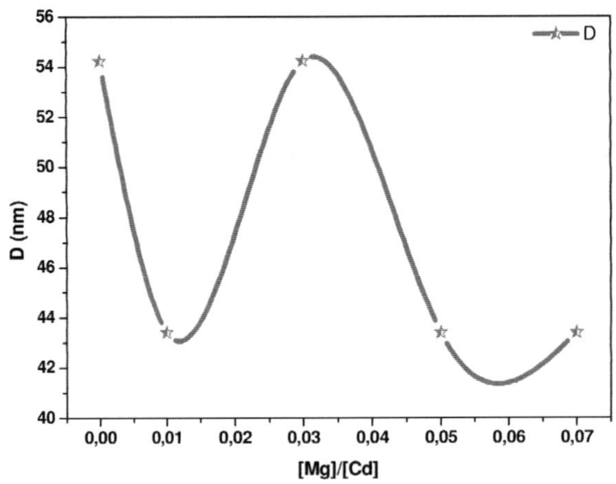

Figure 5. Average crystallites size of Mg and Ni co-doped CdS.

IOP Conf. Series: Materials Science and Engineering **948** (2020) 012019 doi:10.1088/1757-899X/948/1/012019

2.2 Surface morphology

Plan view SEM images of all the samples are presented Figure 7(a)-(c) (with magnification of x40000). Figure 7(a) display the morphology of $Cd_{0.98}Ni_{0.02}S$ thin films the size of the grains is not homogenously distributed over the surface, Figure 7(b) show the presence of homogeneously distributed grains having a relatively smooth surface, indicating that the films are homogeneous and show good substrate adhesion. As the Mg doping take the value 3 % in the spray solution, the average grains size increases which is in agreement with X-ray diffraction (XRD) results.

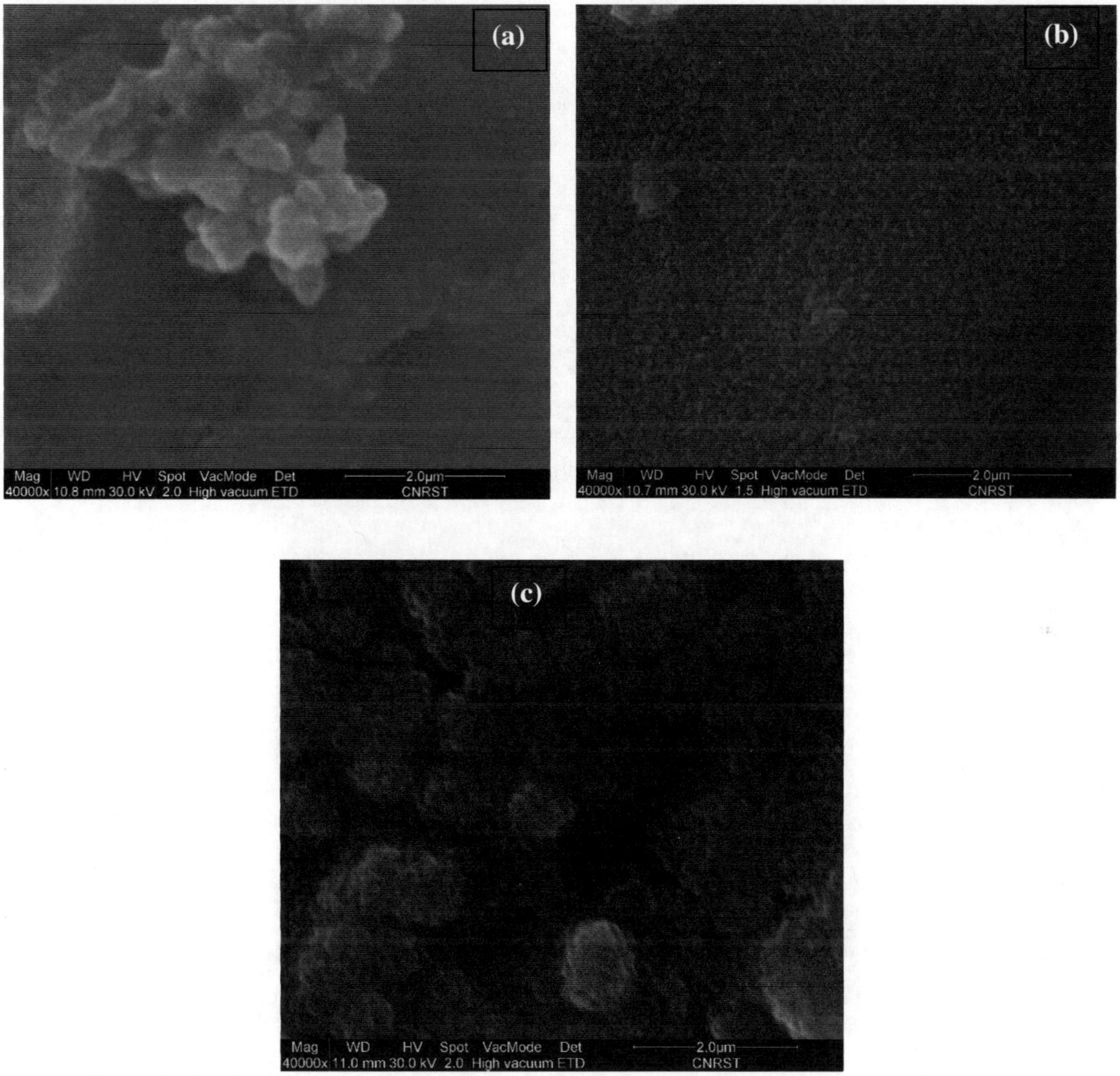

Figure 3. SEM plan views of (a) $Cd_{0.98}Ni_{0.02}S$, (b) $Cd_{0.95}Ni_{0.02}Mg_{0.03}S$, (c) $Cd_{0.91}Ni_{0.02}Mg_{0.07}S$

2.3 Composition analysis

The composition analysis of all films was made by energy dispersive X-ray spectroscopy. Figure 8 presents the EDAX spectra of Mg and Ni co-doped CdS films. EDAX spectra reveal that deposited CdS layers contain cadmium elements and sulfur, with low percentages of nickel and magnesium; the existence of the chlorine (Cl) as a source of Cd ($CdCl_2$), and the carbon (C) is perhaps belongs to the source of sulfur CS (NH_2). When the Mg doping rate exceeds 7%, the oxygen appears with a small amount, which explains the formation of the CdO phase in the deposited films. Other elements such as the Si detected in our layers probably come from the glass substrate.The distribution of Cd, S, Ni and Mg elements on the surface is studied by EDAX mapping and the results are given in Figure 8 (a) and the individual elemental maps of Cd, S, Ni and Mg are also shown in Figure 8 (b, c, d and e) showing the presence and homogeneous distribution of the elements mentioned on the surface.

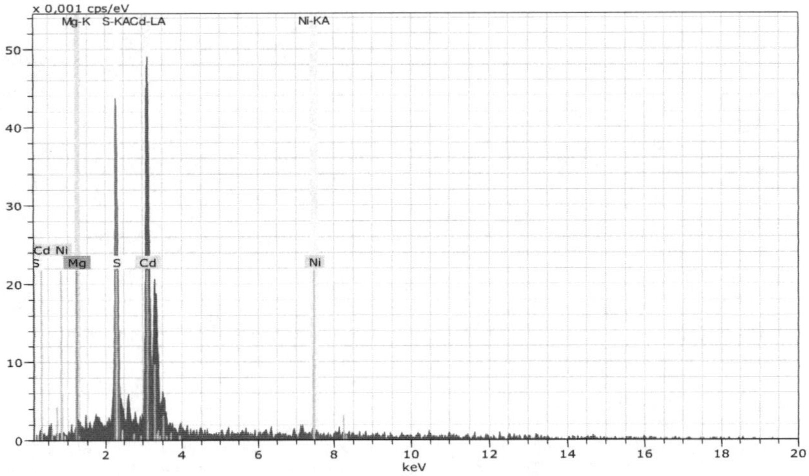

Figure 8. EDS spectrum of Cd0.95Ni0.02Mg0.03S thin films.

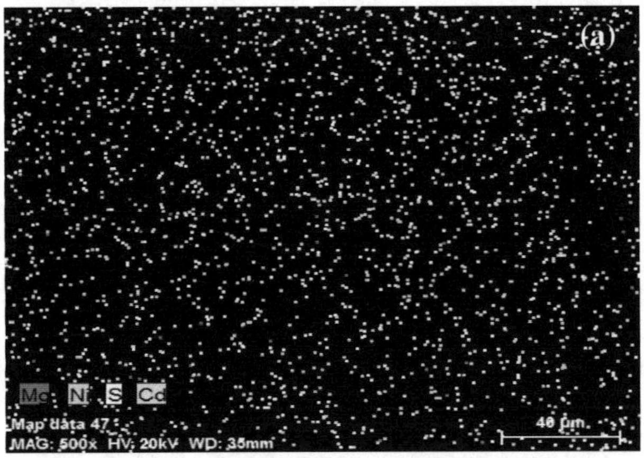

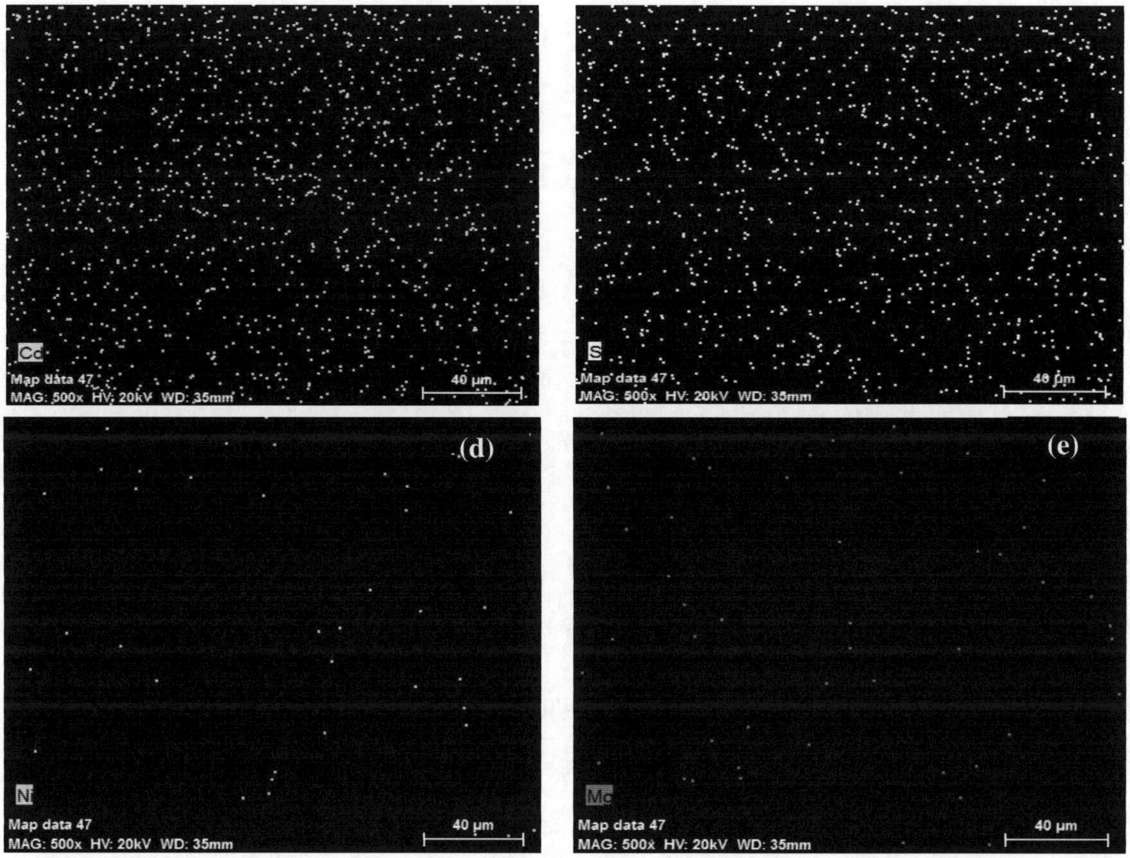

Figure 9. (a) SEM plan view of Cd0.95Ni0.02K0.03S thin films including mapping of Cd, S, Mg and Ni elements on the surface, individual mapping results of (b) Cd, (c) S, (d) Ni and (e) Mg elements.

2.4 Optical properties

Optical properties such as transmission, absorption coefficient and band gap energy of Mg and Ni co-doped CdS thin films are determined from the variation in optical transmission as a function of the wave length (λ) in the range of 300 to 1000 nm. The results show that CdS cadmium sulphide thin films have good transmittance in the visible region of solar spectrum, which allows them to be used as transparent windows or buffer layers in solar cells. Optical transmission curves of the thin films of $Cd_{0.98}N_{0.02}S$, $Cd_{0.97}Ni_{0.02}Mg_{0.01}S$, $Cd_{0.95}N_{0.02}Mg_{0.03}S$, $Cd_{0.93}Ni_{0.02}Mg_{0.05}S$ and $Cd_{0.91}Ni_{0.02}Mg_{0.07}S$ are shown in Figure 9. The CdS sample doped with 2 % Ni shows practically 65 % of transparency in the visible region, after the doping of 1% with Mg, the transmittance of thin films $Cd_{0,\,98}Ni_{0,\,02}S$ decreases until at the wavelength of 700 nm and then increases to reach a transmission value of about 75%. When thin layers $Cd_{0.98}Ni_{0.02}S$ doped with 3, 5 and 7 at % of Mg, the transmission decreases remarkably up to a value of 50% for the ratio of concentration [Mg] / [Cd] equal to 7 %. Films co-doped with nickel and magnesium show a slight decrease in transmission compared to undoped films in the visible and near infra-red regions. This can be explained by the increase in the optical absorption of the charge carriers, following an increase in their density by magnesium doping effect. The same behavior has been reported by T. Sivaraman et al. [31] for thin layers of CdS doped with Mg.

The optical band gap of (Ni, Mg) co-doped CdS has been evaluated using the Tauc's relation [32]:

IOP Conf. Series: Materials Science and Engineering **948** (2020) 012019 doi:10.1088/1757-899X/948/1/012019

$$(\alpha h\nu) = A(h\nu - Eg)^2 \quad (6)$$

α is calculated initially by the relation :

$$\alpha = \frac{\ln(T^{-1})}{d} \quad (7)$$

Where T is the optical transmittance, α is the absorption coefficient, A is a constant and hν is the photon energy and d is the film thickness.

As for pure cadmium sulphide, the optical gap value of Mg and Ni co-doped CdS thin layers is estimated by extrapolation of the straight line of the curve that gives $(\alpha h\nu)^2$ as a function of photon energy (hυ), shown in the Figure 10. The optical gap values of the nickel and magnesium doped thin films found by extrapolation of the absorption coefficient curves as a function of energy are given in Table 3, We have found that the Eg values varied from 2.460 to 2.398 eV, we can explain it as follows: the nickel and magnesium atoms are donor atoms, so incorporation of Mg^{2+} ions create degenerate donor levels in the CdS conduction band, leading to the extension of the conduction band in the band gap region. This results in the reduction of the band gap of the co-doped layers. This is consistent with the results previously obtained by T. Sivaraman et al. [31] for the Mg doped CdS thin films, and C. Lokhande et al. [33] for Al doped CdS thin films.

Table 2: Variation in the value of the optical gap as a function of doping.

Sample	$Cd_{0.98}Ni_{0.02}S$	$Cd_{0,91}Ni_{0.02}Mg_{0.07}S$	$Cd_{0,91}Ni_{0.02}Mg_{0.07}S$	$Cd_{0,91}Ni_{0.02}Mg_{0.07}S$	$Cd_{0,91}Ni_{0.02}Mg_{0.07}S$
E_g (eV)	2,460	2.455	2,440	2,440	2,398

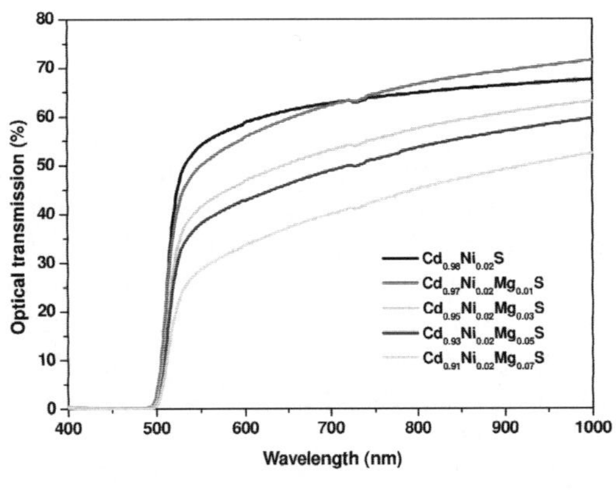

Figure 4 . Transmission plots of $Cd_{0.98}Ni_{0.02}S$, $Cd_{0.97}Ni_{0.02}Mg_{0.01}S$, $Cd_{0.95}Ni_{0.02}Mg_{0.03}S$, $Cd_{0.93}Ni_{0.02}Mg_{0.05}S$ and $Cd_{0.91}Ni_{0.02}Mg_{0.07}S$ thin films.

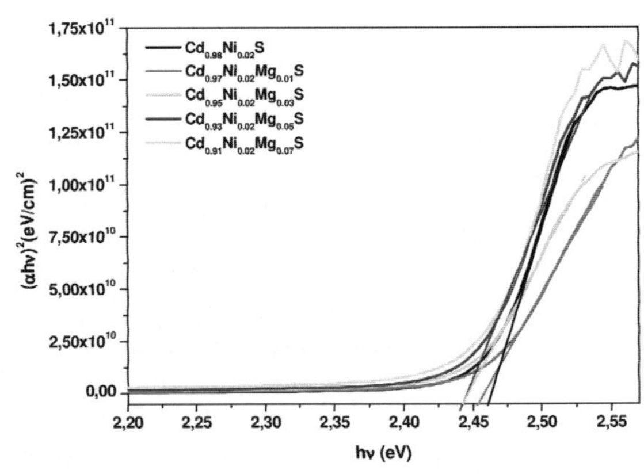

Figure 11. Tauc's plot results of $Cd_{0.98}Ni_{0.02}S$, $Cd_{0.97}Ni_{0.02}Mg_{0.01}S$, $Cd_{0.95}Ni_{0.02}Mg_{0.03}S$, $Cd_{0.93}Ni_{0.02}Mg_{0.05}S$ and $Cd_{0.91}Ni_{0.02}Mg_{0.07}S$ thin films.

3.Conclusions

Mg doped ($Cd_{0.98}Ni_{0.02}S$) films with different Mg concentrations (0, 1, 2, 4 and 7%) were synthesized by a simple Spray Pyrolysis technique. The effect of Mg doping on the structural and optical properties of $Cd_{0.98}Ni_{0.02}S$ films were studied. The structural studies revealed that all the films exhibit hexagonal structure (wurtzite) with preferential orientation along the (101) from 3 % up to 7 % Mg-doping. The SEM analysis shows that the Cd0.95Ni0.02Mg0.03S sample has a homogeneous morphology with a grain allocation. The optical study showed that all the films were highly transparent. The transmittance in the visible region varies between 50% and 75% depending on the Mg-doping. The gap energy decreases with the Mg doping increase. The good structural and optical properties of (Mg, Ni) co-doped CdS make it a potential candidate for window material in solar cell applications.

References

[1] Castillo R H, Acosta M, Riech I, Rodríguez G S, Mendez-Gamboa J, Acosta C, Zambrano M, *Optik*, **148** 95-100.
[2] Yu M, Kenny S D, Section B: Interactions of Beams with Materials and Atoms **412** 66-70.
[3] Olusola O I, Madugu M L, Dharmadasa I M, *J. Mater. Chem. Phys.* **191** 145-150.
[4] Niu X, Zhu H, Liang X, Guo Y, Li Z, Mai Y, *J. Appl. Surf. Sci.* **426** 1213-1220.
[5] Ali A M, Yusoff Y, Ali L M, Misran H, Akhtaruzzaman M, Alghoul M A, Sopian K, Radiman S, Amin N, *J. Sol. Energy* **173** 120-125.
[6] Ojeda-Barrero G, Oliva-Avilés A I, Oliva AI, Maldonado R D, Acosta M, Alonzo-Medina G M, *J. Mat. Sci. Semicon. Proc* **79** 7-13.
[7] Tao H, Jin Z, Wang W, Yang J, Hong Z, *J. Mater.* **65** 1340–1343.
[8] El Kissani A, Ait Dads H, Oucharrou S, Welatta F, Elaakib H, Nkhaili L, Narjis A, Khalfi A, El Assail K, Outzourhit A, *J. Thin Solid Films* **664** 66-69.
[9] Shen X P, Yuan A H, Wang F, Hong J M, Xu Z, *J. Solid State Commun.* **133** 19-22.
[10] Yang X, Bao Z, Luo R, Liu B, Tang P, Li B, Zhang J, Li W, Wu L, Feng L, *J. Mat. Sci. Semicon. Proc.* **48** 27–32.
[11] Sasagawa M, Nosaka Y, *J. Electroanal. Chem.* **536** 141-144.
[12] Jack L. Boone Scott A. Don H D. Cantwell M G, *J. Thin Solid Films* **176** 143-150.
[13] Bairy R, Jayarama A, Shivakumar G K, Kulkarni S D, Maidur S R, Patil P S, *Physica B: Condensed Matter* **555** 145-151.
[14] Butt S, Shah N A, Nazir A, Ali Z, Maqsood A, *J. Alloys Compd.* **587** 582-587.
[15] Soto E, Vaquero F, Mota N, Fateixa S, Trindade T, Navarro R M, Fierro J L G, *J. Mater. Today Energy* **9** 345-358.
[16] Challa K K, Magnone E, Kim E T, *J. Mater* **85** 135–137.
[17] Yılmaz S, Atasoy Y, Tomakin M, Bacaksız E, *J. Superlattices and Microstructures* **88** 299-307.
[18] Mageswari S, Dhivya L, Palanivel B, Murugan R, *J. Alloys Compd.* **545** 41–45.
[19] Kumar V, Kumar K, Jeon H C, Kang T W, Lee D, Kumar S, *J. Phys. Chem. Solids***124** 1-6.
[20] Yahia I S, El Radaf I M, Salem AM, Sakr G B, *J. Alloys Compd.* **776** 1056-1062.
[21] Chand P, Ghosh R, Sukriti, *Optik* **161** 44-53.
[22] Veerathangam K, Pandian M S, Ramasamy P, *J. Mater. Res.* **94** 371-377.
[23] Wang C, Wang H M, Fang Z Y, *J. Alloys Compd.* **486** 702-705.
[24] Thambidurai M, Muthukumarasamy N, Agilan S, Murugan N, Arul N S, Vasantha S, Balasundaraprabhu R, *J. Solid State Sci.***12** 1554-1559.
[25]Standard JCPDS Data Card N° 00-001-0783.
[26] Sivarama T, Nagarethinam V S, Balu A R *J. Mat. Sci. Semicon. Proc.* **2** 6-15.
[27] Rmili A, Ouachtari F, Bouaoud A, Louardi A, Chtouki T, Elidrissi B, Erguig H, *J. Alloys Compd.*, **557** 53-59.

[28] Benkhettou N, Rached D, Soudini B, Driz M, Phys. Stat. Solid ***241*** 101-107.
[29] Balu A R, Nagarethinam V S, Suganya M, Arunkumar N, Selvan G, *J. Electron Devices*, **12** 739 – 749.
[30] Yakuphanoglu F, Ilican S, Caglar M, Caglar Y, *J. Optoelectron. Adv. Mater* **9** "2180''
[31] Sivaraman T, Balu A R, Nagarethinam V S, *J. Mat. Sci. Semicon. Proc.***27** 915–923.
[32] Tauc J, "Amorphous and Liquid Semiconductors" *Plenum Press*, New York (1974).
[33] Lokhande C, Pawar S, *Solid State Commun.* **44** 1137 – 1139.

IOP Conf. Series: Materials Science and Engineering **948** (2020) 012020 doi:10.1088/1757-899X/948/1/012020

Electric and dielectric behavior of carbonate intercalated CoAl-Layered Double Hydroxide (LDH) investigated by impedance spectroscopy

F Z Bouragba[1,*], W Elhatimi[1], R Lahkale[1], E Moujahid[1], E Sabbar[1]

[1]Faculty of Science – Chouaib Doukkali University, Chemistry Departement, Laboratory of Physico -Chemistry of Materials, El Jadida –Morocco

*Corresponding author: bouragba.f@ucd.ac.ma

Abstract- Cobalt-based layered double hydroxides (LDH) have invoked a great deal of interest in the field of energy storage due to their unique properties and low cost. Herein, carbonate intercalated CoAl-LDH was prepared in order to investigate the electrical and dielectric properties of this material. This polycrystalline sample was synthesized by the conventional co-precipitation method, characterized by PXRD, FT-IR and TGA, and analyzed by ICP. Ac impedance was measured by the complex impedance spectroscopy (CIS) method using pressed pellet sample over a frequency range of (1Hz-1MHz) and room temperature, to separate contribution of grains and grain boundaries to capacitance, resistance and conductivity of the sample. The adjustment of conductivity parameters by the double power law allowed the determination of σ_{dc}. The relaxation mechanism is dominated by long range movement of charge carriers: carbonate ions and H^+ protons of water molecules in the interlayer space. This material exhibits high dielectric constant values and a low loss factor at high frequencies which makes the valorization of this material may be interesting in the area of dielectrics.
Keywords- Layered Double Hydroxide; Impedance Spectroscopy; Dielectric Properties; Equivalent Electrical Circuit; Ac Conductivity.

1. Introduction

LDH are a kind of (2D) anionic clays with positively charged brucite-like layers and intercalated charge balancing interlayer anions. LDH have the general formula $[M^{II}_{1-x}M^{III}_{x}(OH)_2]^{x+}[(A^{n-})_{\frac{x}{n}}.mH_2O]^{x-}$, where M^{II} (Mg^{2+}, Cu^{2+}, Ni^{2+}, Co^{2+}, Zn^{2+}...etc.) and M^{III} (Al^{3+}, Fe^{3+}, Cr^{3+}...etc.) are respectively divalent and trivalent metal ions. A^{n-} is the exchangeable hydrated anions located in the interlayer gallery for charge balance [1, 2]. LDH have been intensively studied recently in various applications such as: environment [3], catalysis [4], medicine [5], supercapacitors [6], electrode modified materials [7] , batteries [8] , etc. Transition-metal-based LDH have been identified as efficient and stable with additional advantages, including composition flexibility, low cost, ease of preparation and light weight. Among them, the Co-based LDH has drawn much attention due to its electrochemical performances, not only in the layered form but they were regarded as ideal precursors to the preparation of the corresponding oxide nanoplates, and this by thermal transformation from the corresponding LDH precursors [9]. This kind of studies can be useful to designing functional nanodevices from spinel nanostructures. Indeed, many investigators have been attracted especially by CoAl-LDH. For example, CoAl-LDH nanosheets were in-situ incorporated with reduced graphene oxide to improve the electrochemical performance of lithium-oxygen battery

IOP Conf. Series: Materials Science and Engineering **948** (2020) 012020 doi:10.1088/1757-899X/948/1/012020

[10] . Also, Amperometric sensors were fabricated based on CoAl-LDH for hydrogen peroxide detection [11].

Since the CoAl-LDH has already been used as the active material for capacitors, electrochemical studies are reported on this compound. Although, so far the knowledge of the present author is concerned no significant reports are available on impedance spectroscopy study, dielectric study and ac conductivity.

In order to use LDH based materials in electric/dielectric fields, it is necessary to know their dynamic properties such as resistance, conductivity, time relaxation, permittivity, etc. by a technique which is used for simultaneous electric and dielectric characterization of samples. This is why complex impedance spectroscopy measurement was applied to LDH [12–15] and revealed interesting properties of these materials. Recent works interested to the study of electric and dielectric properties of LDH were focused on the effect of layer charge [16, 17], pH of the synthesis [5, 18], intercalated anions [19, 20], measurements conditions (humidity, temperature, pressure…) [21] on the ionic conductivity and the dielectric parameters.

In the present work, we describe the synthesis and characterization of carbonates intercalated CoAl-LDH by PXRD, FT-IR, TGA and ICP. We investigate the electric properties using modelization of the sample by equivalent electrical circuit. The Ac conductivity, residual conductivity, and the dielectric parameters for the prepared sample were studied.

2. Experimental

2.1. Materials

The starting chemicals of cobalt nitrate ($Co(NO_3)_2{\bullet}6H_2O$) and aluminum nitrate ($Al(NO_3)_3{\bullet}9H_2O$) (Fluka Analytical ≥98%), sodium hydroxide (NaOH) (Fluka analytical > 97%) and sodium carbonate (Na_2CO_3) (Polysciences.INC, 99%) were used without further purification. Distilled water was used as solvent throughout this study.

2.2. Synthesis of sample

The preparation of the precursor at pH=9±0.5 with molar ratio of Co^{2+}/Al^{3+} =1.5 has been done using the co-precipitation method [22]. The synthesis was carried out by mixing 8.43g and 7.56g respectively of $Co(NO_3)_2{\bullet}6H_2O$ and $Al(NO_3)_3{\bullet}9H_2O$ salts dissolved in 100 ml of distilled water. A volume of distilled water was placed in a reactor vessel heated to 70°C and adjusted with an excess of Na_2CO_3 solution to the desired pH. The two metal nitrates solutions were slowly added to the reactor with vigorous stirring. The pH value of the sample was controlled by dropwise addition of aqueous solution of NaOH (1.0M). The resulting slurry was aged at 70°C for 1h then at room temperature for 24h. The precipitate was washed with distilled water several times to remove sodium and nitrate ions, then filtered and dried in an oven at 50°C for 16h. The resultant nanostructure was ground into fine powder. The sample will be denoted as CoAl-CO_3-LDH.

2.3. Characterization

Powder X-ray Diffraction (PXRD) pattern of the samples was recorded on X-ray diffractometer (BRUKER-BINARY) using CuKα (λ = 1.54187Å). Fourier transform infrared spectroscopy (FT-IR) spectrum was carried out on a Thermo scientific NICOLET Is10. Thermogravimetric analysis (TGA) was carried out on instrument from Shimadzu model DTG 60H, air flow rate of 200cm^3/min. Inductively coupled plasma (ICP) analysis was recorded using VARIAN ms/13/35. The sample dissolution is carried out in nitric acid. The dielectric and electric measurements for the sample were recorded using a MTS Modulab (Solartron Analytical). The impedance was measured within a frequency range of 1Hz to 1MHz at room temperature with relative humidity of 74.1% and under low voltage (0.8V). The CoAl-CO_3 sample was prepared in pellet form (diameter = 10mm, thickness = 1.136mm).

3. Results and Discussion

3.1. Powder X-ray diffraction (PXRD) analysis

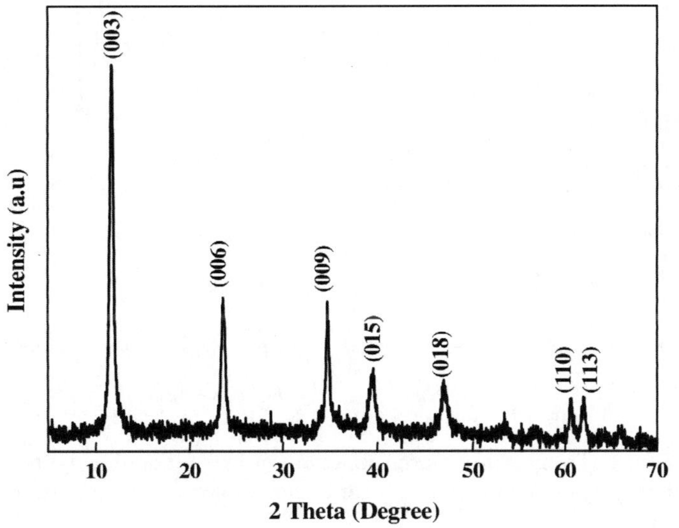

Figure 1. Powder X-ray Diffraction pattern of layered double hydroxide based on cobalt and aluminum, and intercalated by carbonate ions showing the reflections indexed to a hexagonal lattice with rhombohedral symmetry.

The PXRD pattern of CoAl-CO_3-LDH sample is represented in **figure 1**. The intense diffraction lines located at 2θ = 11.5°; 24°; 35°; 60.5°and 62° can be indexed using the hexagonal cell with rhombohedral symmetry as (003), (006), (009), (110) et (113) respectively, which is a characteristic of a well-crystallized hydrotalcite-like phase with carbonate anions in the interlayer space. The lattice parameters $a=2\times d_{110}$ and $c=3\times d_{003}$ are calculated and seems to be in a good agreement with those reported in literature for the same phases and the same molar ratio (**Table 1**) [1, 22].

Table 1. Lattice parameters of layered double hydroxide based on cobalt and aluminum, and intercalated by carbonate ions.

sample	d_{110}	a	d_{003}	c
CoAl-CO_3	1.52	3.04	7.57	22.72

3.2. Fourier transform-infrared spectroscopy (FT-IR)

In order to better study the different functional groups of precursor and intercalated anion, FT-IR measurement has been carried out. The FT-IR spectra represented in **figure 2**, display a large band in the 3400 cm^{-1} region, attributed to the water molecules and hydroxyl groups in the interlayer space. The band situated around 1630 cm^{-1} can be attributed to the bending vibration δ (H_2O) of water molecules. CoAl-CO_3-LDH displays a strong vibration around 1350 cm^{-1} due to the presence of interlayer carbonate anions. It could also be seen that the band between 810 and 400 cm^{-1} could be attributed to the stretching vibration of M–O and O–M–O bending vibration [23, 24].

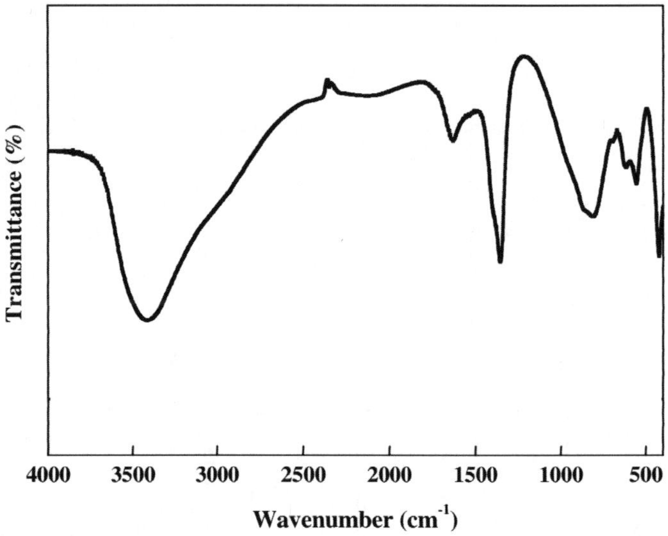

Figure 2. Fourier Transform-Infrared spectrum of layered double hydroxide based on cobalt and aluminum, and intercalated by carbonate ions. (%): Percent, (cm^{-1}): Centimeter^{-1}.

3.3. Thermogravimetric analysis (TGA)

The thermal decomposition of the samples was monitored by thermogravimetric analysis. As seen in **figure 3**, the thermal decomposition of CoAl-CO_3-LDH sample occurs in three well-defined steps. Physisorbed and intercalated water are completely removed near 200°C (about 3% of mass loss below 100°C representing physisorbed water and 13% of mass loss between 100°C and 200°C representing interlayer water).

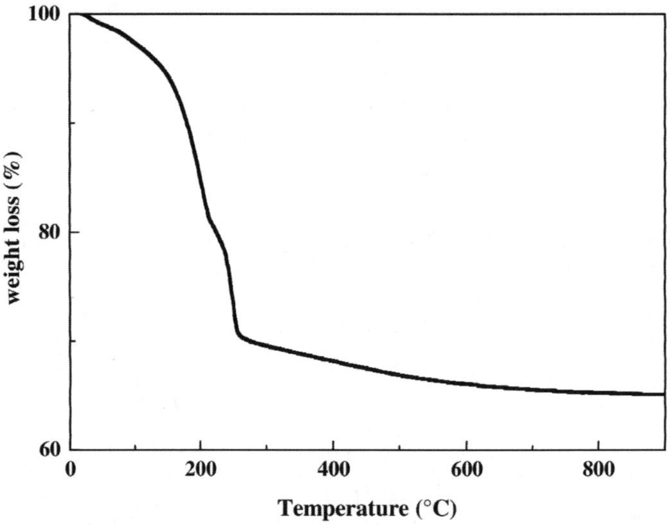

Figure 3. Thermogravimetric Analysis curve of layered double hydroxide based on cobalt and aluminum intercalated by carbonate ions, showing three steeps of thermal decomposition.(dehydration, dehydroxylation and departure of intercalated anions (carbonates in this case), (°C): Celcius degree).

The second step which is in the temperature range of above 200°C and below 260°C is ascribed to the dehydroxylation of brucite-like layers and the beginning of decomposition of interlayer anions. The third process which is anions decomposition is finished at about 550°C [1].

3.4. *ICP*

The result of the elemental chemical analysis (**Table 2**) shows as expected that the experimental ratio $R_{ex}= Co^{2+}/Al^{3+}$obtained is very close to the theoretical ratio R_{th}. The formula is determined using ICP and TGA results.

Table 2. Chemical analysis and formula of the synthesized LDH

Sample	% Co^{2+}	% Al^{3+}	% H_2O^a	R_{ex}	R_{th}	Chemical formula
CoAl-CO_3	41.61	12.44	15.16	1.51	1.5	$Co_{0.605}\,Al_{0.395}\,(OH)_2\,(CO_3^{2-})_{0.197}$•$0.913H_2O$

[a] % H_2O determined by TG analysis

3.5. *Electric and dielectric properties.*

3.5.1. *Complex impedance*

The electrical properties of the present sample have been investigated. The first and the most significant step in this regard is the Nyquist plot between the real and the imaginary part of impedance. Nyquist diagram allows the separation of the effect of grain, grain boundary and electrode interface on the relaxation process. Each contribution is represented by a semicircular arc, and the intersection of each arc with the (Z′) axis gives an estimation of the associated resistance value.
The Nyquist diagram (**Figure 4**) corresponding to CoAl-CO_3-LDH sample shows the adjustment of impedance data by an equivalent circuit consisting on three blocs serially arranged and describing grains, grain boundaries and electrode-LDH interface. Each bloc is a parallel arrangement of a resistor R and a constant element phase CPE.
The arcs associated to the grains and the grain boundaries are confused and seem to be one semicircular arc, thing which is difficult to find out without the adjustment of impedance data and the modelization by an equivalent circuit. The low frequencies straight line forming a 45° angle with (Z′) axis represents the electrode blocking, which is a characteristic of ionic conductors [25]. The complex impedance in the case of three blocs equivalent circuit is given by:

$$Z^*(\omega) = Z^*_{ei}(\omega) + Z^*_{gb}(\omega) + Z^*_{g}(\omega) \tag{1}$$

$$Z^*(\omega) = \frac{R_{ei}}{1+(j\omega\tau_{ei})^{P_{ei}}} + \frac{R_{gb}}{1+(j\omega\tau_{gb})^{P_{gb}}} + \frac{R_{g}}{1+(j\omega\tau_{g})^{P_{g}}} \tag{2}$$

The real and the imaginary parts of complex impedance in this case are described as:

$$Z'(\omega) = \frac{R_{ei}\,(1+\tau_{ei}^{P_{ei}}\omega^{P_{ei}}\cos(\frac{\pi P_{ei}}{2}))}{1+2\tau_{ei}^{P_{ei}}\omega^{P_{ei}}\cos(\frac{\pi P_{ei}}{2})+(\tau_{ei}^{P_{ei}}\omega^{P_{ei}})^2} + \frac{R_{gb}\,(1+\tau_{gb}^{P_{gb}}\omega^{P_{gb}}\cos(\frac{\pi P_{gb}}{2}))}{1+2\tau_{gb}^{P_{gb}}\omega^{P_{gb}}\cos(\frac{\pi P_{gb}}{2})+(\tau_{gb}^{P_{gb}}\omega^{P_{gb}})^2} + \frac{R_{g}\,(1+\tau_{g}^{P_{g}}\omega^{P_{g}}\cos(\frac{\pi P_{g}}{2}))}{1+2\tau_{g}^{P_{g}}\omega^{P_{g}}\cos(\frac{\pi P_{g}}{2})+(\tau_{g}^{P_{g}}\omega^{P_{g}})^2} \tag{3}$$

$$Z''(\omega) = \frac{R_{ei}\,(1+\tau_{ei}^{P_{ei}}\omega^{P_{ei}}\sin(\frac{\pi P_{ei}}{2}))}{1+2\tau_{ei}^{P_{ei}}\omega^{P_{ei}}\sin(\frac{\pi P_{ei}}{2})+(\tau_{ei}^{P_{ei}}\omega^{P_{ei}})^2} + \frac{R_{gb}\,(1+\tau_{gb}^{P_{gb}}\omega^{P_{gb}}\sin(\frac{\pi P_{gb}}{2}))}{1+2\tau_{gb}^{P_{gb}}\omega^{P_{gb}}\sin(\frac{\pi P_{gb}}{2})+(\tau_{gb}^{P_{gb}}\omega^{P_{gb}})^2} + \frac{R_{g}\,(1+\tau_{g}^{P_{g}}\omega^{P_{g}}\sin(\frac{\pi P_{g}}{2}))}{1+2\tau_{g}^{P_{g}}\omega^{P_{g}}\sin(\frac{\pi P_{g}}{2})+(\tau_{g}^{P_{g}}\omega^{P_{g}})^2} \tag{4}$$

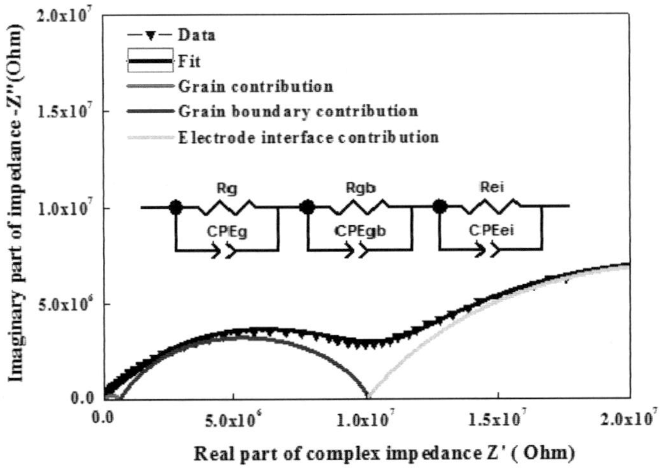

Figure 4. Nyquist diagram of experimental and fitted impedance data of layered double hydroxide based on cobalt and aluminum and intercalated by carbonate ions, measured at room temperature, showing deconvolution of contributions and equivalent circuit used for the fitting.

The values of R_{ei}, T_{ei}, P_{ei}, τ_{ei}, R_{gb}, T_{gb}, P_{gb}, τ_{gb}, R_g, T_g, P_g and τ_g are grouped in **table.3**. Where R_i, T_i, P_i and τ_i are respectively resistance, pseudo-capacitance, dispersion constant and relaxation time, and i for electrode interface (ei), grain boundary (gb) or grain (g). Relaxation time can be confirmed by the following expression: $\tau_i = (R_i \times T)^{P_i^{-1}}$.

Table 3. Fitting parameters of the equivalent circuit for layered double hydroxide based on cobalt and aluminum, and intercalated by carbonate ions.

	Resistance (*R*) (Ohm)	Pseudo-capacitance (*T*) (Farad)	Dispersion Constant (*P*)	Relaxation time (τ) (second)	Relaxation frequency (Hertz)
Inerface (ei)	2.40×10^{7}	1.27×10^{-8}	0.67	0.17	0.93
Grain boundary (gb)	9.50×10^{6}	3.06×10^{-10}	0.76	4.60×10^{-4}	345.95
Grain (g)	6.27×10^{5}	3.21×10^{-10}	0.76	1.37×10^{-5}	11612.06

The grain boundaries are founding to be more insulating owing to their high resistance. It acts as thin traps for charge carriers with high capacitance. Since the grain resistance is smaller compared to the grain boundaries, the predominant contribution to the sample conduction comes from the grains. This is confirmed in the case of the present sample.

In the case of an ideal behavior, the polarization mechanism exhibits a Debye relaxation consisting on equation with a single relaxation time. In polycrystalline materials case, the electrical response confirms the non-ideal behavior proved by the existence of a distribution of relaxation times. The frequency dependence of imaginary part Z″ (***Figure 5***) shows the existence of two relaxation frequencies, one for the electrode interface and another for the grain boundaries, while the grain relaxation is masked. Then, CoAl-CO_3-LDH exhibits a non-Debye type relaxation characterized by a distribution of relaxation times [26]. The two relaxations are found at 1Hz and 345.08 Hz respectively for electrode-LDH interface and grain boundaries, which confirms the previous results.

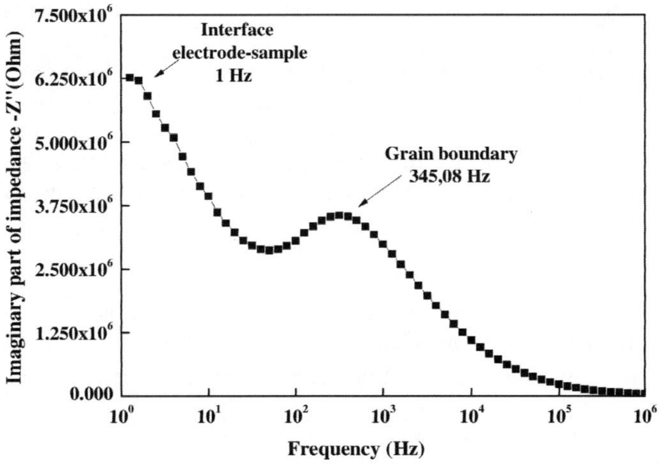

Figure 5. Frequency dependence of imaginary part of complex impedance of layered double hydroxide based on cobalt and aluminum, and intercalated by carbonate ions, showing two relaxation peaks.

3.5.2. Electric modulus

In order to magnify any other contribution present in the sample and maybe masked or unidentifiable on the CIS formalisms, it is important to adopt an alternative approach which can discriminate against extrinsic relaxation effect. Hence the necessity of applying complex modulus analysis. It suppresses the unwanted contributions of electrode polarization and grain boundaries and highlights the intrinsic properties.

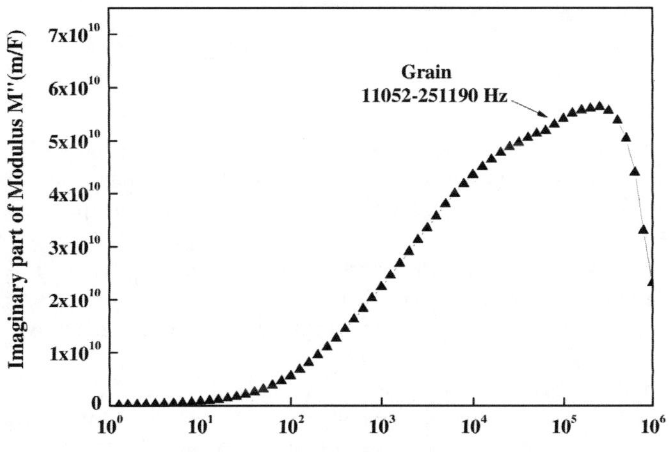

Figure 6. Frequency dependence of imaginary part of complex modulus of layered double hydroxide based on cobalt and aluminum, and intercalated by carbonate ions, showing one relaxation peak related to the grain contribution. (m/F): meter per Farad, (Hz): Hertz.

Figure 6 shows the frequency dependence of imaginary part of complex modulus M″ .The asymmetry of the curve is a consequence of the non-Debye type relaxation in the CoAl-CO_3-LDH sample, and the clear peak at high frequencies confirms the presence of grain contribution which was hidden in

complex impedance formalism. This contribution is previously found to be the predominant one in the material conduction.
According to the representation of imaginary part of modulus M″, probably a dispersion of grain relaxation time occurs at high frequencies causing an enlargement of the peak over a frequency range of 11052 Hz to 251190 Hz. The relaxation in this frequency range confirms the results of grains relaxation time obtained previously by the three blocs equivalent circuit. The nature of modulus spectrum suggests the existence of hopping mechanism of electrical conduction in the material [27].

3.5.3. Ac conductivity study

In polycrystalline materials, Ac conductivity analysis is the best way to understand conduction mechanism. For that, the frequency-dependent Ac electrical conductivity (σ_{ac}) of the CoAl-CO_3-LDH sample was investigated and obtained from the following equation: $\sigma_{ac} = \frac{1}{Y'} k$. Where Y' is the real part of complex admittance, k is the cell constant ($k = \frac{e}{s}$ with e is the sample pellet thickness and s is the electrode surface)

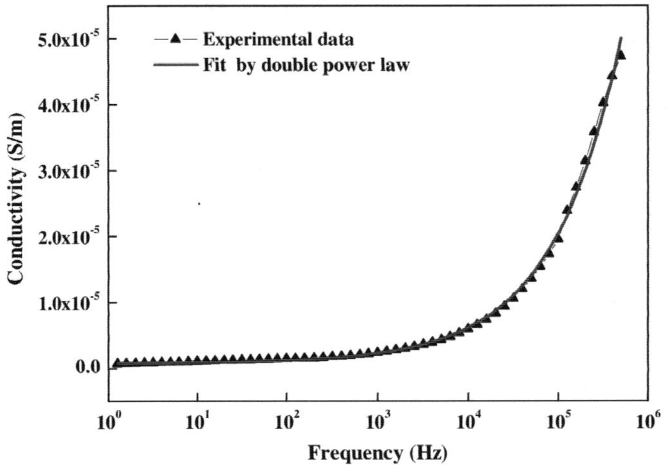

Figure 7. Frequency dependence of alternative current conductivity of layered double hydroxide based on cobalt and aluminum, intercalated by carbonate, showing the best fit of curve to a double power law by Origin 8 software. (S/m): Siemens per meter. (Hz): Hertz.

As shown in **Figure 7**, the Ac conductivity increases with increasing frequency, which is attributed to the gradual decrease of the sample resistance. It is also worth mentioning that a frequency independent region appears at low frequencies which smoothly relaxes to a strong dispersive region situated at high frequencies.
In order to analyze the Ac conductivity, experimental data has been fitted with Origin software using nonlinear curve fitting. A convenient formalism to investigate the frequency dependence of conductivity at constant temperature is based on the double power law: $\sigma_{ac} = \sigma_{dc} + A\omega^n + B\omega^m$. Where σ_{dc} is the Dc conductivity for a particular temperature, the terms $A\omega^n$ and $B\omega^m$ denotes the frequency dependence and characterize dispersion phenomenon [28–30]**,** and n and m are the power law exponents which generally vary between 0 and 2 it represents the degree of interaction between mobile ions and the lattices around them [31]**.** The obtained values of σ_{dc}, A, B, n and m are grouped in **table.3**.

Table 4. Fitting parameters values of alternative current conductivity (σ_{ac}) for layered double hydroxide based on cobalt and aluminum, and intercalated by carbonate ions. (R^2): correlation coefficient, (σ_{dc}):direct current conductivity, (s/m): (siemens per meter), (A) and (B): polarizability strength , (n) and (m) : the fractional exponents.

Sample	σ_{dc} (S/m)	*A*	*B*	*n*	*m*	R^2
CoAl-Co_3	6.75×10^{-7}	1.15×10^{-7}	2.61×10^{-8}	0.12	0.57	0.99

The value of σ_{dc} determined by the Jonscher law was verified and compared to the one calculated using data obtained from the fitting by the equivalent electrical circuit. For this, σ_{dc} was calculated using the following equation: $\sigma_{dc} = \frac{1}{Rs} k$, with $k = \frac{e}{s}$ (m^{-1}) (*k*: cell constant; *e:* The thickness of sample pellet; *s* is the contact surface between the sample and the electrode and R_s is sample resistance). The σ_{dc} obtained in some other works dealing with the same kind of materials is varying between 10^{-6} S/m and 10^{-8} S/m [12, 13, 15, 32]. So the value found in the present work is included in this range. The exponent *n* and *m* give an idea about the kind of hopping motion in the corresponding region, and this according to its value. Indeed, $n < 1$ means that the hopping movement of charge carriers involves long-range movement with a sudden hopping (a translational motion), whereas $n \geq 1$ means that the movement involves localized hopping [33, 34].
In this work, the obtained values of *n* and *m* are lower than 1 (**Table.3**). It means that the charge carriers undergo the same hopping process in both frequency-independent region and dispersive region which is hopping through translational motion. Since the cluster formed by the intercalated species (carbonate ions and H^+ protons of water molecules) is the only origin of charge carriers responsible of the hopping process, the conduction within the present material is of ionic type [12, 15].

3.5.4. Dielectric constant

The dielectric relaxation can be analyzed in terms of dipolar and conductivity mechanisms. In the case of CoAl-CO_3-LDH sample, **Figure 8** is showing the logarithmic frequency dependence of real part of dielectric constant. A relaxation is observed as a gradual decrease in ε_r'. This can be explained by the frequency dependent orientational polarization. In fact, permanent dipoles can align themselves following the applied field when frequencies are low. In this case they contribute fully to the total polarization. But with the increase of frequencies, dipoles are not able to follow the field because its variation is too rapid, so their contribution to the dielectric constant becomes negligible causing the decrease of ε_r' with the increasing frequency. CoAl-CO_3-LDH exhibits interesting values of dielectric constant for low and medium frequencies then decreases gradually to attain 6.88 at high frequencies.

Figure 9 shows the value of tanδ with frequency, measured at room temperature and calculated according to the following formula: $\tan\delta = \varepsilon_r'' / \varepsilon_r'$. Maximum of dispersion in the loss tangent is observed at low frequencies (10Hz to 100Hz) due to polarization phenomenon, leading to a maximum of energy dissipation. Then loss tangent decreases as the frequency of the alternating field increases. Beyond a frequency of 10^4Hz, the tangent of dielectric loss decreases to attain a value of 0.09 where the loss is minimal.

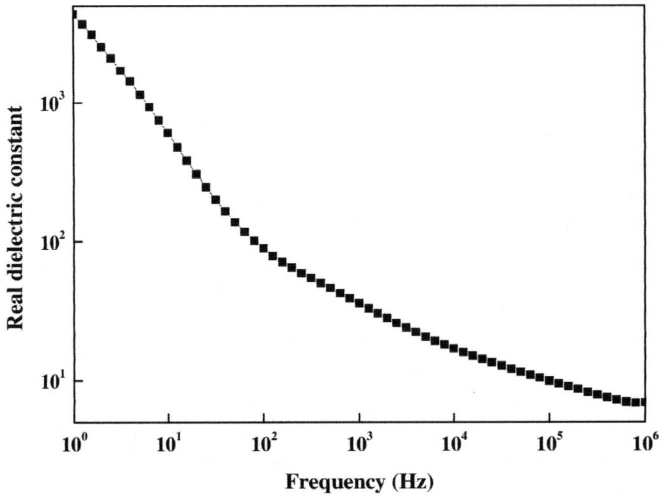

Figure 8. Frequency dependence of real dielectric constant of layered double hydroxide based on cobalt and aluminum, and intercalated by carbonate ions. (Hz): Hertz.

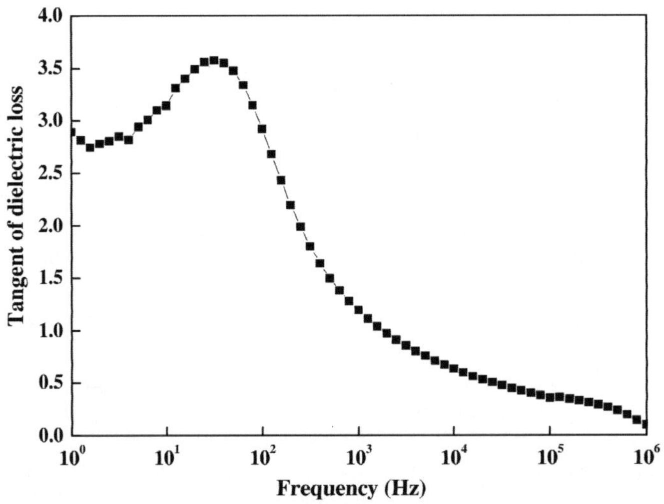

Figure 9. Frequency dependence of tangent of dielectric loss of layered double hydroxide based on cobalt and aluminum, intercalated by carbonate ions.

4. Conclusion

Carbonate intercalated CoAl-LDH sample was prepared by the conventional co-precipitation method. According to the PXRD pattern, the sample exhibit a well crystalized hydrotalcite-like phase with carbonate ions in the interlayer space. FT-IR analysis was carried out to confirm the presence of gallery anions and the forming of LDH layers. Using TG analysis coupled to ICP analysis the chemical formula of the sample was determined.

Electric and dielectric properties of the LDH phase were systematically investigated by the complex impedance spectroscopy. The Nyquist diagram corresponding to CoAl-CO_3-LDH sample was adjusted by an equivalent electrical circuit consisting on three blocs serially arranged. Each bloc is a parallel arrangement of resistor (R) and constant phase element (CPE), allowing the determination of the resistance and the relaxation time of each contribution. The imaginary parts of impedance and Modulus analysis magnified all contributions and confirmed relaxation times calculated previously. To have an idea about the conduction mechanism within the present material, σ_{ac} was investigated by the adjustment by the double power law. The adjustment parameters allowed the determination of σ_{dc} and gave an idea about the movement of charge carriers (H^+ protons of water molecules and gallery anions) under the applied field, which is found to be a long-range movement according to the value of the fractional exponents of the double power law. The movement of charge carriers is similar in both frequency independent region and dispersive region of σ_{ac}. The representation of frequency dependence of ε_r' showed an interesting dielectric constant value at low frequencies with a gradual decrease at high frequencies. Moreover, the maximum of energy dissipation was observed at low frequencies, and then tan δ decreased to attain 0.09 beyond 10^4 Hz. According to these results, this material may be promising in the field of dielectrics and energy storage, especially for the assembly of nanocomposites and hybrid materials with interesting properties.

Acknowledgement

This work was supported by the National Center of Scientific and Technical Research of Morocco.

References

[1] Białas A, Mazur M, Natkański P, Dudek B, Kozak M, Wach A and Kuśtrowski P 2016 Hydrotalcite-derived cobalt–aluminum mixed oxide catalysts for toluene combustion *Appl. Surf. Sci.* **362** 297–303
[2] Mahjoubi F Z, Khalidi A, Abdennouri M and Barka N 2017 Zn–Al layered double hydroxides intercalated with carbonate, nitrate, chloride and sulphate ions: Synthesis, characterisation and dye removal properties *J. Taibah Univ. Sci.* **11** 90–100
[3] Laipan M, Fu H, Zhu R, Sun L, Steel R M, Ye S, Zhu J and He H 2018 Calcined Mg/Al-LDH for acidic wastewater treatment: Simultaneous neutralization and contaminant removal *Appl. Clay Sci.* **153** 46–53
[4] Prevot V and Tokudome Y 2017 3D hierarchical and porous layered double hydroxide structures: an overview of synthesis methods and applications. *J. Mater. Sci.* **52** 11229–50
[5] Ahmed A A A, Talib Z A and Hussein M Z 2016 Synthesis and optimization of electric conductivity and thermal diffusivity of zinc-aluminum hydroxide (Zn–Al–NO_3–LDH) prepared at different pH values *Mater. Today Proc.* **3** 130–144
[6] Wang W, Zhang N, Shi Z, Ye Z, Gao Q, Zhi M and Hong Z 2018 Preparation of Ni-Al layered double hydroxide hollow microspheres for supercapacitor electrode *Chem. Eng. J.* **338** 55–61
[7] Eshghi A and kheirmand M 2017 Graphene/Ni–Fe layered double hydroxide nano composites as advanced electrode materials for glucose electro oxidation *Int. J. Hydrog. Energy* **42** 15064–72
[8] Pan X, Ma J, Yuan R and Yang X 2017 Layered double hydroxides for preparing $CoMn_2O_4$ nanoparticles as anodes of lithium ion batteries *Mater. Chem. Phys.* **194** 137–141
[9] Hu L, Wu L, Liao M, Hu X and Fang X 2012 Electrical transport properties of large, individual $NiCo_2O_4$ nanoplates *Adv. Funct. Mater.* **22** 998–1004
[10] Wang X, Wang Q, Hou X, Liu Y, Zheng P, Huo J, Yin L and Guo S 2018 Facile fabrication of two-dimensional reduced graphene oxide/CoAl-layered double hydroxides nanocomposites for lithium-oxygen battery with improved electrochemical performance *J. Alloys Compd.* **744**, 196–203
[11] Rashid M, Hussain A, Shoaib M, Basit M A, Khan H and Kim Y S 2019 An amperometric hydrogen sensor based on Pt nanoparticles supported multi-wall carbon nanotubes *J. Electroanal. Chem.* **833** 173–180
[12] Elmelouky A; Mortadi A; Elmoznine R, Lahkale R, Sabbar E and Chahid E 2015 Dielectric spectroscopy studies in Zn_2FeNO_3, Zn_2AlNO_3 and $Zn_2Fe_{0.5}Al_{0.5}NO_3$ *Int. J. Adv. Res.***3** 465-475

[13] Elhatimi W, Lahkale R, Bouragba F Z, Sadik R, Lebbar N, Elmelouky A, Mortadi A, Siniti M and Sabbar E 2017. Electric and dielectric properties of copper-chromium layered double hydroxide intercalated with chloride ion using impedance spectroscopy *Spectrosc. Lett.* **50** 189–195

[14] Ahmed A A A, Talib Z A, Hussein M Z bin and Zakaria A 2012 In situ dielectric measurements of Zn–Al layered double hydroxide with anionic nitrate ions *Solid State Sci.* **14** 1196–1202

[15] Elmelouky A, Mortadi A, Moznine R E, Chahid E, Lahkale R and Sabbar E 2015 Analysis of conduction mechanism in Zn_2FeNO_3 using impedance spectroscopy *Int. j. adv. res. phys. sci.* **2** 1-10

[16] Ahmed A A A, Talib Z A, Hussein M Z bin and Zakaria A 2012 Zn–Al layered double hydroxide prepared at different molar ratios: Preparation, characterization, optical and dielectric properties *J. Solid State Chem.* **191**, 271–278

[17] Lahkale R, Elhatimi W, Sadik R, Bouragba FZ, Lebbar N, Elmelouky A, Mortadi A and Sabbar E 2018 Nitrate-intercalated Mg_{1-x} Al_x -layered double hydroxides with different layer charges (x): Preparation, characterization, and study by impedance spectroscopy *Appl. Clay Sci.* **158** 55–64

[18] Tamaki T, Nakanishi N, Ohashi H and Yamaguchi T 2012 The effect of particle size and surface area on the ion conductivity of layered double hydroxide *Electrochem. Commun.* **25** 50–53

[19] Furukawa Y, Tadanaga K, Hayashi A and Tatsumisago M 2011 Evaluation of ionic conductivity for Mg–Al layered double hydroxide intercalated with inorganic anions *Solid State Ion.* **192** 185–187

[20] Kubo D, Tadanaga K, Hayashi A and Tatsumisago M 2012 Hydroxide ion conduction in Ni–Al layered double hydroxide *J. Electroanal. Chem.* **671** 102–105

[21] Kim H S, Yamazaki Y, Kim J D, Kudo T and Honma I 2010 High ionic conductivity of Mg–Al layered double hydroxides at intermediate temperature (100–200°C) under saturated humidity condition (100% RH) *Solid State Ion.* **181** 883–888

[22] Gabrovska M, Edreva-Kardjieva R, Tenchev K, Tzvetkov P, Spojakina A and Petrov L 2011 Effect of Co-content on the structure and activity of Co–Al hydrotalcite-like materials as catalyst precursors for CO oxidation *Appl. Catal. Gen.* **399** 242–251

[23] Wang H, Yi H, Tang X, Ning P, Yu L, He D, Zhao S and Li K 2012 Catalytic hydrolysis of COS over calcined CoNiAl hydrotalcite-like compounds modified by cerium *Appl. Clay Sci.* **70** 8–13

[24] Feng Z, Yang Z, Yang B, Zhang Z and Xie X 2014 The application of Co–Al-hydrotalcite as a novel additive of positive material for nickel–metal hydride secondary cells *J. Power Sources* **266** 22–28

[25] Mouahid F Z 2001 Cristallochemistry and ionic conductivity of nasicon type composed with formula: $A_{1+X}M_{2-X}Al_x(PO_4)_3$ (A=Li, Na; M=Ti, Ge) Thesis, University of Chouaîb Doukkali, *El Jadida, Morocco*

[26] Xu Q, Lanagan M T, Huang X, Xie J, Zhang L, Hao H and Liu H 2016 Dielectric behavior and impedance spectroscopy in lead-free BNT–BT–NBN perovskite ceramics for energy storage *Ceram. Int.* **42** 9728–36

[27] Kathayat K, Panigrahi A, Pandey A and Kar S 2012 Characterization of Electrical Behavior of $Ba_5HoTi_3V_7O_{30}$ ceramic using impedance analysis *Mater. Sci. Appl.* **03** 390–397

[28] Peláiz-Barranco A, Gutiérrez-Amador M P, Huanosta A and Valenzuela R 1998 Phase transitions in ferrimagnetic and ferroelectric ceramics by ac measurements *Appl. Phys. Lett.* **73** 2039–2041

[29] Mahboub S, Prasad G and Kumar G S 2006 Electrical conduction in $(Na_{0.125}Bi_{0.125}Ba_{0.65}Ca_{0.1})(Nd_{0.065}Ti_{0.87}Nb_{0.065})O_3$ ceramic *Bull Mater Sci* **29** 35–41

[30] Das S, Banerjee S and Sinha T P 2016 Structural and AC conductivity study of CdTe nanomaterials *Phys. E Low-Dimens. Syst. Nanostructures* **78** 73–78

[31] Jlassi I, Sdiri N and Elhouichet 2017 Electrical conductivity and dielectric properties of MgO doped lithium phosphate glasses *J. Non-Cryst. Solids* **466–467** 45–51

[32] Ivanov M, Klemkaite K, Khinsky A, Kareiva A and Banys J 2011 Dielectric and Conductive Properties of Hydrotalcite. Ferroelectrics **417** 136–142
[33] Funke K 1993 Jump relaxation in solid electrolytes *Prog. Solid State Chem.* **22** 111–195
[34] Izquierdo J L, Astudillo A, Bolaños G, Zapata V H, Morán O 2015 Dielectric response of Al-substituted multiferroic $TbMnO_3$ at high temperatures *Ceram. Int.* **41** 1285–96

IOP Conf. Series: Materials Science and Engineering **948** (2020) 012021 doi:10.1088/1757-899X/948/1/012021

Optical properties of Red pigment for Dye Sensitized Solar Cells

Rania Anoua1, 2*, Samira Touhtouh1, Mohamed El Jouad1, Abdelowahed Hajjaji1, Mina Bakasse3, Przemysław Płociennik2 and Anna Zawadzka2

[1] *Laboratory of Engineering Sciences for Energy, National School of Applied Sciences, Chouaib Doukkali University, El Jadida, Morocco,*

[2] *Institute of Physics, Faculty of Physics, Astronomy and Informatics, Nicolaus Copernicus University,*
Grudziądzka 5, PL 87-100 Torun, Poland.

[3] *Laboratory of Organic Bioorganic Chemistry and Environment, Faculty of Science, Chouaib Doukkali University, El Jadida, Morocco*

**Corresponding author: anoua.r@ucd.ac.ma*

Abstract: This work contains a description of the optical properties of Lithol Bordeaux dye thin films. The main use of this dye is for Dye Sensitized Solar Cells (DSSCs). Hence the need to deposit this dye as a thin layer in order to know its optical properties. The films were successfully grown by physical vapor deposition (PVD) technique in a high vacuum on transparent (glass) kept at room temperature during the deposition process. Optical properties were investigated by transmittance and photoluminescence measurements. Strong luminescence was observed around 1,68×107 CPS and a reaching an absorbance band about 100 nm in the visible range.

KEYWORDS: Physical Vapor Deposition (PVD), Pigments, Dye Sensitized Solar Cell (DSSC), Transmittance, Photoluminescence.

1. Introduction:

Nowadays, photovoltaic represents an important stake in renewable energies and more generally in energies in general [1] [2].
However, the real challenge for photovoltaic is to significantly reduce the cost [3] per kilowatt-hour produced. Silicon currently offers the best performance, but its purification and treatment lead to very high costs as well as a manufacturing process very [4] toxic to the environment and very expensive. All this is a major obstacle to the development of photovoltaic.
A more original solution is to take inspiration from nature (biomimicry) in the case of photosynthesis of plants. Dye Sensitized Solar Cells (DSSCs) are developed for this purpose. In the future, they could be a reliable solution with low cost and equivalent or better performance.
As its name suggests, the dye is the key element of the cell that behaves like a kind of artificial chlorophyll, converting the absorbed solar radiation into electricity by a process very different from the usual photovoltaic cells made of silicon.
Unlike the production of silicon photovoltaic panels, the production of Grätzel cells requires neither much energy nor delicate operations [5] [6].

IOP Conf. Series: Materials Science and Engineering **948** (2020) 012021 doi:10.1088/1757-899X/948/1/012021

The DSSCs are made of a dye/liquid solvent mixture trapped between two transparent electrodes, which makes it possible to make transparent panels of various colors according to the dye chosen [7] [8] [9] .
For our study, we will rely on a red dye under the industrial name Lithol Bordeaux and the chemical formula is $C_{21}H_{12}N_2O_6SCa$ which is recommended normally for coating and interior painting.
So the purpose of this work is to enhance its use as a coating and paint pigment and apply it to DSSCs.
In this paper, we will present the finding of the optical properties of the red dye via transmission and photoluminescence deposited by Physical Vapor Deposition (PVD) technique [10].

2.Technique of deposition:

Lithol Bordeaux dye was deposited as a thin film on glass substrates by means of Physical Vapor Deposition (PVD) technique [11] [12] using equipment shown on Figure 1.
Cleaning substrates are important in the results of the optical properties. So the procedure of cleaning is as follows: acetone bath in ultrasonic followed by the ethanol for 15 min each. After that, cleaned by the isopropanol using the flow synthesis [13] [14].
The technique of evaporation was performed by the heat of the material by a tungsten resistance coil in a vacuum chamber under pressure of about 2×10^{-5} Tr. Once the evaporation is done, the thin film is created.

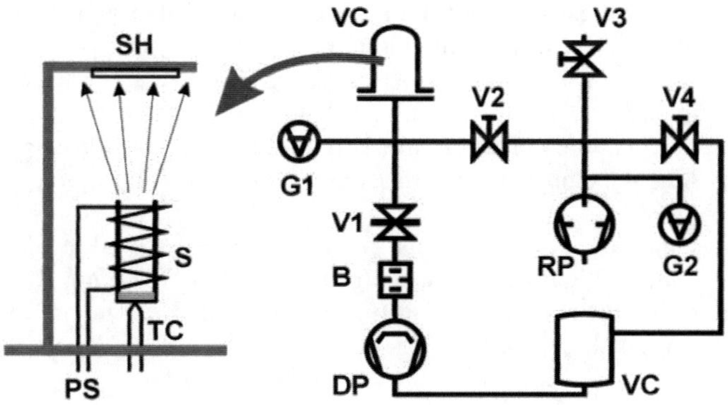

Figure 1: PVD apparatus. Left part – evaporation equipment: S – source (effusion cell and heater), SH – sample (substrate) holder, TC – K-type thermocouple, PS – power supply. Right part – scheme of vacuum system: VC – vacuum chamber, V1-4 – manual valves, B – b baffle, G1-2 – vacuum gauges, VC – equalization vacuum chamber, RP – sliding vane rotary vacuum pump, DP – diffusion pump.

3.Transmittance:

The transmittance (λ) of the film was measured at normal incidence in the spectral range 190–1100 nm using a double–beam spectrophotometer (Perkin Elmer Lambda 2Spectrophotometer). The spectral distribution of the transmittance for the red dye thin film is shown in Figure 2 .
Figure 2 shows that in the visible range the material absorbs the light. Nevertheless, from 490 nm the transmittance increases and after reaches 0.6 in the infrared. In brief, we can say that it absorbs in visible range.

IOP Conf. Series: Materials Science and Engineering **948** (2020) 012021 doi:10.1088/1757-899X/948/1/012021

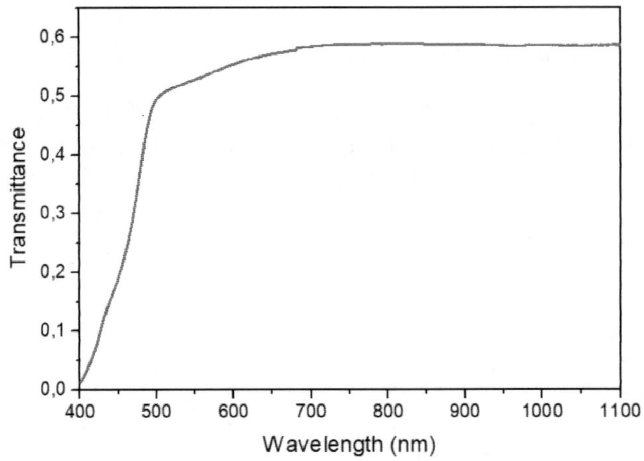

Figure 2: Transmittance $T\ (\lambda)$ of the red dye

4.Photoluminescence:

Photoluminescence was investigated by the FluoroMax 4P apparatus and the excitation source is a xenon lamp. For our material, we excited with 480 nm.

As shown in Figure 3, the intensity luminescence is strong and reaches exactly 1, 67754×10^7 CPS at the emission wavelength 587 nm. It is interesting to not have only one maximum emission wavelength but to have a broad emission in the maximum intensity. It means, that if we exit with 480 nm as in our case, the material will shin and emit in many emission wavelengths for 10 nm and from 584 nm to 593 nm.

Thus, Figure 4 shows the 3D excitation-emission fluorescence spectrum [15] and exhibits three maximum intensity of 1,68×10^7 CPS corresponding to 355(396)(478) nm / 587 nm. This means that not only 480 nm can excite our material, but also it excites with 355 nm and 396 nm for getting the maximum intensity photoluminescence.

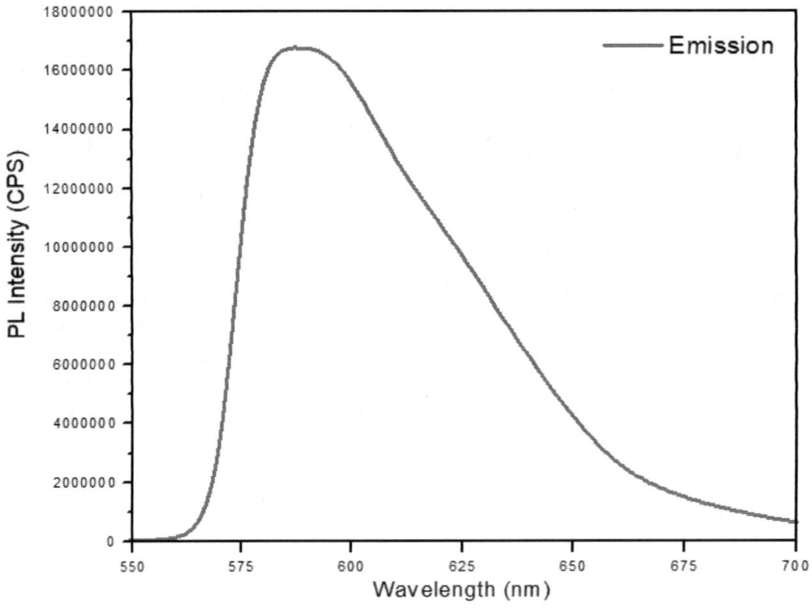

Figure 3: Emission spectrum of the red dye

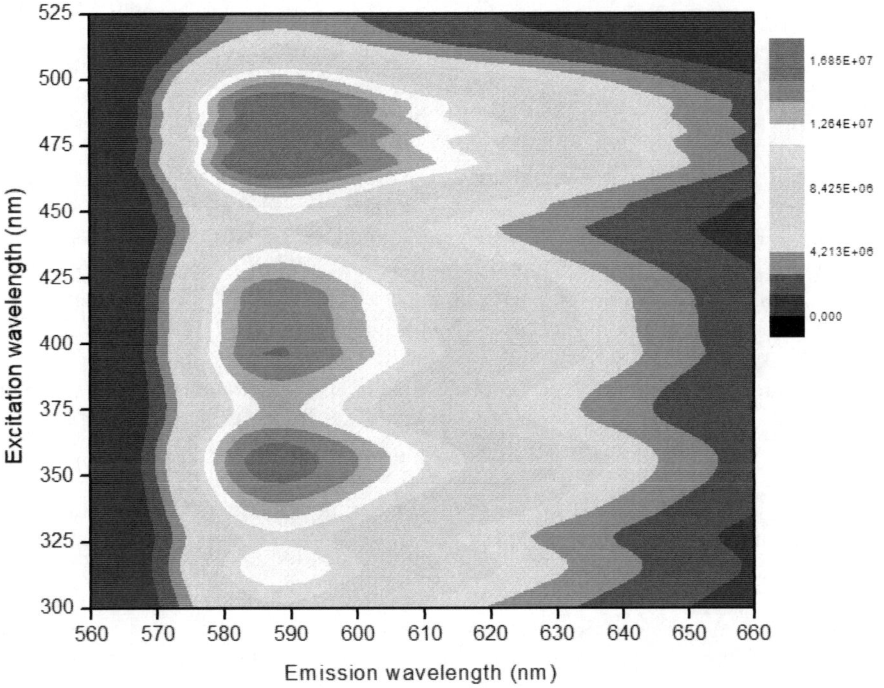

Figure 4: 3D excitation-emission fluorescence spectrum of the red dye

5.Conclusion:

The present study was designed to determine the optical properties of Lithol Bordeaux dye thin films. The Physical Vapor Deposition (PVD) was adopted as a technique of deposition in a glass substrate where we obtained a very homogeneous thin layer. This work then go to the optical characterization by transmission and it reaches 60% and also investigated by photoluminescence. The results of this investigation show a strong luminescence from 584 nm to 593 nm in the average of 1, 665×10^7 CPS, and a second major finding was in 3D excitation-emission fluorescence spectrum which presents a three maximum intensity corresponding to the same emission wavelength but in three different excitation wavelength. In conclusion, since our material is shiny and presents strong luminescence with only a small slit of the measurement about 2 nm, and also absorbs in the visible range, it is encouraging to apply this dye in solar cells.

ACKNOWLEDGEMENTS

This research has been financed from the funds of the Polish National Science Centre (grant no. 2017/25/B/ST7/02124). The films used in this paper were obtained using Interdisciplinary Centre for Modern Technologies facilities, NCU, Torun, Poland.

REFERENCES:

[1] A. Louwen et W. van Sark, « Photovoltaic solar energy », in Technological Learning in the Transition to a Low-Carbon Energy System, Elsevier, 2020, p. 65□86.
[2] A. Bartolotta et G. Calogero, « Dye-sensitized solar cells: from synthetic dyes to natural pigments », in Solar Cells and Light Management, Elsevier, 2020, p. 107□161.
[3] A. Goetzberger et C. Hebling, « Photovoltaic materials, past, present, future », Solar Energy Materials and Solar Cells, vol. 62, n° 1□2, p. 1□19, avr. 2000.
[4] A. Amkassou et H. Zgou, « New dyes for DSSC containing thienylen-phenylene: a theoretical investigation », Materials Today: Proceedings, vol. 13, p. 569□578, 2019.

IOP Conf. Series: Materials Science and Engineering **948** (2020) 012021 doi:10.1088/1757-899X/948/1/012021

[5] M. Grätzel, « Dye-sensitized solar cells », Journal of Photochemistry and Photobiology C: Photochemistry Reviews, vol. 4, n° 2, p. 145□153, oct. 2003.
[6] K. Tennakone, A. R. Kumarasinghe, G. R. R. A. Kumara, K. G. U. Wijayantha, et P. M. Sirimanne, « Nanoporous TiO2 photoanode sensitized with the flower pigment cyanidin », Journal of Photochemistry and Photobiology A: Chemistry, vol. 108, n° 2□3, p. 193□195, août 1997.
[7] S. Ito et al., « Fabrication of thin film dye sensitized solar cells with solar to electric power conversion efficiency over 10% », Thin Solid Films, vol. 516, n° 14, p. 4613□4619, mai 2008.
[8] D. Sinha, D. De, D. Goswami, et A. Ayaz, « Fabrication of DSSC with Nanostructured ZnO Photo Anode and Natural Dye Sensitizer », Materials Today: Proceedings, vol. 5, n° 1, Part 2, p. 2056□2063, janv. 2018.
[9] K. Subalakshmi, K. A. Kumar, O. P. Paul, S. Saraswathy, A. Pandurangan, et J. Senthilselvan, « Platinum-free metal sulfide counter electrodes for DSSC applications: Structural, electrochemical and power conversion efficiency analyses », Solar Energy, vol. 193, p. 507□518, nov. 2019.
[10] K. Medjnoun, « Etude et réalisation de semi conducteurs transparents ZnO dopé vanadium et oxyde de vanadium en couches minces pour des applications photovoltaïques », thesis, Perpignan, 2015.
[11] A. Zawadzka, P. Płóciennik, I. Czarnecka, J. Sztupecka, et Z. Łukasiak, « The effects of annealing process influence on optical properties and the molecular orientation of selected organometallic compounds thin films », Optical Materials, vol. 34, n° 10, p. 1686□1691, août 2012.
[12] P. Plociennik, A. Zawadzka, R. Frankowski, et A. Korcala, « Selected methods of thin films deposition and their applications », in 2016 18th International Conference on Transparent Optical Networks (ICTON), Trento, Italy, 2016, p. 1□4.
[13] R. Anoua et al., « Physical Properties of Yellow Natural Moroccan Dye », in 2019 21st International Conference on Transparent Optical Networks (ICTON), Angers, France, 2019, p. 1□3.
[14] A. Korcala, P. Plociennik, M. Lougdali, et R. Anoua, « Admittance Spectroscopy of Organic and Metal-Organic Complexes », in 2019 21st International Conference on Transparent Optical Networks (ICTON), Angers, France, 2019, p. 1□3.
[15] A. Zawadzka, P. Płóciennik, A. Korcala, et P. Szroeder, « Optical properties of chiral single-walled carbon nanotubes thin films », Optical Materials, vol. 96, p. 109295, oct. 2019.

ORIENTED MEMBRANE PROCESSES FOR EXTRACTION OF METHYLENE BLUE AND BLUE P3R DYES ACROSS POLYMER INCLUSION MEMBRANEs CONTAINING CHITIN AS NEW EXTRACTIVE AGENT

Y. CHAOUQI[1,2,*], R. OUCHN [1,2], M. EL BOUCHTI[2], O. CHERKAOUI[2], M. HLAIBI[1]

[1] Laboratoire Génie des Matériaux pour Environnement et Valorisation (GeMEV), Faculté des Sciences Aïn Chock, Université Hassan II, Casablanca, Maroc
[2] Laboratoire de Recherche sur les Matériaux Textiles (REMTEX), ESITH Casablanca, Maroc
Email: youssef.chaouqi-etu@etu.univh2c.ma
Project PPR2 (MESRSFC-CNRST)

Abstract: In the present work, according to the phase inversion technique, we have developed polymer inclusion membranes (PIMs) based on polymer supports polyvinylidene fluoride (PVDF) and polyvinyl pyrrolidone (PVP), with chitin as extractive agent (EA). These PIMs have been used to study the facilitated extraction and recovery of the Methylene Blue (MB) and Blue P3R dyes from simulated aqueous textile solutions. FT-IR spectra and SEM micrographs of these membranes were recorded to explore composition, structure and morphology of the elaborated PIMs. The adoption of the kinetic and thermodynamic models based on the Fick's laws, and the saturation law of the EA by the substrate (dyes) were tested to explain the performance of each membrane and the mechanism of the studied process for the recovery of dyes from water. The overall data indicate that the used extractive agent is effective for this oriented process, and the adopted membranes can remove about 87% of dyes from wastewater, without applying any external pressure on the polluted aqueous medium.

Keywords: Textile industry, polymer inclusion membranes, Methylene Blue, Bleu P3R dye, facilitated extraction process.

1. Introduction

Dyes are widely used in various industries such as textile, paper, leather and plastic. Textile industries produce large amounts of liquid wastes that contain organic and inorganic compounds. At present, 100000 different types of dyes exist, with an annual production of 735 new species. Among them textile industries consume about 36000 ton/year, 10 to 20 percent of which remains in wastewater [1].

During the dyeing process it has been estimated that the losses of colorants to the environment can reach 10–50% [2,3]. In addition, the effects caused by other pollutants, it is noteworthy that some dyes are highly toxic and mutagenic, and also decrease light penetration and photosynthetic activity, causing oxygen deficiency and limiting downstream beneficial uses such as recreation, drinking water and irrigation [4,5].

To solve the problems related to textile effluents, several studies have focused on the origin of pollution (integrated approach) [6,7] and at the treatment of the final effluents (end-of-pipe approach) [8–10].

IOP Conf. Series: Materials Science and Engineering **948** (2020) 012022 doi:10.1088/1757-899X/948/1/012022

The Membrane separation was selected as an appropriate treatment tool, characterized by their ability to clarify, concentrate, and continuously separate, are potentially interesting for effluent treatment through recycling. In the last few years, the treatment of industrial waste effluents by membrane processes has gained more and more interest because of technical, economical, and political reasons. Technically, the basic technology of process engineering has been developed. Economically, the shortage of water and the increasing cost of auxiliary chemicals and energy have pushed this technology. Politically, the increasing interest of government and people to environmental problems has forced industries to observe severe environmentally safe procedures [11,12].
The goal of this work is to adopt polymer inclusion membrane type (PIM) for the extraction and recovery of Methylene Blue and Blue P3R dyes from simulated aqueous textile solutions. The PIM membrane based on polymers Polyvinyllidene fluoride (PVDF) and polyvinylpyrrolidone (PVP), containing chitin as a new extractive agent was prepared by phase inversion method. Elaborated PIM has been characterized by two methods (Fourier transform infrared attenuated total reflection and SEM), to confirm the inclusion of the extractive agent (carrier) into polymeric support. Kinetic and thermodynamic models were developed, based on the formation of unstable entity (**TS**) by the low interactions of the substrate **S** (dyes) with the immobilized extractive agent **T** (chitin), necessary for the diffusion of the substrate molecules through the membrane phase. These models were used to determine the ***macroscopic*** and ***microscopic*** parameters, respectively: the permeability ***P*** and the initial flux J_0 related to the performance of elaborated PIM membrane for the facilitated extraction process of the dyes. Then the association constants K_{ass} related to the formed entities (Chitin-EA), and the apparent diffusion coefficients ***D**** related to the diffusion of the substrate **S** through the membrane phase of the adopted PIM.

2. Materials and methods

2.1 Chemicals:

All chemicals, reagents, and solvents were pure commercial products (Aldrich, Fluka) of analytical grade. The dye solutions were prepared by dissolving dye (as supplied) in distilled water without adding any auxiliary compounds. The used polymeric support is “Poly(vinyllidene fluoride” supplied by Alfa Aesar.

2.2 Membrane preparation:

The membrane was prepared by phase inversion method, the mixture of PVDF/PVP was completely dissolved in Dimethylformamide as solvent and Chitin as extractive agent and as the water bath step of phase inversion at room temperature.

2.3 IR and SEM observations:

The Fourier transform infrared spectroscopy was used to characterize the elaborated PIM membrane. The prepared membrane was previously dried in vacuum desiccators at room temperature. Samples shaped square were cut from the PIM. The samples were fractured in liquid nitrogen and coated with gold–palladium and then observed by a scanning electron microscope (ZEISS EVO40EP).

2.4 Facilitated extraction experiments:

Extraction experiments were carried out in a cell that contains two compartments, the source phase and the receiving phase, the system is immersed in a thermostatic bath and agitation provided by a multi station magnetic stirring, to follow the extraction evolution of dyes substrate, we took samples from the receiving phase as a function of time, and we used *THERMO SCIENTIFIC* UV adsorption spectroscopy to analyze the samples [13,14].

3. Results and discussion

3.1 Characterization of the membrane:

We developed under the same conditions two PIMs membranes PVDF/PVP without and with Chitin as extractive agent, and we made the characterization of these prepared membranes:

3.1.1 Fourier transforms infrared-attenuated total reflection:

For the developed support and PIM membrane, the following Figure 1 illustrate the infrared spectra of polymeric support PVDF/PVP, before and after the addition of extractive agent. The two new peaks at 1653 cm^{-1} and at 3996 cm^{-1} in the spectrum of the PVDF+PVP+Chitin membrane are attributed respectively to the vibration of C =O carbonyl group of the ester function, and the vibration of O-H groups. Those new peaks indicate that the Chitin extractive agent is inserted correctly in the PVDF+PVP support.

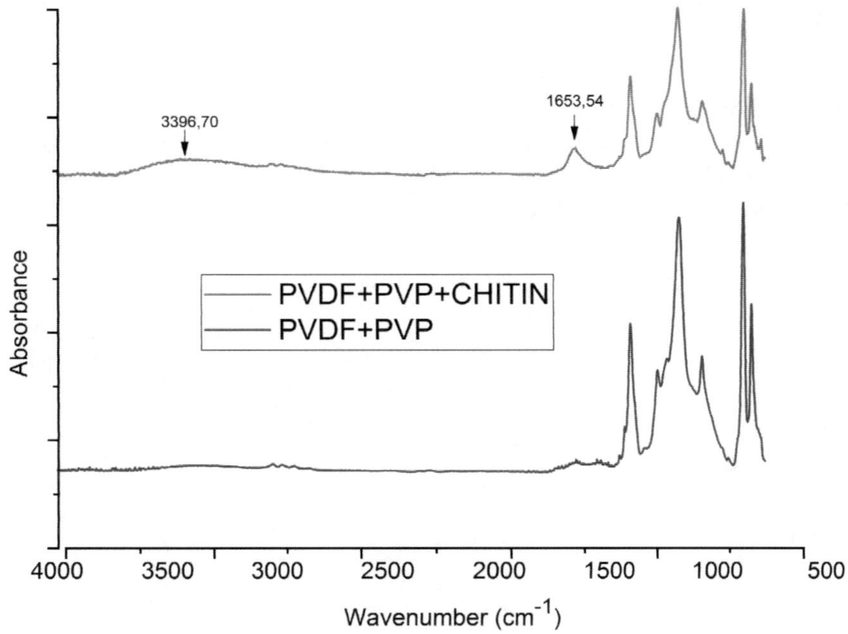

Figure 1. Fourier transform infrared-attenuated total reflection spectra of the PVDF/PVP and PVDF/PVP/Chitin membranes.

3.1.3 Membrane morphology:

The SEM images in Figure 2 show that has asymmetric membranes, and there are homogeneous and contain cavities of different sizes, communicative through pores.

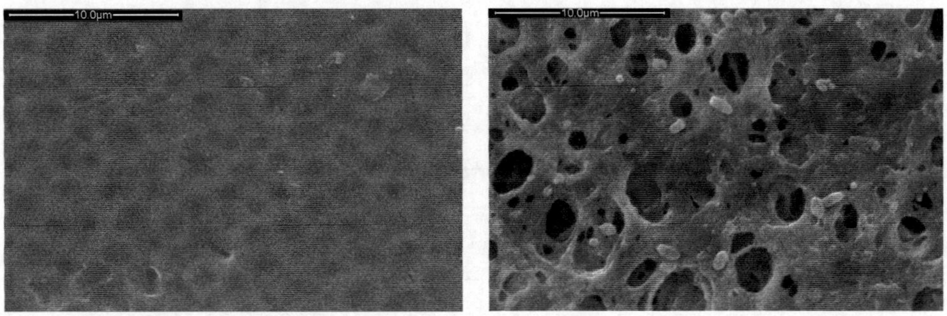

Figure2. Micrographs SEM of the surface of polymer support and PVDF/PVP/Chitin PIM membrane

3.2 Facilitated extraction of Bleu P3R dye:

We have conducted the facilitated extraction process of dyes using the PIM membrane PVDF+PVP+Chitin at T = 298 K and pH = 4 for initial concentration C_0 (0.00014 M to 0.0011 M) for Blue P3R dye, and C_0 (0.00078 M to 0.00009 M) for Methylene Blue dye. Several studies conducted by Hlaibi et al [15], indicated that the macroscopic parameters, permeability ***P*** and initial flux J_0 are calculated from the following equations (1) and (2):

$$P(t-t_L) = (\ell \times V/2S) \ln(C_0 / C_0 - 2C_r) \quad (1)$$

$$P = a \times V \times \ell / 2*S \quad \text{and} \quad J_0 = P \times C_0/\ell \quad (2)$$

a: the slope experimental values of the straight lines $-Ln(C_0-2C_r) = f(\mathbf{t})$. ℓ: the membrane thickness. **S:** the membrane active area in contact with the aqueous solutions. **V**: the receiving phase volume.

3.2.1 Macroscopic parameters *P* and J_0:

Initially we studied the influence of the substrate (Blue P3R dye) initial concentration $\mathbf{C_0}$ on the evolution of the parameters for this facilitated extraction process. Our experimental results for the facilitated extraction process of Blue P3R dye, verify the developed kinetic model. The study of the function $-Ln(\mathbf{C_0-2C_r}) = f(\mathbf{t})$ for the different studied concentrations $\mathbf{C_0}$, provides straight lines shown by the graph in the Figure 3, then we can calculate the macroscopic parameters ***P*** and J_0, from the slopes of these straight line and the obtained values are grouped in table 1.

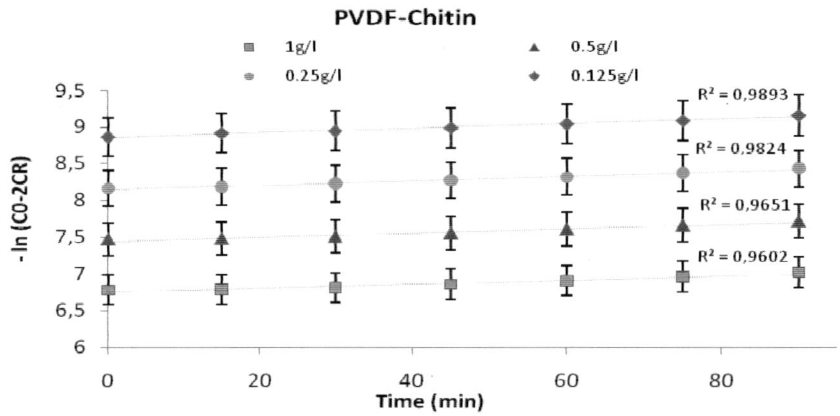

Figure 3: Evolution of the kinetic function $-Ln(\mathbf{C_0-2C_r}) = f(\mathbf{t})$ for the facilitated extraction of Bleu P3R dye at different initial substrate concentrations.
$[Chitin]_0 = 0.0007035$ M, pH = 4, and T = 298 K

Table 1: Influence of initial concentration $\mathbf{C_0}$ on the evolution of ***P*** and J_0 parameters for the facilitated extraction process of the dyes

Substrate	C_0 (mol L^{-1})	$P\times10^6$ (cm^2 s^{-1})	$J_0\times10^5$ (mmol cm^2 s^{-1})
Blue P3R dye	0.00113	5.66	0.022
	0.00056	6.06	0.011
	0.00028	6.27	0.006
	0.00014	6.52	0.003
Methylene Blue dye	0.00078	12.60	0.685
	0.00039	13.83	0.298
	0.00019	15.15	0.163
	0.00009	16.20	0.087
$[Chitin]_0 = 0.0007035$M, pH = 4, and T = 298 K, PIM=PVDF+PVP+Chitin			

The results show a very good performance of the developed membrane for the oriented process of the facilitated extraction of studied dyes.

3.2.1 Microscopic parameters *D** and K_{ass}:

The same studies conducted by Hlaibi et al [15–17]. and the developed thermodynamic model, based on a saturation law of the extractive agent by the substrate, show that the microscopic parameters are determined from the expressions in the following relationships.

$$1/J_0 = 1/D^*((1/[T]_0 * K_{ass} * C_0) + 1/[T]_0) \quad (3)$$

$$K_{ass} = \text{intercept } (oo) / \text{slope } (p) \quad \text{and} \quad D^* = (1/oo) * (1/[T]_0) \quad (4)$$

The experimental results verify the thermodynamic model, and make it possible to obtain the line segments presented by the graph of the figure 4 for the presentation of function $1/\boldsymbol{J_0} = f(1/\mathbf{C_0})$. The intercepts and the slopes of these segments make it possible to calculate $\boldsymbol{K_{ass}}$ and $\boldsymbol{D^*}$ parameters according to the expressions in equation (4).

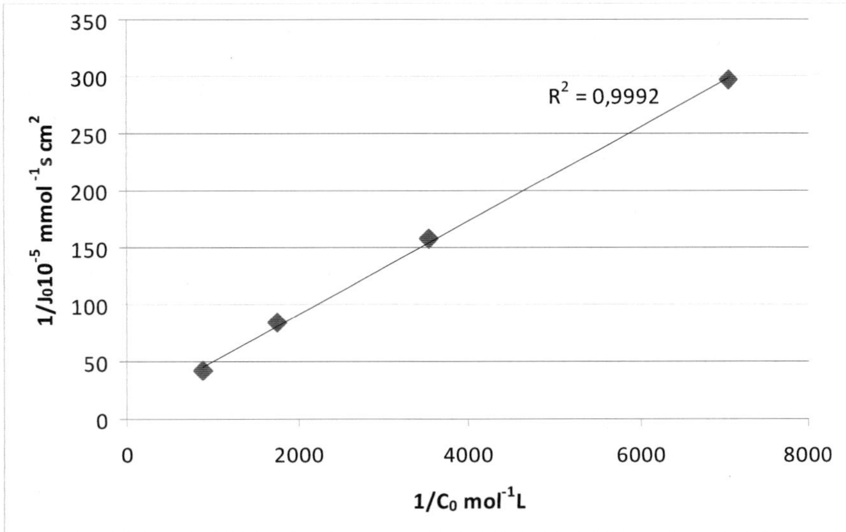

Figure 4: Line weaver - Burk straights for the facilitated extraction phenomenon of Bleu P3R dye by the membranes: PVDF+PVP+Chitin.
[Chitin]$_0$ = 0.0007035M, pH = 4, and T = 298 K

Table 2: Evolution of the parameters $\boldsymbol{K_{ass}}$ and $\boldsymbol{D^*}$ according to the nature of the used substrate

Substrate	**D×10^{-5} (cm^2 s^{-1})**	**Kass (L. mol^{-1})**
Blue P3R dye	4.15	242.42
Methylene Blue dye	22.49	7.74
[Chitin]$_0$ = 0.0007035 M, pH = 4, and T = 298 K		

All obtained results in Table 2 show an inverse evolution of these parameters with low values of the association constant $\boldsymbol{K_{ass}}$, which suggests a ***jumping mechanism on fixed-sites*** for the facilitated diffusion process of the studied dyes through this PIM membrane type.

4. Conclusions

In this work, we prepared by phase inversion method new PIM membrane based on PVDF+PVP as supports and Chitin as extractive agent. We characterized the obtained membrane by SEM and IR, and we found that insertion of the extractive agent Chitin in the membrane phase of the PIM allows the passage of dyes. On the other hand, we have verified our models and we were able to calculate all parameters related to facilitated extraction process of dyes through the elaborated PIM membrane

Finally, we have elucidated a mechanism by successive jumps on fixed sites for the studied process through this PIM membrane type, and we can conclude that the elaborated membrane is efficient for a possible application relating to the extraction and the recovery of the studied dyes from the waters waste.

Acknowledgments
The works in this manuscript are conducted within the framework of a PPR2 project, funded by the Ministry of Higher Education and Scientific Research (MESRSFC) and the National Center of Scientific and Technical Research (CNRST). We would like to thank both organizations for their financial support.

REFERENCES

[1] Contreras S, Pascual E, Esplugas S, Gim J, Rodrı M. Comparison of different advanced oxidation processes for phenol degradation. Water Res. 2002;36:1034–42.
[2] Ghaly AE, Ananthashankar R, Alhattab M, Ramakrishnan V V. Chemical Engineering & Process Technology Production , Characterization and Treatment of Textile Effluents□: A Critical Review. J Chem Eng Process Technol. 2014;5(1):1–18.
[3] Katheresan V, Kansedo J, Lau SY. Efficiency of Various Recent Wastewater Dye Removal Methods: A Review. Biochem Pharmacol . 2018;
[4] Al-alwani MAM, Ludin NA, Bakar A, Amir A, Kadhum H, Mukhlus A. Spectrochimica Acta Part A□: Molecular and Biomolecular Spectroscopy Application of dyes extracted from Alternanthera dentata leaves and Musa acuminata bracts as natural sensitizers for dye-sensitized solar cells. Spectrochim Acta Part A Mol Biomol Spectrosc . 2018;192:487–98.
[5] Liu L, Zhang J, Tan Y, Jiang Y, Hu M, Li S, et al. Rapid decolorization of anthraquinone and triphenylmethane dye using chloroperoxidase□: Catalytic mechanism , analysis of products and degradation route. Chem Eng J. 2014;244:9–18.
[6] Hubbe MA, Hasan SH, Ducoste JJ. Cellulosic Substrates for Removal of Pollutants. BioResources. 2011;6(1998):2161–287.
[7] Lopez A, Ricco G, Ciannarella R, Rozzi A, Di Pinto AC, Passino R. Textile wastewater reuse: Ozonation of membrane concentrated secondary effluent. Water Sci Technol. 1999;40(4–5):99–105.
[8] Forgacs E, Cserháti T, Oros G. Removal of synthetic dyes from wastewaters: A review. Environ Int. 2004;30(7):953–71.
[9] Robinson T, McMullan G, Marchant R, Nigam P. Remediation of dyes in textile effluent: A critical review on current treatment technologies with a proposed alternative. Bioresour Technol. 2001;77(3):247–55.
[10] Przystaś W, Zabłocka-Godlewska E, Grabińska-Sota E. Biological Removal of Azo and Triphenylmethane Dyes and Toxicity of Process By-Products. Water, Air, Soil Pollut. 2012 May;223(4):1581–92.
[11] Askari N, Farhadian M, Razmjou A. Environmental Nanotechnology , Monitoring & Management Simultaneous e ff ects of pH , concentration , pressure on dye removal by a polyamide nano fi lter membrane□; optimization through response surface methodology. Environ Nanotechnology, Monit Manag . 2018;10:223–30.
[12] Abid MF, Zablouk MA, Abid-alameer AM. ENVIRONMENTAL HEALTH Experimental study of dye removal from industrial wastewater by membrane technologies of reverse osmosis and nanofiltration. Iranian J Environ Health Sci Eng. 2012;9(1):1.
[13] Kamal O, Eljaddi T, Atmani EH EL, Touarssi I, Lebrun L, Hlaïbi M. Grafted polymer membranes with extractive agents for the extraction process of VO_2^+ ions. Polym Adv Technol. 2016;(September):76821.
[14] El Atmani EH, Benelyamani A, Mouadili H, Tarhouchi S, Majid S, Touaj K, et al. The oriented processes for extraction and recovery of paracetamol compound across different affinity polymer membranes. Parameters and mechanisms. Eur J Pharm Biopharm. 2018;126:201–10.
[15] Hlaïbi M, Tbeur N, Benjjar A, Kamal O, Lebrun L. Carbohydrate-resorcinarene complexes involved in the facilitated transport of alditols across a supported liquid membrane. J Memb Sci. 2011;377(1–2):231–40.
[16] Benjjar A, Eljaddi T, Kamal O, Touaj K, Lebrun L, Hlaibi M. The development of new supported liquid membranes (SLMs) with agents: Methyl cholate and resorcinarene as carriers for the removal of

dichromate ions (Cr2O72−). J Environ Chem Eng. 2014;2(1):503–9.
[17] Hor M, Riad A, Benjjar A, Lebrun L, Hlaïbi M. Technique of supported liquid membranes (SLMs) for the facilitated transport of vanadium ions (VO2+). Parameters and mechanism on the transport process. Desalination. 2010;255(1–3):188–95.

SYNTHESIS AND CHARACTERIZATION OF NICKEL-MOLYBDENUM CATALYSTS SUPPORTED ON COPPER ORTHOPHOSPHATES

OUCHABI Mbarka[1,*], BAALALA Mohammed[1], ELAISSI Abdelkrim[1], BENSITEL Mohammed[1,].**
[1] Laboratoire de Catalyse et Corrosion des Matériaux, Université Chouaib Doukkali, Faculté des Sciences d'El Jadida, BP. 20, El Jadida 24000, Morocco.
*Corresponding author: m.ouchabi@yahoo.fr ; **bensitel.m@ucd.ac.ma

Abstract. Synthesis of a pure copper orthophosphate **(CuP)** prepared by Coprecipitation, and **CuP** modified by impregnation of NiMo (2-14 wt % of Ni-Mo oxide) have been carried out. The solids obtained were investigated as synthesized or after calcination by various physico-chemical techniques such as X-Ray Diffraction (XRD), Infrared Spectroscopy (IR), Thermogravimetric analysis (TGA), and differential thermal analysis (DTA). The results revealed that the solids **NiMo/CuP** consisted of copper orthophosphate $Cu_3(PO_4)_2$ as major phases, together with $NiMoO_4$ as minor phase. The diffraction lines of $NiMoO_4$ increase by increasing the nickel-molybdenum content.
Keywords: Copper orthophosphate **CuP**, nickel-molybdenum catalysts supported on copper orthophosphates **NiMo/CuP**, XRD, DTA-TGA.

1.Introduction

The numerous investigations of the systems Cu–P–O and Cu–P–O–H have revealed their rich crystal chemistry. Copper can accommodate various coordinations from tetrahedral or square to pyramidal andoctahedral. Thirteen different structures of copper phosphates or hydroxyphosphates have been discovered to date [1]. It is well accepted that there is a close relationship between the morphology and their properties of inorganic materials, that is, morphologies determine their properties since the crystal shape dictates the interfacial atomic arrangement of the material [2-7]. Acidic solids derived from metal orthophosphates are widely used to catalyze various reactions occurring at the gas/solid interface. Acid–base properties and redox properties are amongst important types of surface chemical properties of metal orthophosphate catalysts.

A number of NiMo catalysts have been tested in a number of heterogenous catalytic oxidation reactions such as: Deoxygenation of vegetable oils over sulfided Ni, Mo and NiMo [8]; The Mo_3NiS_4/NaY catalyst showed much higher activity for HDS of benzothiophene than the Mo_3S_4/NaY catalyst [9]; NiMo/Al_2O_3 catalyst containing nano-sized zeolite Y for deep hydrodesulfurization and hydrodenitrogenation of diesel [10]; Recently Sadiq tested the Influence of acid–base properties of cobalt–molybdenum catalysts supported on magnesium orthophosphates in isomerization of 3,3-dimethylbut-1-ene [11].

The most of nickel-molybdenum oxide based catalysts consist of a $NiMoO_4$ phase deposited on the surface or in pores of an oxide support. However, numerous studies in the last decade indicated that the activity and selectivity of supported metal oxide catalysts are significantly affected by the properties of the support oxide material, and that the dispersed phase has a particular structure and properties different from that of the bulk oxide.
The attempt of this work is to prepare the pure copper orthophosphate **CuP** by coprecipitation, and **CuP** modified by impregnation of Nickel-molybdenum. Additionally, the studied powders were characterized by thermogravimetry–differential thermogravimetry (TG/DTG), X-ray diffraction (XRD), Fourier transform infrared (FTIR) and UV-Visible. The data obtained will be important for further studies of the compound.

2. Experimental

1.1. Sample preparation

Copper orthophosphate **CuP** were synthesized upon evaporation to dryness from aqueous solutions of $Cu(NO_3)_2.3H_2O$ and $(NH_4)_2HPO_4$. The resulting solids were dried for 12h at 120°C and were finally calcined in air at 450°C for 10h.
A series of nickel–molybdenum catalysts NiMo/CuP, were prepared using successive impregnation method, over an CuP support. A mass of nickel nitrate $[Ni(NO_3)_2,6H_2O]$ calculated to yield the desired percentage of nickel on the support, was dissolved in a predetermined volume of distiled water. This volume calculation is based on the pore volume of the support. The impregnated catalysts Ni/CuP were dried at 120°C K for 12 h and calcined under air flow at 450°C during 4 h. A second impregnation over Ni/CuP by an aqueous solution of ammonium heptamolybdate $[(NH_4)6Mo_7O_{24},4H_2O]$. Finally, the impregnated catalysts are dried at 120°C during 12 h and calcined under air flow at 450°C for 4 h, finally at this solids calcined at 650°C for 10h to clearly identify the phase obtained.

1.2.Characterization techniques

The X-ray diffraction (XRD) analysis was performed for the fresh catalyst using Bruker-eco D8 Advance diffract meter with Cu-Kα radiation (λ = 1.5418 Å). The XRD patterns were scanned from 5 to 70 θ.
TGA–DTA analysis has been recorded with a TA Instruments balance, model SDT 2966 (TGA–DTA). It allows obtaining simultaneous DTA and TGA diagrams under similar experimental conditions. Experiments were performed under an N_2 flow, from 298 K to 1073 K using a heating rate of 5°C/min.
Diffuse reflectance infrared spectra for the synthesized solids were recorded from 4000 to 400 cm^{-1} on a Nicolet 460 spectrophotometer.
UV–Visible spectra in diffuse reflectance mode were recorded by Shimadzu UV-2400PC Series spectrometer, from 200 to 1000 nm.

3.Results and discussion

3.1. DTA and TG analyses

The TG/DTA curves of CuP are shown in Figure 1. The TG curve of CuP shows the three weight loss steps in the range of 50–670°C. The weight loss steps in the TG curve were observed in ranges of 50–150°C and 150–320°C. Correspond respectively to 4.372% of water and 41.524% of ammonium nitrate and water. These two stages appear in the corresponding DTA curve as two endothermic peaks at 130 and 290°C.

The third stage of the weight loss was related to the start of dehydroxylation of libenethite $Cu_2(PO_4)$ (OH) which leads to an orthophosphate $Cu_3(PO_4)_2$, corresponding an one endothermic peaks at 330°C. The endothermic peak appeared at 605°C corresponds to the better crystallization of Orthophosphate $Cu_3(PO_4)_2$ [12-13].

The TGA curve of the solid NiMo(14%)/CuP dried at 110 ° C (Figure 2) has two mass loss between room temperature and 670°C. These weight losses are due to the removal of water and ammonia from the structure ammonium heptamolybdate. The endothermic peak appeared at 605°C corresponds to the crystallization of Orthophosphate $Cu_3(PO_4)_2$. While No peak observed after 605°C attributed to crystallization of the $NiMoO_4$ phase. May be the Nickel-Molybdenum oxide is well dispersed on the surface of the copper orthophosphate or the amount of Nickel-Molybdenum oxide is too small.

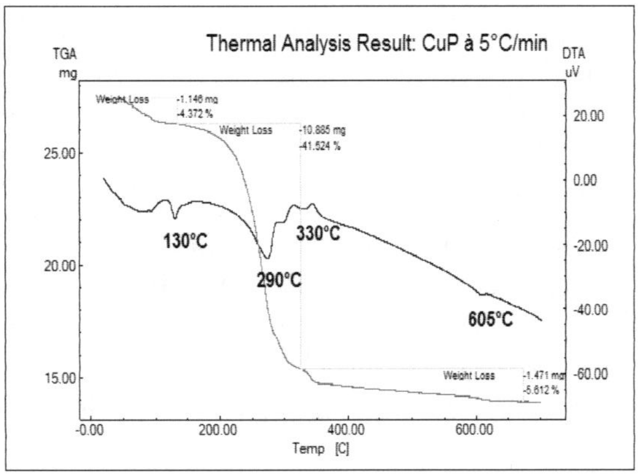

FIGURE 1. TGA-TDA CURVES FOR PURE SOLID CuP

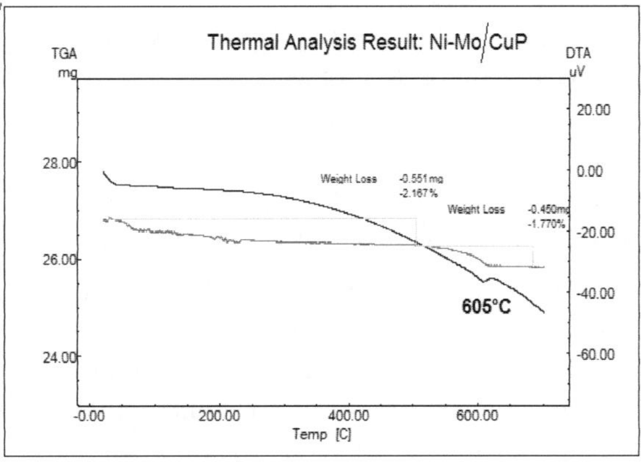

FIGURE. 2. TGA-TDA CURVES FOR SOLID NiMo(12%)/CuP

3.3. XRD results

The XRD pattern of the nickel-molybdenum oxide catalysts supported on copper orthophosphate NiMo/CuP calcined at 450°C and 650°C are presented in Figure 3 and figure 4, respectively.

The pure CuP solid treated at 450°C exhibits broad diffraction lines corresponding $Cu_3(PO_4)_2$ (PDF Index Name : Copper Phosphate .ref cod: 01-070-0494), and also some weaker diffraction stripes (2□ = 15°) have been assigned to libethenite $Cu_2(PO_4)(OH)$(PDF Index Name : Copper Phosphate .ref cod: 01-083-1557). Progressively with the increase in temperature at 650°C (Figure 4), diffraction lines of the copper orthophosphate sharpen indicating a better crystallization of $Cu_3(PO_4)_2$ (PDF Index Name :Copper Phosphate .ref cod: 01-070-0494) [12-13].

From figure 4,The solids impregnated with nickel-molybdenum treated at 450°C lead only to the previous phase identified for $Cu_3(PO_4)_2$.

At high temperature treatment (650°C), all the catalysts of the series showed lines characteristic for Copper orthophosphate (figure 4) and all crystalline phases of these solids belong to the anorthic system and crystallize in the P-1 (2) spatial group.
The appearance of peaks at 2□ = 26° and 33° indicates the presence of Nickel Molybdenum oxide $NiMoO_4$ (ref cod: 00-045-0142) this phase belong to the monoclinic, spatial group C2/m (12). This suggests that if Nickel-molybdenum-containing phases are present, they are small and poorly crystalline. The formations of $Cu_3(PO_4)_2$ and $NiMoO_4$ oxide is well confirmed by XRD.
However, Average crystallite size of NiMo/CuP calcined at 450°C and 650°C calculated by applying the debye -scherrer's equation from the high score software and reported in table 2. The compositions of these solids are nanocrystalline with an average crystallite size of up to about 77 nm. The crystallite size of all solids calcined at 650°C remains stable which confirms the better crystallization of catalysts.

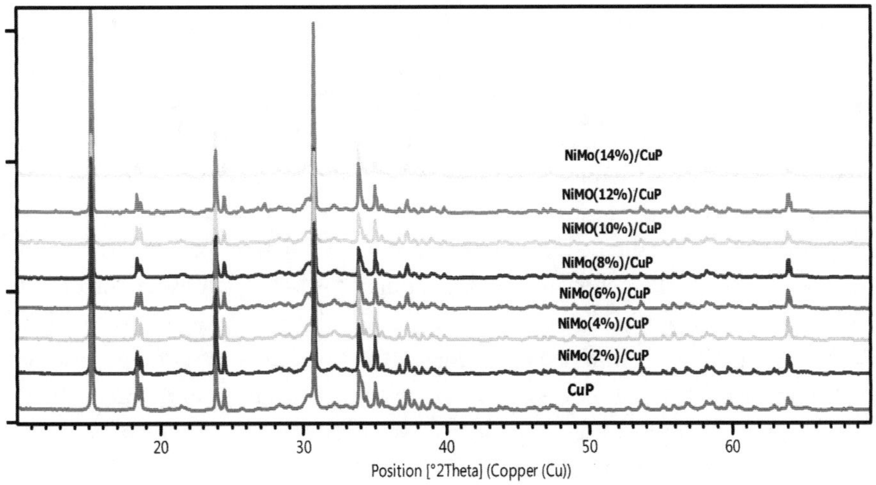

FIGURE 3. XRD PATTERNS OF NiMo/CuP (450°C).

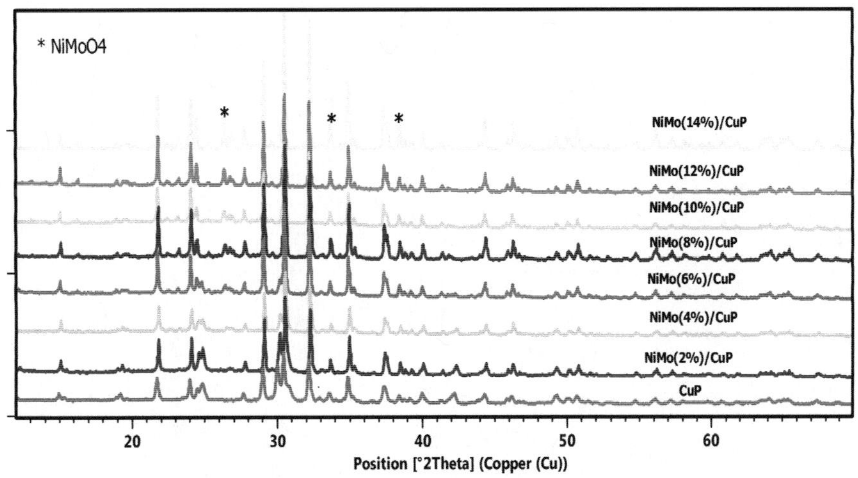

FIGURE 4. XRD PATTERNS OF NiMo/CuP (650°C).

TABLE 2: AVERAGE CRYSTALLITE SIZE of NiMo/CuP solids CALCULATED FROM XRD DATA .

Catalysts	CuP	NiMo(2 %)	NiMo(4 %)	NiMo(6 %)	NiMo(8 %)	NiMo(10 %)	NiMo(12 %)	NiMo(14 %)
crystallite size(nm) (450°C)	40.12	40.12	40.12	77.03	40.88	77.03	40.12	77.03
crystallite size(nm) (650°C)	40.86	40.86	40.86	40.86	40.86	40.86	40.86	40.86

3.3. FTIR results

The FTIR spectra of the synthesized NiMo/CuP are shown in Figure 5. For characteristic vibration of CuP calcined at 650°C, FTIR bands are assigned according to the literature [14-15] based on the fundamental vibrating unit PO_4^{3-} anion. One of the most noteworthy features of the spectrum is the presence of the strong bands at 1120, 965-935, 608-527 and 406 cm^{-1}. These bands can be assigned to the $\upsilon_{as}(PO_4^{3-})$, $\upsilon_s(PO_4^{3-})$, $\delta_s(PO_4^{3-})$ and δ_s(POP), respectively. The characteristic bands of groups (PO_4^{3-}) refine, divide, which reflects a structural evolution of crystalline compound.

For FT-IR spectra of NiMo/CuP (figure 5) and comparing with the CuP samples, the appearance of the bands at 738, 820 and 870 cm^{-1} indicates the presence of weakly crystallized $NiMoO_4$. From literature the spectrum of molybdate is characterized by appearance the three strong absorption bands at 566, 762 and 848 cm^{-1}. The first two are assigned to Mo-0 stretching vibrations with Mo in octahedral coordination. The band at 848 cm^{-1} indicates that part of the Mo is in a tetredral environment [16].

In our case, the broad band centered around 738 cm^{-1} assigned to Mo-0 stretching vibrations with Mo in octahedral coordination. The band at 566 (due to Mo-0 octahedral vibration) is absent, perhaps they may be masked by the bands characteristic of phosphates.

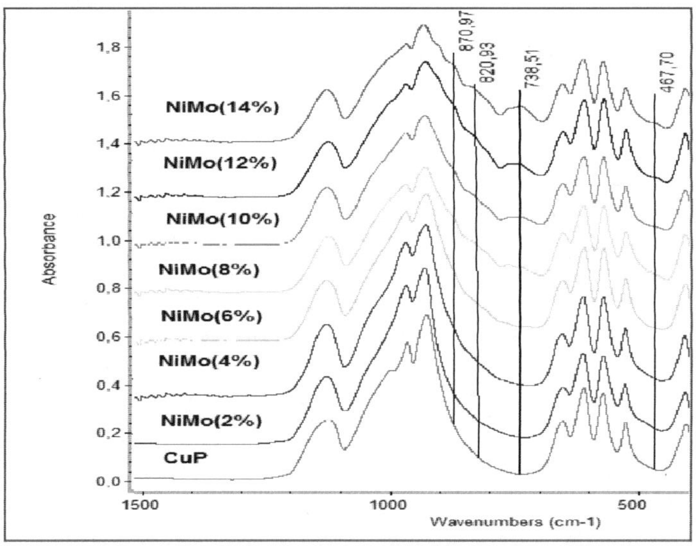

FIGURE 5. FTIR SPECTRA OF NiMO/CuP SOLIDS

IOP Conf. Series: Materials Science and Engineering **948** (2020) 012023 doi:10.1088/1757-899X/948/1/012023

3.4. UV-Visible results

The UV–Vis spectra of the NiMo/CuP samples are compared in Figure 6. All spectra of NiMo/CuP samples exhibit two major CTLM (charge transfer ligand-metal) bands centered at about 289 and 294 nm of P–O and Cu–O. In addition, the NiMo/CuP samples exhibit a small and broad visible absorption band with a maximum at 800 nm corresponding to d–d transitions ($d^{10} \rightarrow 3d^{9}4s^{1}$) of Cu^{2+}, this results suggested that the Cu^{2+} ions must be present in approximately octahedral coordination [17-18].

For NiMo/CuP samples a broad band between 700 and 800 nm is also probably attributed to reduced Mo ions Mo^{5+} and/or Mo^{4+}. M.C. Abello et al. observed for MoO_3 / Al_2O_3 catalysts two absorption bands at 220 and 260 nm and a broad band in the 280–350 nm region, suggesting that Mo^{6+} octahedral species are mainly present [19-22]. For NiMo/CuP the band around 300 nm probably attributed to Mo^{6+} was observed.

The Ni^{2+} ions, in NiMo/CuP sample (Figure 6), were located in oxygen octahedra. According to the literature data a band around 740 nm indicates the appearance of Ni^{2+} ions in octahedral coordination. Several papers describe the nickel catalysts supported on niobia, titania, magnesia and alumina which were characterized by UV-VIS spectroscopy. The appearance of a band around 660-670 nm noticed for NiO/TiO_2 [23] and NiO/MgO [24] catalysts was attributed to the dispersed octahedrally coordinated Ni^{2+} surface species. Similar results were obtained by Y. Chen et al. (~665 nm) [25] and M.A. Zanjanchi (~660 and 730 nm) [19]. For NiO/Nb_2O_5 well defined bands attributed to octahedrally coordinated Ni^{2+} were observed at ca. 840 nm [26].

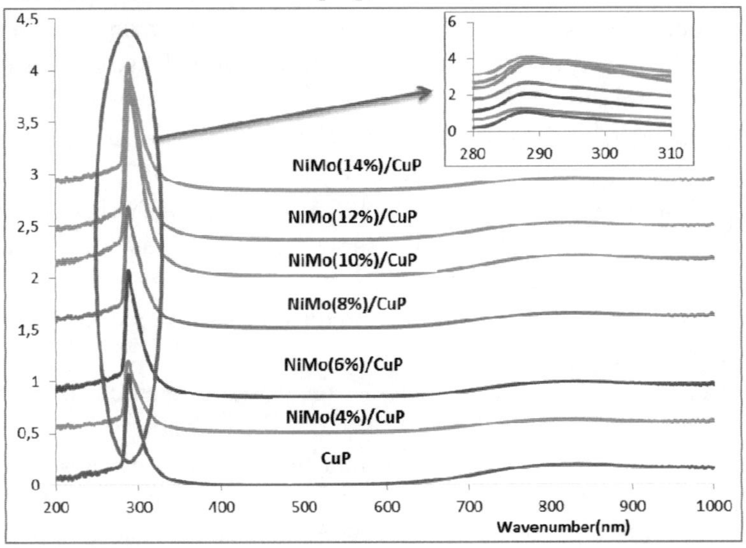

FIGURE 6. UV–VIS SPECTRA OF NiMo/CuP SOLIDS.

4. Conclusions

The In this work we have reported the synthesis of a pure copper orthophosphate (CuP) and CuP modified by introduction of nickel-molybdenum through an easy route, in comparison with previous synthesis of other metal orthphosphates. The XRD and FTIR results confirmed the formation of $Cu_3(PO_4)_2$ product and we have demonstrated also the formation of the $NiMoO_4$ on the surface of support. The solids obtained are necessary for theoretical study, applications development, and catalysis reaction.

In perspective, these catalysts NiMo/CuP will be used for hydrodesulfuration reaction. The Influence of acid–base properties of nickel-molybdenum catalysts supported on copper orthophosphates in this reaction will be studied by Catalytic decomposition of isopropanol.

REFERENCES

[1] Baies R., Caignaert V., Pralong V., Raveau B., *Inorg.Chem.* 44 (2005) 2376–2380.
[2] Chen S., Yu S., Jiang J., Li F., Liu Y., *Chem. Mater.* 18 (2006) 115–122.
[3] Pan C., Gu Z., Dong L., *Acta Chim. Sinica* 17 (2009) 1981–1986.
[4] Akla A., Howari H., *J. Phys. Chem. Solids* 70 (2009) 1337–1343.
[5] Khalavka Y., Sonnichsen C., *Adv. Mater.* 20 (2008) 588–591.
[6] Xu J., Xue D., *J. Phys. Chem. Solids* 67 (2006) 1427–1431.
[7] Huang X., Guo C., Zuo J., Zheng N., Stucky G.D., *Small* 5 (2009) 361–365.
[8] David Kubička[a,] , Luděk Kaluža[b] Applied Catalysis A: General Volume 372, Issue 2, 15 January 2010, Pages 199–208.
[9] MitsuguTaniguchia, Daichi Imamuraa, Hiromi Ishigea, Youichi Ishiib,Takashi Muratab, Masanobu Hidaib, Takashi Tatsumia1, Hydrodesulfurization of Benzothiophene over Zeolite-Supported Catalysts Prepared from Mo and Mo–Ni Sulfide Clusters, Journal of Catalysis 187(1):139-150 • October 1999.
[10] Hailiang Yin∗, Tongna Zhou, Yunqi Liu, Yongming Chai, Chenguang Liu∗NiMo/Al2O3 catalyst containing nano-sized zeolite Y for deep hydrodesulfurization and hydrodenitrogenation of diesel Journal of Natural Gas Chemistry 20(2011)441–448
[11] M. Sadiq a, A. Sahibed-dine a, M. Baalala a, K. Nohair a, M. Abdennouri a, M. Bensitel a,*, C. Lamonier b, J. Leglise c Arabian Journal of Chemistry (2011) 4, 449–457
[12] M. Ouchabi, M. Baalala, A. Elaissi, M. Bensitel, J. Mater. Environ. Sci. 7 (4) (2016) 1417-1424.
[13] M. Ouchabi, M. Baalala, A. Elaissi, M. Bensitel, 2017, IOP Conf. Ser.: Mater. Sci. Eng. 186 012013.
[14] Colthup N.B., Daly L.H., Wiberley S.E., Introduction to Infrared and Raman Spectroscopy, Academic Press, New York, 1964.
[15] Steger E., Ka¨ßner B., Die infrarotspektren von wasser freienschwer metall-Diphosphaten, Spectrochimica Acta, 24A (1968) 447–456.
[16] Anagha A. Belhekar, Subbiah Ayyappan & Arumugamangalam V. Ramaswamy., J . Chem. Tech. Biotechnol. 1994, 59, 395-402
[17] Morsi M.M., Metwalli E.S., Mohamed A.A., 1999 Phys. Chem. Glasses 40 p314.
[18] Y. Ywasawa, M. Yamagishi, J. Catal., 82, (1983), 373.
[19] H. Aritani, T. Tanaka, T. Funabiki, S. Yoshida, K. Eda and N. Sotani, M. Kudo, S. Hasegawa, J. Phys. Chem., 100, (1996), 19495.
[20] M.C. Abello,., M.F. Gomez, M. Casella, O.A. Ferretti, M.A. Bañares, J.L.G. Fierro, Appl. Cat. A: Gen., 251, (2003) 435.
[21] M.C. Abello, M.F. Gomez, O. Ferretti, Appl. Catal. A: Gen. 207 (2001) 421.
[22] B. Xu, L. Dong, Y. Fan, Y. Chen, J. Catal., 193, (2000), 88.
[23] J.D. Dunitz, L.E.Orgel, J. Phys. Chem. Solids, 3, (1957), 318
[24] J. Wang, L. Dong, Y. Hu, G. Zheng, Z. Hu, Y. Chen, J. Solid State Chem. 157, (2001), 274.
[25] M.A. Zanjanchi, L. Abdollahi, J. Inclus. Phen. Macrocyc. Chem., 39 (1-2), (2001), 151.
[26] N. Rajic, I. Arcon, V. Kaucic, A. Kodre, Croatica Chemica Acta, 72 (2-3), (1999) 645

Piezoelectric β-polymorph enhancement in graphene oxide - PLA nanocomposite films

K. Oumghar [1,3], N. Chakhchaoui [1,2,6,*], R. Farhan [3], A. Eddiai [3,], M. Meddad [4], O. Cherkaoui [1], M. Mazroui [3], Lieva Van Langenhove[5]**

[1] REMTEX Laboratory, Higher School of Textile and Clothing Industries (ESITH), Casablanca, Morocco
[2] BGIM Laboratory, Higher Normal School (ENS), Hassan II University, Casablanca, Morocco
[3] Laboratory of Physics of Condensed Matter (LPMC), Faculty of Sciences Ben M'Sik, Hassan II University, Casablanca, Morocco
[4] LAS Laboratory of Setif, Mohamed el Bachir el Ibrahimi BBA University, Algeria
[5] Centre for Textile Science and Engineering, Department of Materials, Textiles and Chemical Engineering, Ghent University, Belgium
[6] Author to whom any correspondence should be addressed

*chakhchaoui.nabil@gmail.com ; **Adil.eddiai@univh2c.ma

Abstract: The flexible piezoelectric nanocomposite films have attracted considerable interest for use in autonomous nanosystems, sensors and portable electronics. The effect of graphene oxide (GO) nanoparticles on the crystal structure, unique crystallization, mechanical behavior and significant improvement in piezoelectricity of polylactic acid (PLA) has been reported. Polylactic acid /graphene oxide (PLA/GO) nanocomposite films were prepared by the solvent casting technique using different percentages of GO. The different samples were characterized by Polarized optical microscope (POM), Fourier transformed infrared (FTIR) and X-ray diffraction analysis. The presence of β phase in nanocomposites and the enhancement of their piezoelectric properties were confirmed by FTIR and XRD analysis. These electroactive and biodegradable polymer nanocomposite thin films may be a potential candidate for the piezoelectric nanogenerators, energy storage devices and energy harvesting applications.
Keywords: PLA; Graphene oxide; piezoelectric properties; β-phase

1. Introduction

The use of mechanical energy wasted as a useful source of energy was honored. In the recent years, several studies have focused on the themes of harvesting energy [1-6]. There is a growing demand for the recovery of the mechanical energy west from our daily life for autonomous sensors and implantable energy devices [7, 8]. The electroactive polymers are one of the most important materials that have received considerable attention as electronic systems, thermal and mechanical detectors and sensors [9].

The piezoelectric nanocomposite films have attracted a growing amount of interest that its energy can be valuable for sensing and energy harvesting application.

In this paper, different amounts of graphene oxide nanosheet (GO) were added into the polylactic acid (PLA) matrix in order to enhance the crystalline β-phase and to improve the piezoelectric properties of the PLA/GO nanocomposite. The nanosheets of graphene oxide (GO) can also be used as a nucleating agent for the PVDF to produce high-performance nanocomposite. GO is a single layer of graphite oxide (GO) obtained by a process of oxidation/exfoliation natural graphite [10]. GO are essentially graphene sheets on which oxygen-containing functional groups are thought to be present in the form of carboxyl, hydroxyl and epoxy groups [11-13].

The graphene oxide was prepared by the Hummers method and the nanocomposite films were made by the solvent casting-evaporation technique. The characterization of graphene oxide are observed by Fourier transform infrared and X-ray diffraction studies, and the nanocomposite films were characterized by polarized optical microscope (POM), Fourier transform infrared and X-ray diffraction analysis, to confirm the enhancement of piezoelectric crystalline β-phase and barrier properties of the nanocomposite to can be valuable for energy harvesting application.

2. Experimental section:

2.1. Materials

The PLA (pellets) used has a density of ρ = 1.78 g/cm3, Mn = 130,000 g/mol and Mw = 400,000 g/mol, was obtained from Sigma Aldrich. Tetrahydrofuran (THF) is used as the organic polar solvent often used to dissolve PLA. It has a density of 0.949 g/cm3 at 20 °C and boiling temperature of 163 to 165 °C, was obtained from LabChem.

2.2. Graphene oxide synthesis

The synthesis of graphene oxide was made by the method of Hummers change. This method of synthesis involves both oxidation and exfoliation of graphite sheets due to the heat treatment of the solution.

An amount of 2 g of graphite powder with 2 g of sodium nitrate ($NaNO_3$) dissolved in 90 ml of H_2SO_4 (98%) and maintained under an ice bath (0-5 ºC) with continuous stirring. After 4 hours of stirring, 12 g of potassium permanganate ($KMnO_4$) was added to the suspension very slowly. The rate of addition is carefully controlled to keep the reaction temperature below 15 ºC. The solution was diluted with a large amount of distilled water and kept stirring for 2 hours. The ice bath was then removed, and the mixture was stirred at 35 ° C for 2 hours. The mixture is maintained in a reflux system at 98 ° C for 10-15 min. After 10 min, the temperature was changed to 30 °C which gives the solution a brown color. Again after 10 min, the temperature was lowered to 25 °C and held for 2 hours. The solution is finally treated with 40 ml of H_2O_2 to reduce the residual permanganate to soluble manganese ions until the gas evolution ceases, giving a yellow color. 200 ml of water was added to the prepared solution and stirred for 1 hour. It is then maintained without agitation for 3-4 hours. The latter was washed several times with HCl (10 %) and then with deionized water several times until neutralization (neutral pH). After centrifugation, the gel-like substance was dried under vacuum at 60 ° C overnight to obtain graphene oxide (Figure 1).

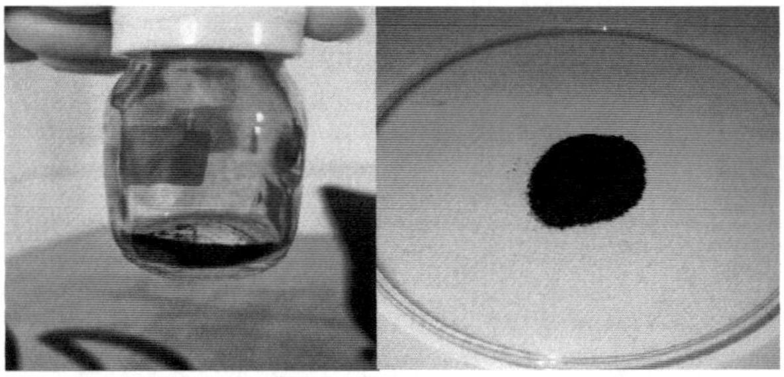

Figure 1. Photograph of graphene oxide nanosheets

2.3. PLA/GO based nanocomposite synthesis

The nanocomposite PLA / GO films were developed by the solution-mixing technique (Figure 2).

1 g of PLA was dissolved in 20 ml of THF using the mechanical stirring at 71 ° C for 2 hours, as soon as a dissolution of the known GO masses (0.1, 0.3, 0.5, 0.8, 1 and 2 wt.%) in 5 ml of THF by ultrasonication. Then the solution of the sonified nanosheet is added in order to prepare a premixed PLA-GO-THF which the mixture is brought to ultrasonication for 5 min followed by stirring for 1 hour. Then the PLA-GO solution is left to dry in the oven at 66 ° C for 15 min to form a nanocomposite film.

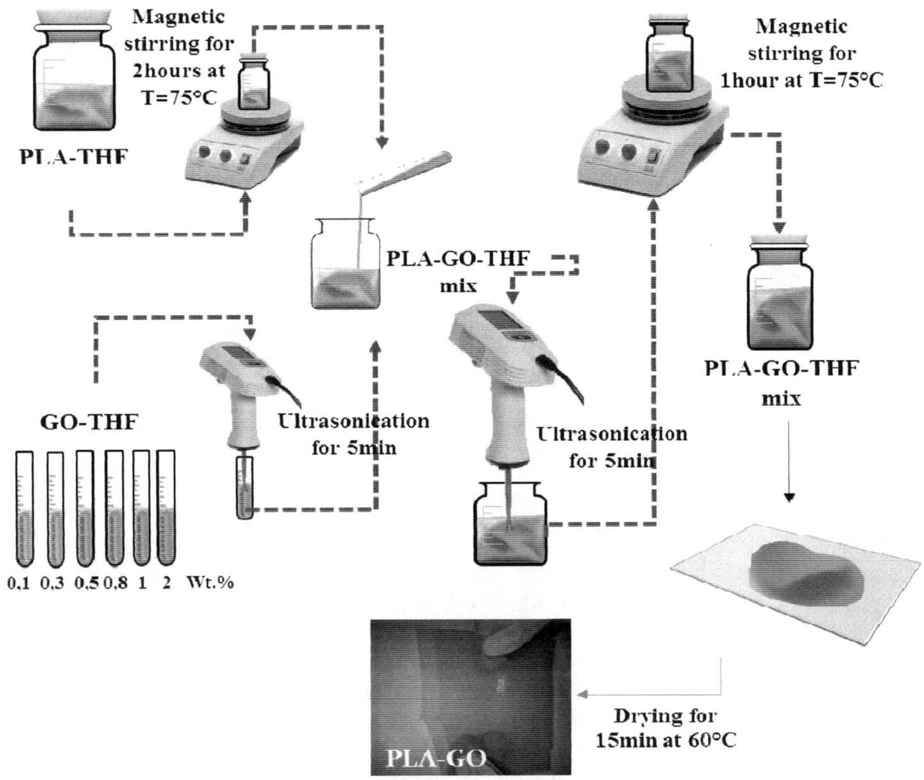

Figure 2. PLA/GO based nanocomposite preparation steps.

3. Results and discussions

3.1. Structural analysis of graphene oxide:

Graphite was used as the starting material in the process of preparing graphene oxide. Graphene oxide powder was examined by XRD and the result obtained was compared with that of natural graphite (Figure 3 (a)). The X-ray spectrum of graphite powder shows the hexagonal structure by the single peak at $2\theta = 26.23$ ° according to the orientation (200), which corresponds to a spacing of ~ 0.34 nm between the graphite planes [14]. In the graphene oxide spectrum, we observe the disappearance of this structure which confirms the formation of graphene oxide.

The infrared analysis (FTIR) revealed that the mixture of potassium permanganate, sulfuric acid and sodium nitrate, used during the chemical oxidation of graphite, results in the formation of groups of epoxide, carboxyl, carbonyl and also of hydroxide [15]. The FTIR spectra obtained for graphite and graphene oxide are shown in (Figure 3 (b)). The existence of several transmission bands confirms the presence of a variety of oxygen groups on the surfaces and edges of each plane of graphene oxide. The bands at 3410 and 1410 cm^{-1} are attributed to the O-H hydroxyl groups. The band at 1720 cm^{-1} is attributed to the elongation vibration of the carbonyl group (= O), the band at 1612 cm^{-1} is attributed to the bond C = C, and the vibrations at 1036 cm^{-1} are attributed to epoxy group (> O). In addition, several vibration bands of the C-H bond can be observed around 2000- 3340 cm^{-1} [16], which confirms the formation of graphene oxide.

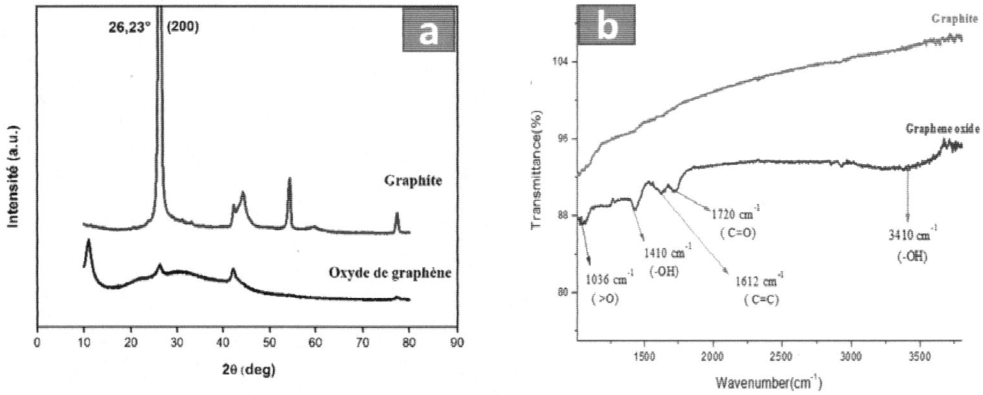

Figure 3. (a) FTIR spectroscopy and (b) XRD spectroscopy of graphene oxide

3.2. PLA/GO based nanocomposite

3.2.1. Polarized optical microscope

Figure 4 presents POM images showing the state of dispersion of the GO nanoparticles in terms of different quantities of nanoparticles (0.1 and 1% by weight) within the PLA matrix. The neat PLA matrix shows a porous surface. It can generally be seen in the POM images that the GO layers have been dispersed homogeneously in the matrix (Figure 4(b)). In Figure 4(c) i.e. when 1wt.% of graphene oxide is added, the appearance of a few agglomerations is observed, which is probably linked to the rheological aspects which explain the percolation phenomenon [17].

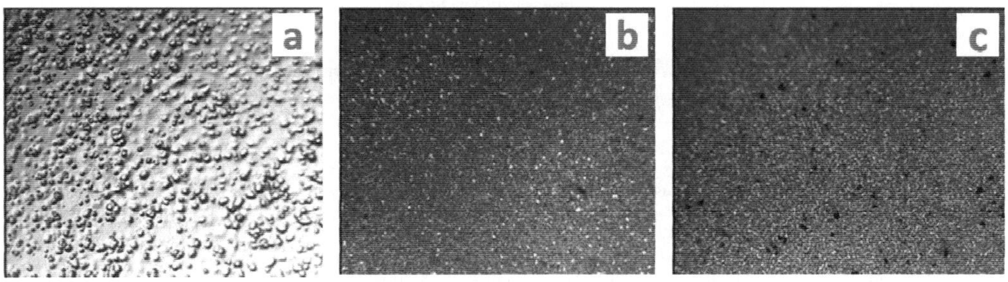

Figure 4. Polarized optical microscope of (a) Neat PLA, (b) PLA- 0.1wt.% GO and (c) PLA- 1 wt.% GO.

3.2.2. Fourier transform infrared spectroscopy

Figure 5 show the FTIR spectroscopy of different PLA phases (α, β and γ) which identified by excitation of the molecular bonds of the PLA/GO samples. At 1384 cm^{-1}, we observe a weak presence of polymorphs of the α phase which hardly changes with the increase in the fraction of GO nanofillers in the PLA matrix and which disappears under the effect of the addition of nanofillers [17]. The β phase is strongly identified by the absorption bands 872 and 1212 cm^{-1} [18] and becomes more dominant by the addition of only a small amount of GO (0.1wt.%). Thus, bands 1130 cm^{-1} show the presence of the semi-polar γ phase which remains insignificant compared to the strong dominance of the β phase. This shows the influence of the incorporation of a small amount of GO on the enhancement of the β phase of PLA and all the di-/ piezo-/pyro-electric properties that this phase can bring to the nanocomposite.

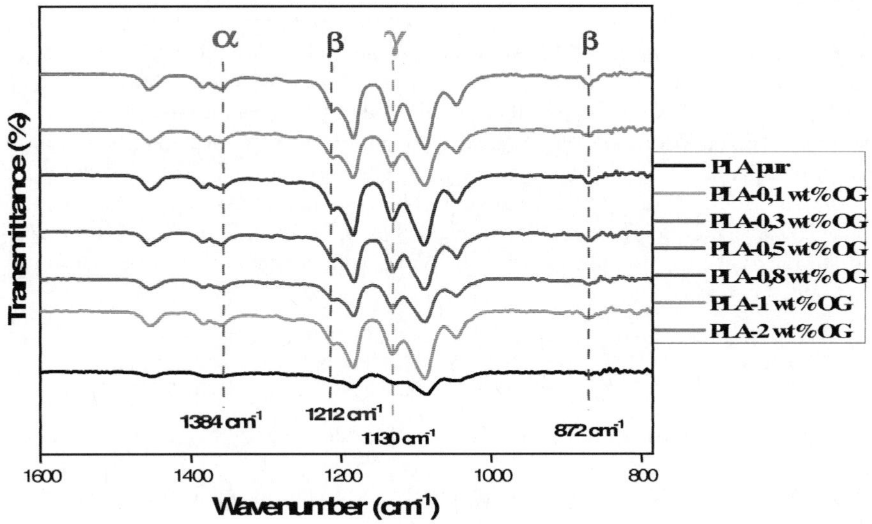

Figure 5. FTIR spectra of PLA/GO nanocomposite films with 0.1; 0.3; 0.5; 0.8; 1 and 2wt.% of GO

3.2.3. X-ray diffraction spectroscopy

Figure 6 show the X-ray diffraction spectroscopy of neat PLA and its nanocomposites. According to the spectra of the PLA/GO nanocomposite series with different mass fractions of graphene oxide (0.1; 0.5 and 1 wt.%) and in comparison with those of neat PLA, no peak corresponds to the graphitic structure was not detected. This indicates that the graphene oxide nanosheets are well dispersed in the PLA matrix without the formation of stacked structures. From this figure, it can be inferred that PLA exhibits a strong reflection at 2θ = 16,6 ° and 22,5 ° corresponding to the piezoelectric β-phase [13] which becomes more and more intense with the increase in GO percentage.

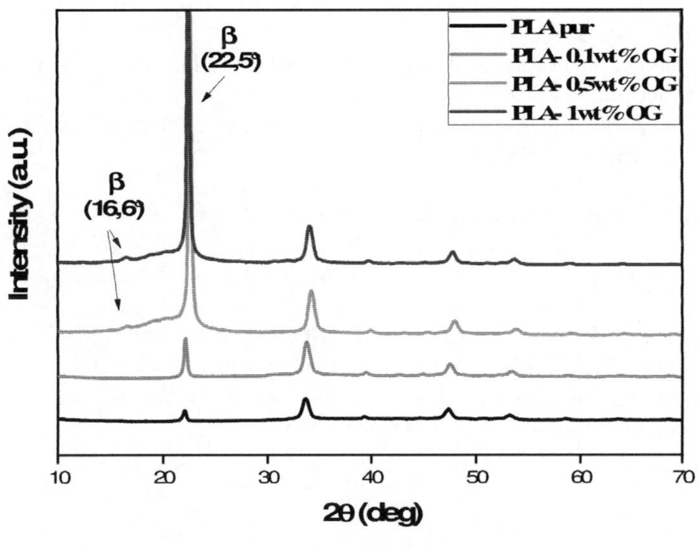

Figure 6. X-ray diffraction spectra of PLA/GO nanocomposite films with 0.1; 0.5; and 1 wt.% of GO

4. Conclusion

In this work, the crystallization of PLA/GO as a potential smart piezoelectric nanocomposite, was investigated. The GO nanoparticles were prepared by modified Hummers method and the nanocomposite samples were prepared through the casting-evaporation technic using different amounts of GO. The nanocomposites obtained were analyzed by polarized optical microscope, Fourier transform infrared and X-ray diffraction analysis. The morphological results show that GO is well dispersed and distributed within the PLA matrix. The crystalline structures of samples were studied by FTIR and XRD analysis, and the results showed that the presence of 0.1 wt% of GO nanoparticles enhances the piezoelectric β-phase of PLA than the enhancement in piezoelectric properties of the nanocomposite.

References

[1] R. Farhan, M. Rguiti, A. Eddiai, M. Mazroui, M. Meddad, C. Courtois, J of Comp Mat. 53 (3) 345–352 (2019)

[2] M.Meddad, A.Eddiai, R.Farhan, S.Benahadouga, M.Mazroui, M.Rguiti, Superl and Microst. 127 86-92 (2019)

[3] N.Chakhchaoui, H. Ennamiri, A. Hajjaji, A. Eddiai, M,Meddad,Y.Boughaleb, Polymer advanced and techn. 28,1170-1178 (2017)

[4] R. Farhan, A. Eddiai, M. Meddad, M. Mazroui, D. Guyomar, Smart Mater. Struct. 28, 035024 (2019)

[5] A. Eddiai, M. Meddad,R. Farhan,M. Mazroui, M. Rguiti,D. Guyomar, Superl. and Microstr. 127, 20-26 (2019).

[6] N. Chakhchaoui, H. Jaouani, H. Ennamiri, A. Eddiai, A. Hajjaji, M. Meddad, L. Van Langenhove, and Y. Boughaleb, J. Compos. Mater. 0021998318808869 (2018).

[7] K. Oumghar, N. Chakhchaoui, R. Farhan, A. Eddiai, M. Meddad, O. Cherkaoui, L. Van Langenhove, Mat sci eng. **827** 012034 (2020)

[8] L. Huang, S. Lin, Z. Xu, H. Zhou, J. Duan, B. Hu, J. Zhou, Adv mater. **32(5)** 1902034 (2020)

[9] M. Choolaei, V. Goodarzi, H A. Khonakdar, S H. Jafari, J. Seyfi, M R. Saeb, R. Boldt, Macromol Chem Phys**.** **218**(19) 1700103 (2017)

[10] S. Korkmaz, İ A. Kariper, J.Energy Storage Mater. **27** 101038 (2020)

[11] D A. Dikin, S. Stankovich, E J. Zimney, R D. Piner, G H. Dommett, G. Evmenenko, R S. Ruoff, Nature **448**(7152) 457-460 (2007)

[12] T. Szabó, O. Berkesi, P. Forgó, K. Josepovits, Y. Sanakis, D. Petridis, I. Dékány, Chem. Mater. **18**(11) 2740-2749 (2006)

[13] N. Wrońska, A. Anouar, M. El Achaby, K. Zawadzka, M. Kędzierska, K. Miłowska, M. Bryszewska, *Materials* **13**(4) 998 (2020)

[14] F Y. Liu, Y M. Dai, F H. Chen, C C. Chen, Colloid interf sci. **562** 112-124 (2020)

[15] M. Muniyalakshmi, K. Sethuraman, D. Silambarasan Mater. Today. **21** 408-410 (2020)

[16] N. Sharma, M. Arif, S. Monga, M. Shkir, Y K. Mishra, A. Singh, Appl surf sci. 513 145396 (2020)

[17] N. Chakhchaoui, R. Farhan, A. Eddiai, M. Meddad, O. Cherkaoui, Y. Boughaleb, L. Van Langenhove, Mat sci eng. **827** 012034 (2020)

[18] K. Hasheminejad, A. Montazeri, Appl surf sci. **502** 144150 (2020)

[19] M. Koç, L. Paralı, O. Şan, Polym. Test 106695 (2020)

Growth and Characterization of CZTS Thin Films Synthetized by Electrodeposition Method for Photovoltaic Applications

Ahmed ZITI[1,4], Bouchaib HARTITI[1,*], Hicham LABRIM[2], Salah FADILI[1], Abdelkrim BATAN[3], Abderraouf RIDAH[4], Philippe THEVENIN[5]

[1]ERDyS laboratory, MEEM & DD Group, Hassan II University of Casablanca, FSTM BP 146 Mohammedia 20650,Morocco
[2]Materials Science Unit / DERS / CNESTEN, Rabat, Morocco
[3]University Moulay Ismail, FST Errachidia, Department of Chemistry, Equipe Sciences des Matériaux, BP 509 Boutalamine 52000 Errachidia, Morocco.
[4]LIMAT laboratory, Department of Physics FSB, Hassan II University of Casablanca B.P 7955, Casablanca, Morocco
[5]University of Lorraine, LMOPS, METZ, France
*b.hartiti@univh2m.ac.ma

Abstract. In the present work we deposited CZTS absorber material using a one-step electrodeposition method on ITO glass substrates. We studied the effect of sulfurization temperature on structural, morphological and optical properties of CZTS thin films sulfurized in N_2+H_2S atmosphere at 350, 375, 400 and 425°C for 10 min respectively for 10 min. The films have been charaterized by different techniques including X-ray diffractometer (XRD), Raman spectroscopy, scanning electron microscopy (SEM) and UV-visible-NIR spectrophotometer. XRD patterns indicated kesterite CZTS with preferential orientation along (112) plane, the intensity of peaks increased with increase of sulfurization temperature signified the amelioration of the crystallinity with sulfurization temperature increased, the estimated crystallite size are in the range of 14.27-37.32 nm. Raman scattering experiments confirmed the presence of CZTS thin films by the characteristic peak at 336 cm^{-1}. SEM images showed the morphology of CZTS thin films improved with increasing of sulfurization temperature. UV-Visible-NIR spectrophotometer showed that the values of optical absorption coefficient are larger than 10^4 cm^{-1} and optical band gap energy of CZTS thin films decreased with increasing of sulfurization temperature ranging 1.56 and 1.66 eV. These results make the electrodeposited CZTS films a suitable material as absorber material in solar cells applications based thin films.
Keywords : CZTS, electrodeposition, sulfurization, thin films, gap energy.

1. Introduction

The first p Copper ternary and quaternary chalcogenide compounds are the most promising absorber materials for use in solar cells based thin films due to their high conversion efficiency [1, 2]. copper zinc tin sulphur CZTS attracted considerable attention in recent years because there suitable properties for photovoltaic devices such as p-type conductivity direct band gap approximated to 1.5 eV and optical absorption coefficient larger than 10^4 cm^{-1} ranging the visible spectrum [3]. The elements constituents of the quaternary CZTS thin films are non-toxic and abundant in the crust of the earth making them favourable replacements for copper indium gallium sulphur CIGS in which Zn and Sn substitutes In and Ga respectively. Wang et al achieved the champion conversion efficiency of 12.6% in 2014 for solar cells based $Cu_2ZnSn(S,Se)_4$ thin films prepared by spin coating technique using hydrazine based solution process in laboratory conditions [4]. The theoretical conversion efficiency reported by shockley –Queisser of CZTS films is about 32.2% [5]. Several methods employed to elaborate CZTS thin films such as pulsed laser deposition [6], chemical bath deposition [7], co-evaporation [8], electron beam evaporation [9], sputtering [10], spray pyrolysis [11], silar [12], spin coating [13], dip coating [14] and electro-deposition [15]. Among these methods, Electrodeposition is applied alternative for elaborate the thin films for solar cells applications. Non vacuum electrochemical deposition process is low cost and environmentally friendly industrial production, usually electrodeposition method can be devised into two categories are two steps process where the deposition of Cu-Zn-Sn followed by sulfurization for obtain CZTS thin films and single step is the deposition of copper (Cu), zinc (Zn), Tin (Sn) and sulfur (S) directly from electrolyte to prepare CZTS films.

Several groups presented in the literature have attempted the fabrication of CZTS films using electrodeposition method, Ennaoui et al prepared CZTS solar cells by one step electrochemical deposition of Cu-Zn-Sn followed by sulfurization using a gas combination of Argon with H_2S during 2h at 550°C, to show CZTS thin films. Secondary phases were formed with CZTS phase including ternary Cu_2SnS_3 and binary ZnS. The device obtained conversion efficiency of 3.4% [16]. Araki et al have produced CZTS thin films using electrodeposited CZT staked and precursors containing Cu-Zn-Sn. Sulfurization needed at 600°C duration 2h in an N_2 to result CZTS films with the best efficiency of 3.16% [17]. Scargg et al have synthetized thin films solar cells based CZTS absorber material by sequential electrochemical deposition of metallic precursor Cu/Sn/Cu/Zn stack onto Mo-coated soda lime glass substrates followed by annealing the stack with (H_2S + N_2) atmosphere for 2h at 575°C with 100 mg of sulfur (S) to get CZTS thin films. ZnS and Cu_2SnS_3 phases were showed with kesterite CZTS phase. The showed conversion efficiency of solar cell based electrodeposited CZTS of 3.2% [18]. Pawar et al prepared CZTS thin films by single step electrodepostion method followed by annealing at N_2+H_2S (95%+5%) atmosphere at 550°C for 1h [19].

In this work we focus on the synthesis of CZTS thin films by single step electrodeposition method onto ITO coated glass substrates at 50 min to study the influence of sulfurization temperature in the range between 350 and 425°C by step of 25°C in the N_2+H_2S atmosphere during 10 min on the structural, morphological and optical properties of CZTS absorber material by XRD, Raman, SEM and UV–vis spectrophotometry, respectively. The some optical parameters were calculated to ameliorate the photovoltaic devises.

2. Experimental Details

Precursor solution for elaborate CZTS films were prepared from electrolytic solution containing of copper sulfate (0.02 M) $CuSO_4$, zinc sulfate (0.01 M) $ZnSO_4$, tin sulfate (0.02 M) $SnSO_4$ and sodium thiosulfate (0.02M) $Na_2S_2O_3$ using a single-step electrodeposition method. Electrodeposition was carried out potentiostatically in a three-electrode cell configuration where the reference electrode was Ag/AgCl and the counter electrode, a platinum wire, and ITO coated glass substrate were used as a working electrode. The ITO substrates were ultrasonically cleaned with acetone and ethanol for 20 min, rinsed with distilled water and subsequently dried. The 0.2 M trisodium citrate was used as a complexing agent. The PH of the solution was adjusted to 4.5 by adding tartaric acid. The applied potential was used fixed at -1.05 V, during the deposition the bath temperature were maintained 25° C without magnetic stirrer. All the samples were deposited at 50 min, subsequently the as deposited samples were sulfurized in N_2+H_2S atmosphere at 350, 375, 400 and 425°C for 10 min.

The structural properties were determined using X-ray diffractometer (XRD) using CuKα radiation (λ = 1.54060 Å) scanning angle (20°-60°) and Raman scattering experiments in the frequency range of

IOP Conf. Series: Materials Science and Engineering **948** (2020) 012025 doi:10.1088/1757-899X/948/1/012025

0-1000.cm^{-1}. The morphological properties were characterized by field-emission scanning electron microscopy (FESEM), UV-Visible-NIR specrtrophotometer were used employed to control the optical properties.

3. Results and discussion

3.1. Structural properties

Figure 1 shows the X-ray diffraction of annealed CZTS thin films. All samples indicate the peaks corresponding to the CZTS thin films situated about 28.32°, 29.51° and 56.21 respectively for (112), (103) and (312) planes. The preferential direction of (112) plane and peaks corresponding to the CZTS thin films confirmed the existence of kesterite CZTS phase of all films matched by (JCPDS-26-0575). The intensity of (112) peak increased with increasing of annealing temperature which prove that the crystallinity of the films ameliorate when the CZTS is annealed at 425°C. The intensity of peaks is better to the obtained by Chtouki et al [20]. The extra peaks are indicating to presence of secondary phases SnS at 24.15 according to (JCPDS 01-083-1758) and $Cu_2Sn_3S_8$ at 34.33° and 75.45° according to according to (JCPDS 27-0197). Other peaks corresponding to the (ITO) glass substrates observed at various films according to (JCPDS-00-006-0416). Figure 2 indicates the sample annealed at 425°C for clarified the X-ray diffraction peaks.

The crystallites size (D) of electrodeposited CZTS thin films is calculated using Debye Scherrer formula (1) [21], where k is the shape factor (k=0.94), is Bragg diffraction angle, λ is the wavelength of the X-ray diffractometer and β is the full-width at half maximum of the peak. "D" expressed using formula (1). Dislocation density value shows the quantity of defects in the structure. Higher dislocation density values show minor crystallinity for the thin films. The dislocation density (δ) is defined as the dislocation lines per unit volume of the crystal and expressed from the crystallite size using formula (2) [22]. The micro-strain (s) of the films was calculated by using the relation (3) [23]. Crystallite size, dislocation density and micro-strain were tabulated in the Table 1. The crystallite size of the samples annealed at 350, 375, 400 and 425°C is 14.27, 29.52, 30.58 and 37.22 nm respectively, the dislocation density of CZTS thin films is decreased with sulfurization temperature increased from 4.9×10^{16} to 0.72×10^{16} (Lines/m^2), These behaviours indicate the decreasing of the defect in a electrodeposited CZTS absorber layers with sulfurization temperature increased. These values are close to the reported by Rondiya et al for CZTS thin films prepared by RF sputtering method [24].

$$D = \frac{k\lambda}{\beta \cos(\theta)} \quad (1)$$

$$\delta = \frac{1}{D^2} \quad (2)$$

$$S = \frac{\beta\cos(\theta)}{4} \quad (3)$$

Raman scattering measurements was used to confirm the presence of CZTS phase because of benary ZnS and ternary Cu_2SnS_3 have similar two theta values to CZTS phase. Raman spectra of all samples showed in Figure 3. Only single peak located at position 336 cm^{-1} is corresponding to the kesterite CZTS thin films, this Raman peak obtained by several authors reported by various elaboration methods [24, 25]. The intensity of Raman peak increases with sufurization temperature increased, this behavior confirmed the results obtained in the (XRD). No other binary and ternary phases were identified in the Raman shift range from 0 and 1000 cm^{-1}, including ZnS (Raman peak about 356 cm^{-1}) [26], Cu_2SnS_3 (Raman peaks about 297 cm^{-1}, 352 cm^{-1} and 374 cm^{-1}) [27], $Cu_{2-x}S$ (Raman peak about 475 cm^{-1}) [28], SnS_2 (Raman peak about 315 cm^{-1}) [29], Cu_3SnS_4 (Raman peaks about 318 cm^{-1}, 348 cm^{-1}, and 295 cm^{-1}) [30].

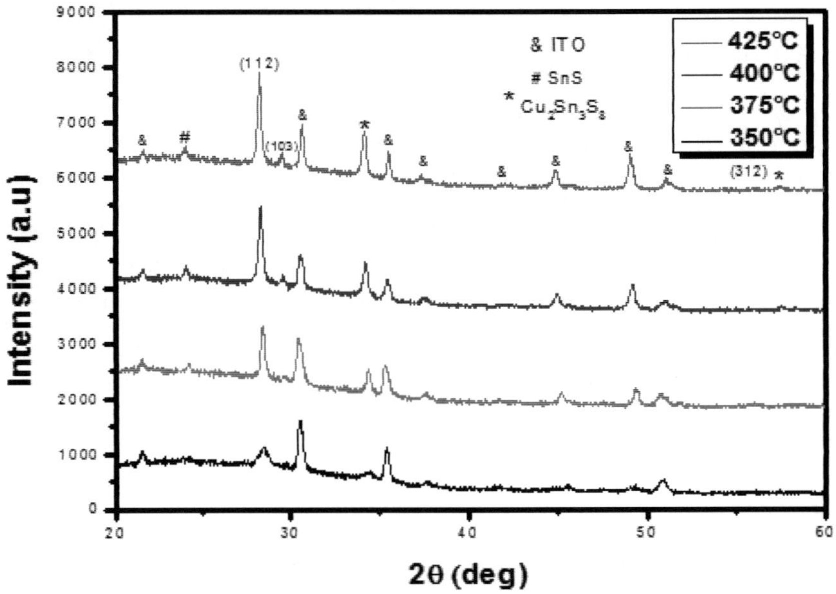

Figure 1. XRD patterns of the electrodeposited CZTS thin films at sulfurized at various temperatures

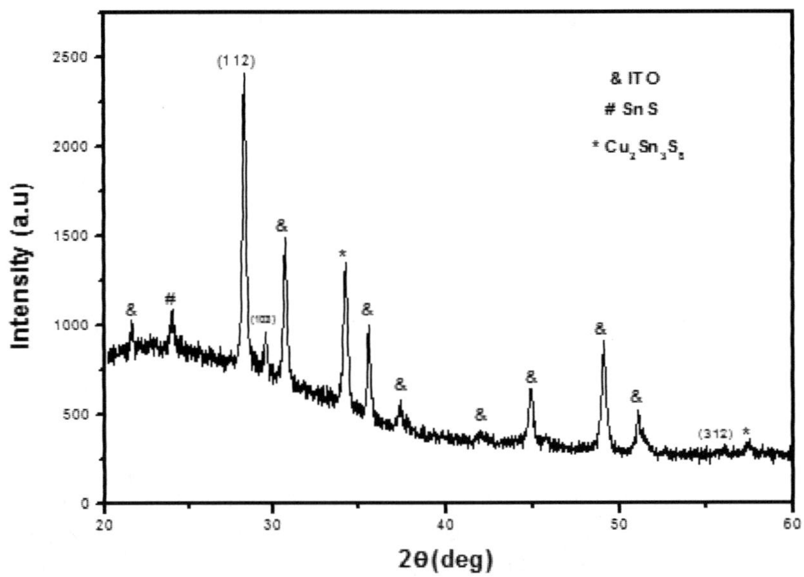

Figure 2. XRD patterns of the electrodeposited CZTS thin films at sulfurized at various temperatures

Table 1. Microstructural information of CZTS thin films with (112) plane

Annealing Temperature (°C)	Crystallite size D (nm)	Dislocation density δ x 10^{16} ($Lines/m^2$)	Micro-strain x10^{-2}ε
350	12.27	4.91	0.74
375	29.52	1.14	0.50
400	30.58	1.06	0.48
425	37.22	0.72	0.40

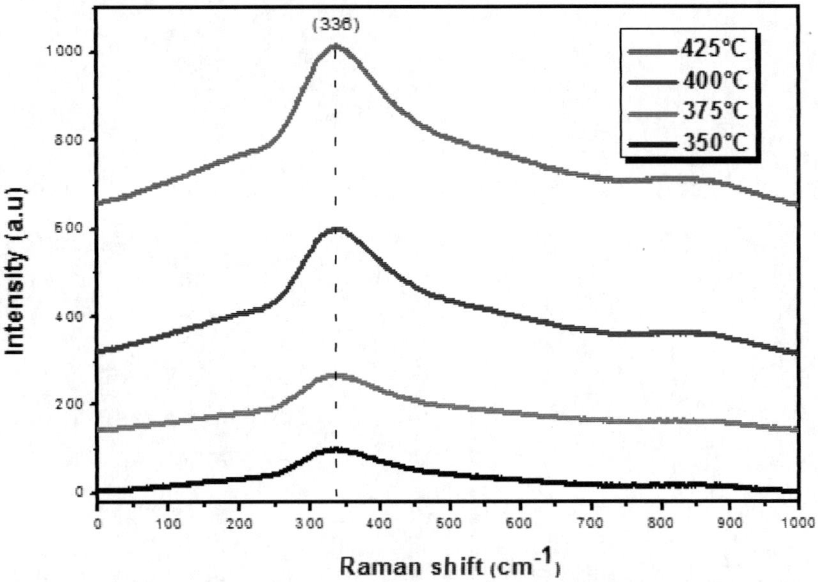

Figure 3. Raman scattering measurements of CZTS thin films sulfurized at different temperatures

3.2. Morphological properties

The CZTS thin films formation rely on nucleation, coalescence and growth phenomena, and this depending on several growth parameters such as Thickness, annealing temperature, deposition time, concentration of precursors. In fact, the temperature plays a major role for CZTS growth, behaving as kenitic energy for the substrates to enhance the ability to diffuse the incident atoms into the substrate surface. Generally, the adatoms (adsorbed atoms) move on the surface until finding the best nucleation centers and creation the clusters, which are developing by the temperature effect.

Figure 4 shows the CZTS film growth process by controlling the sulfurization temperatures. At low temperature the substrate is not totally covering by the material and the shape of clusters are not stable demonstrating the insufficient temperature (energy) for grain formation and crystallization of CZTS phase, by increasing temperature from (a) to (c) the nucleation enhances and the coalescence starts. There where, the produced clusters agglomerate via cluster-cluster interaction, controlling the form of particles and resulting small particles in semi-granular form as seen in the Figure (c). These cleary shows the effect substrate temperature which influences the film morphology by modifying the growing mechanism through a control of the interaction taking place at the CZTS solution–substrate interface. The final state is the crystallization of CZTS films as shown in Figure (d) with the appearance the tetragonal form of particles.

Figure 4. SEM images of CZTS thin films growth at (a) 350°C, (b) 375°C , (c) 400°C and (d) 425°C

3.3. Optical properties

The transmittance and absorbance data of CZTS thin films were examined by UV-visible spectrophotometer. The wavelength of incident light is ranging from 400 and 1050 nm. The Figure 5 exhibits the transmittance as a function of the wavelength λ, the transmittance of the annealed films increase with annealing temperature increased in the visible region, in the wavelength between 400 and 700 nm the transmittance of radiation is lower than 30 % and after 800 nm the transmittance of the samples is shifted, this behaviour shows that the absorption of photon energy is changing in the high wavelength of incident light (800-1050 nm). Figure 6 shows the absorbance spectra of CZTS thin films versus the wavelength λ, the maximum absorbance of the films annealed at 350, 375, 400 and 425°C is 1.45, 1.51, 1.65 and 2.12, respectively. The suitable absorbance of the samples annealed at

425°C is due to more absorption of photon energy from CZTS thin film. These results are near to reported in the literature for an absorber material for photovoltaic applications [31, 32].
Using the measured transmittance data the absorption coefficient (α) of the samples annealed at different temperature was calculated followed the relation (4) [33].

$$\alpha = \frac{1}{d}\ln\left(\frac{1}{T}\right) \quad (4)$$

where d is the thickness of the films. The optical absorption coefficient of all annealed CZTS thin films showed in the Figure 7 present the values great than 10^4 cm^{-1}, this values are in agreement literature for CZTS absorber materiel for solar cell applications [34, 35].
The direct band gap of CZTS absorber layer was calculated using Tauc's relation [36]:

$$(\alpha h\nu) = C(h\nu - Eg)^n \quad (5)$$

where C is a constant, E_g is the optical band gap energy, hν is the incident photon energy and $n = \frac{1}{2}$ because of CZTS thin films has the direct allowed transition. Figure 8 presents the graph of optical band gap energy (Eg) for electrodeposited CZTS thin films and inset Figure 8 indicate the $(\alpha h\nu)^2$ versus hν of the films annealed at different temperature. The band gap was determined by extrapolating the linear part of the plot versus photon energy hν to $(\alpha h\nu)^2=0$. The obtained gap energies are 1.66, 1.64, 1.60 and 1.56 eV for CZTS growth at 350, 375, 400 and 425°C respectively. The gap energy of CZTS thin film annealed at 425°C is suitable for photovoltaic applications [37]. The optimum band gap energy of CZTS thin films is 1.5 eV, the shift of band gap is due existence of secondary phases observed in the XRD patterns. With increase in the sulfurization temperature the optical band gap energy is deviated to low energies, when the sulfurization temperature increased more energy is applied to improve the crystallinity of the material, which naturally decreases the band gap of CZTS absorber layer. The intensity of X-ray diffraction peaks of electrodeposited CZTS thin films increased with increasing of sulfurization temperature leading the enhancement of FESEM images quality and reducing of band gap energies, this behaviour is reported by previously study [38, 39].

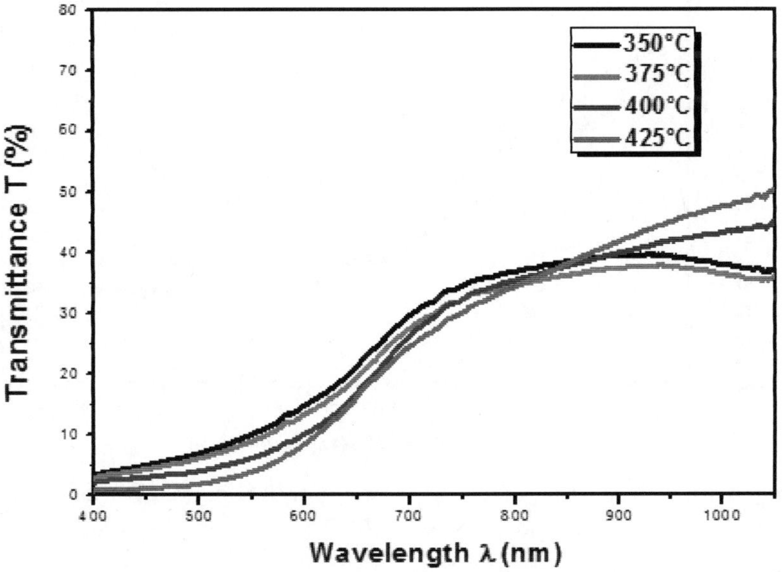

Figure 5. Transmittance spectra versus wavelength λ for CZTS thin films

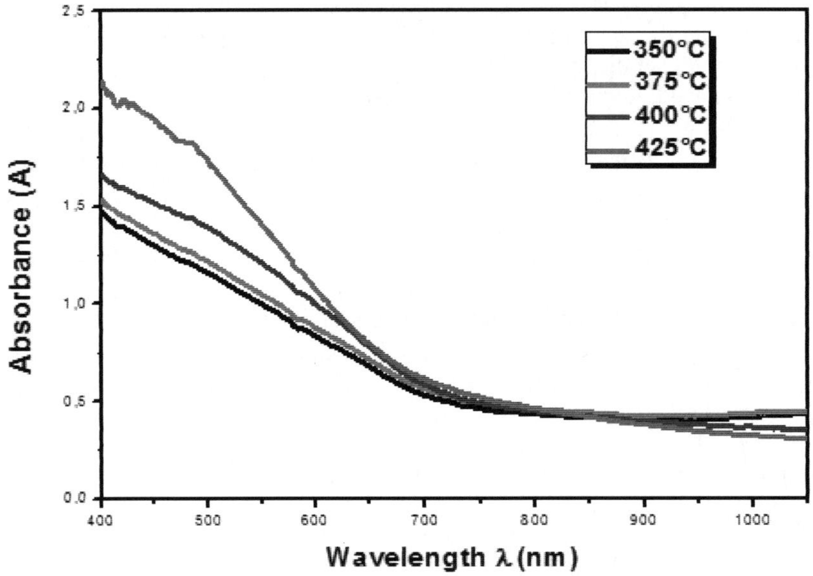

Figure 6. Absorbance data versus wavelength λ for CZTS absorber material

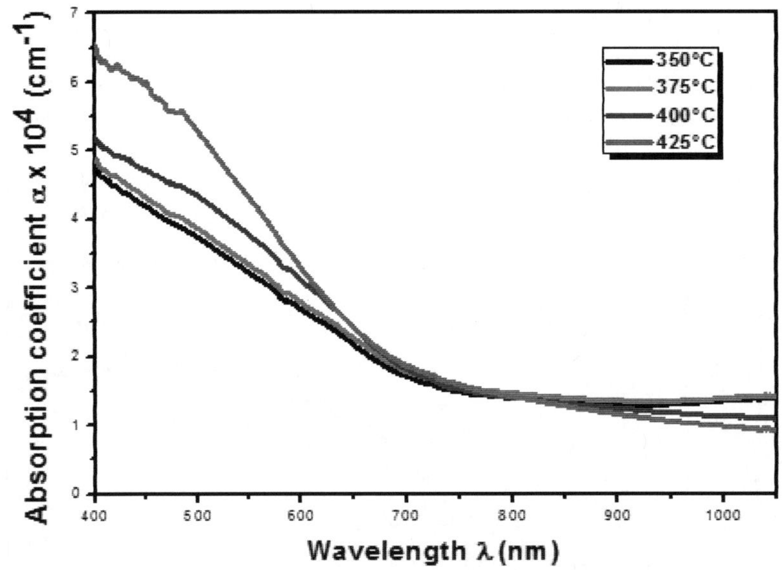

Figure 7. Optical absorption coefficient α of CZTS thin films as a function of wavelength λ

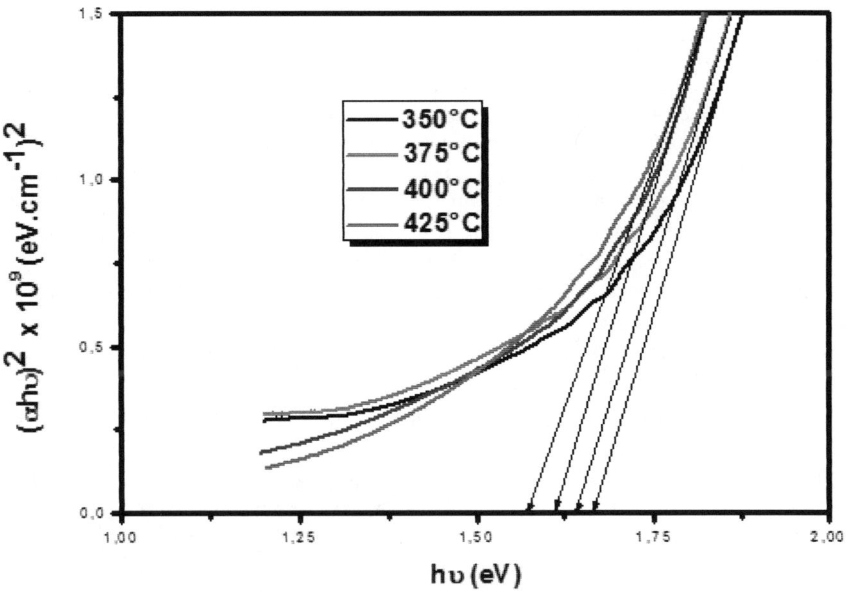

Figure 8. Determination of optical band gap by extrapolation method from the variation of $(\alpha h\nu)^2$ versus hν for CZTS absorber layer

Refractive index (n) is essential parameter in optical communication and representation of the optical devices. The refractive index (n) can be calculated from the reflectance (R) coefficient following the relation (6) [40]. Extinction coefficient (k) signifies the amount of absorption of electromagnetic wave in a semiconductor material due to inelastic scattering phenomena. The extinction coefficient can be determined from the absorption coefficient (α) using the relation (7) [41]. Real part (ε_r) of dielectric constant related to slowing down of speed of light in the medium Imaginary part (ε_i) of the dielectric constant defined by the correlation of the interaction of matter with the electric field vector. Real and imaginary dielectric constants are essential intrinsic properties of a material. The real and imaginary dielectric constants can be calculated using the equations (8) and (9) [42].

$$n = \frac{1+R}{1-R} + \left(\frac{4R}{(1-R)^2} - k^2\right)^{1/2} \quad (6)$$

$$k = \frac{\alpha\lambda}{4\pi} \quad (7)$$

$$\varepsilon_r = n^2 - k^2 \quad (8)$$

$$\varepsilon_i = 2nk \quad (9)$$

where "R" is reflectance, "n" is refractive index "k" is extinction coefficient, "ε_r" is real dielectric constant and "ε_i" is imaginary dielectric constant. The refractive index increases with the sulfurization temperature increase in the wavelength range of 400 and 1050 nm. According to Figures 9 and 10, the high values of refractive index of the electrodeposited CZTS absorber material with different sulfurization temperature can be attributed to the enhancement in the crystallinity of the films [44]. A film with high density gives rise to a high refractive index. The values of refractive index are in the range of 1.51-7.27, these values are comparable by Babichuk et al for CZTS thin films [43]. The decreasing of refractive index with increasing of wavelength λ signifies the normal dispersion behaviour. The extinction coefficient increases with increasing of sulfurization temperature in the wavelength range of 400-800 nm, from the 800 to 1050 nm we observe the deviation of extinction

IOP Conf. Series: Materials Science and Engineering **948** (2020) 012025 doi:10.1088/1757-899X/948/1/012025

coefficient behaviour in which the films annealed at 350 and 375°C have the values greater than the films annealed at 400 and 425°C respectively, this deviation indicate the changing of extinction coefficient behaviour between 800 and 1050 nm. The values extinction coefficient is obtained to be between 0.03 and 0.09 since "k" is directly relational to the absorption coefficient. The found values of extinction coefficient are comparable by Bakr for CZTS deposited by spray pyrolysis method [44]. Figures 11 and 12 showed that the values of the real and imaginary parts dielectric constants. The real dielectric constant increase with increasing of sulfurization temperature, this behaviour is close to the behaviour of the refractive index, because the real part depends a lot on refractive index compared to extinction coefficient. The imaginary dielectric constant increase with sulfurization temperature increased in the visible region (400-800 nm), after 800 nm we indicate the shift in the behaviour of imaginary dielectric constant in which the CZTS thin films annealed at 350 and 375°C have the values superior than the films annealed at 400 and 425°C respectively. The values of real and imaginary dielectric constants are in the range of 2.51- 55.85 and 0.10-1.33 respectively. The optical parameters are used to develop and improve the photovoltaic devices [45].

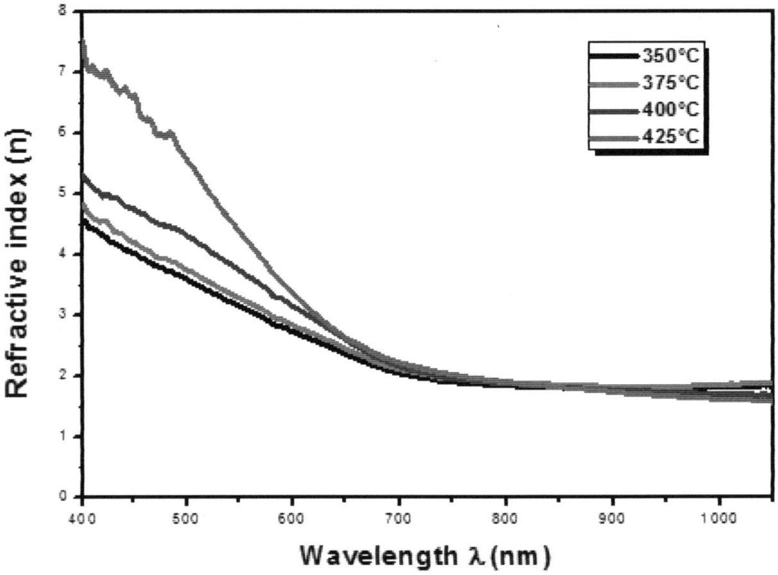

Figure 9. Refractive index as a function of wavelength λ for CZTS thin films with various temperatures

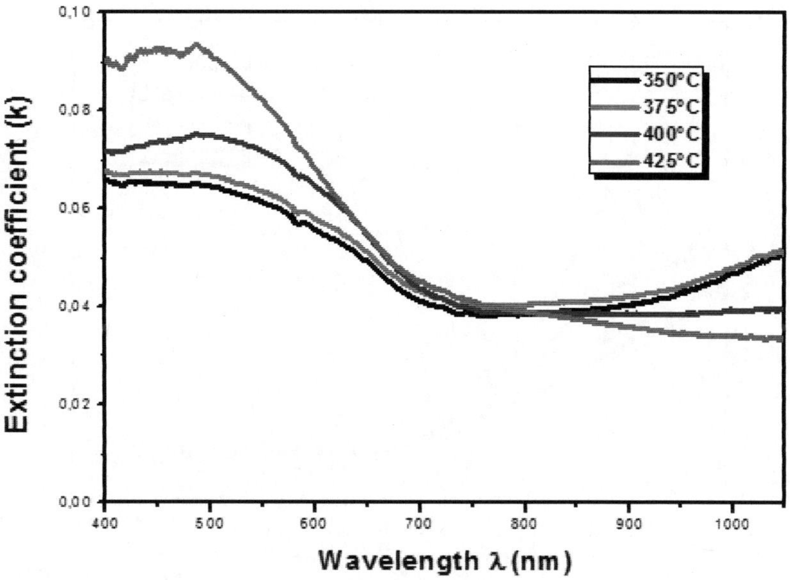

Figure 10. Extinction coefficient as a function of wavelength λ for CZTS thin films with various temperatures

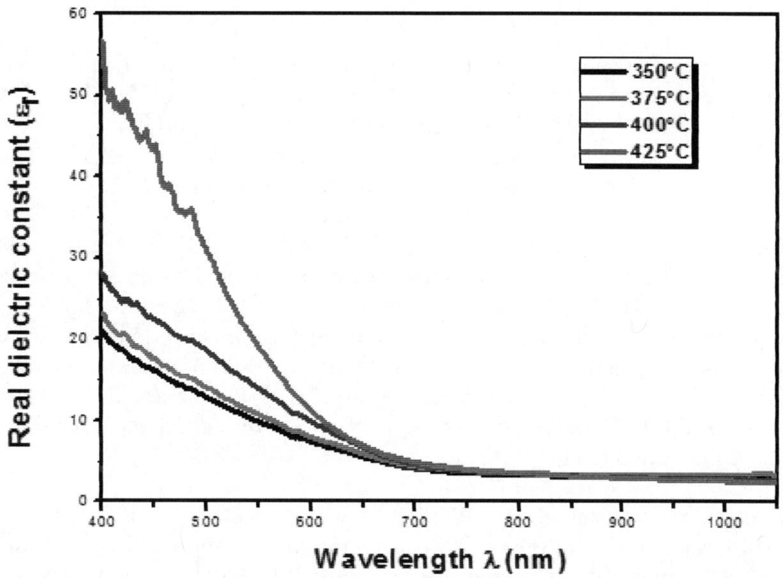

Figure 11. Real dielectric constant as a function of wavelength λ for CZTS absorber material

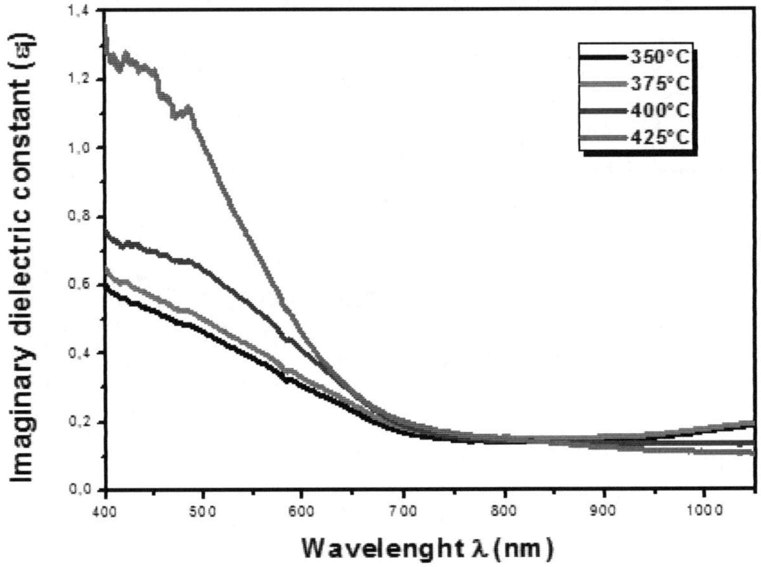

Figure 12. Imaginary dielectric constant as a function of wavelength λ for CZTS absorber material

Conclusion

CZTS thin films were deposited using single step on ITO glass substrates by electrodeposition method. The obtained samples were sulfurized in N_2+H_2S atmosphere at 350, 375, 400 and 425°C for 10 min for study influence of sulfurization temperature on the properties of CZTS thin films. The obtained films show a kesterite phase with the preferential orientation along (112) plane. Crystallites size was found to be between 14.27 and 37.22 nm, the $Cu_2Sn_3S_8$ and SnS secondary phases were observed in X-ray diffraction patterns with CZTS phase. Raman peak of all CZTS thin films was observed at 336 cm^{-1}confirming the existence of the kesterite CZTS structure. FESEM images indicated a surface of CZTS thin films morphology ameliorated with sulfurization temperature increased in the range of 350-425°C. The sample annealed at 425°C has good surface morphology for photovoltaic applications. . The optical parameters are calculated using transmittance and absorbance data. The optical band gap energy was found to decrease with sulfurization temperature increased in the range of 1.56-1.66′eV. The best optical band gap of 1.56 eV is obtained for the sample annealed at 425°C. The values of refractive index, extinction coefficient, real and imaginary dielectric constants are increase with sulfurization temperature increased.

Acknowledgments: Prof. Bouchaib HARTITI, Senior Associate at ICTP (The Abdus Salam International Centre for Theoretical Physics), is very grateful to ICTP for financial support. Technical support from LMOPS (University of Lorraine, France) and LGOPS (University of Valencia, Spain).

IOP Conf. Series: Materials Science and Engineering **948** (2020) 012025 doi:10.1088/1757-899X/948/1/012025

References

[1] G. Heise, A. Heiss , H. Vogt, H. P. Huber "Ultrafast lasers improve the efficiency of CIS thin film solar cells" Physics Procedia 39 702 – 708 (2012).

[2] M. A. Green, K. Emery, Y. Hishikawa, W. Warta and E. D. Dunlop "Solar cell efficiency tables (version 46)" Prog. Photovolt: Res. Appl; 23:805–812 (2015).

[3] M-Y. Yeh, Po-H. Lei, S-H. Lin and C-D. Yang "Copper-Zinc-Tin-Sulfur Thin Film Using Spin-Coating Technology" Materials 9, 526, (2016).

[4] W. Wang, M. T. Winkler , O. Gunawan, T. Gokmen , Teodor K. Todorov ,Y. Zhu , and D. B. Mitzi "Device Characteristics of CZTSSe Thin-Film Solar Cells with 12.6% Effi ciency" Adv. Energy Mater. 2013.

[5] W. Shockley and H. J. Queisser "Detailed Balance Limit of Efficiency of p-n Junction Solar Cells" Journal of Applied Physics Volume 32, Number 3 (1 9 6 1).

[6] S.A. Vanalakar, G.L. Agawane, S.W. Shin, M.P. Suryawanshi, K.V. Gurav, K.S. Jeon, P.S. Patil, C.W. Jeong, J.Y. Kim, J.H. Kim "A review on pulsed laser deposited CZTS thin films for solar cell applications" Journal of Alloys and Compounds 619 (2015) 109–121.

[7] T. R. Rana, N. M. Shinde, J. Kim "Novel chemical route for chemical bath deposition of Cu_2ZnSnS_4 (CZTS) thin films with stacked precursor thin films" Materials Letters Volume 162, Pages 40-43 (2016).

[8] R Moreno1 , E A Ramirez1 and G Gordillo Guzmán, Study of optical and structural properties of CZTS thin films grown by co-evaporation and spray pyrolysis, Journal of Physics: Conference Series 687 (2016) 012041.

[9] H. Katagiri, K. Saitoh, T. Washio,H. Shinohara, T. Kurumadani, S. Miyajima "Development of thin films solar cell based on Cu_2ZnSnS_4 thin films" Solar Energy Materials & Solar Cells 65 141148 (2001).

[10] N. Song, M. A. Green, J. Huang, Y. Hu, X. Hao, Study of sputtered Cu_2ZnSnS_4 thin films on Si, Applied Surface Science 459 700–706 (2018).

[11] Y. Arba, J. H. Nkuissi Tchognia, B. Hartiti, A. Ridah and P. Thevenin, "Determination of the optimal conditions for the deposition of Cu_2ZnSnS_4 (CZTS) thin films by spray pyrolysis using Taguchi method" 2014 International Renewable and Sustainable Energy Conference (IRSEC), Ouarzazate, 2014, pp.705-710.

[12] A. Krishnan, K. Rishad Ali, G. Vishnu, P. Kannan, Towards phase pure CZTS thin flms by SILAR method with augmented Zn adsorption for photovoltaic applications, Materials for Renewable and Sustainable Energy 8:16 (2019).

[13] S. B Patel, J. V Gohel, Effect of type of solvent on the sol-gel spin coated CZTS thin films, Phys Astron Int J,1(4):126-129 2017.

[14] A. Ziti, B. Hartiti, H. Labrim, S. Fadili, M. Tahri, A. Belafhaili, A. Ridah, P. Thevenin, Application of Taguchi method to optimize the sol–gel dip-coating process of the semiconductor

IOP Conf. Series: Materials Science and Engineering **948** (2020) 012025 doi:10.1088/1757-899X/948/1/012025

Cu_2ZnSnS_4 with good optical properties, Journal of Sol-Gel Science and Technology 91:364–373 (2019).

[15] S Sugiarti, A Sjahriza, T Juniarti, H Firmansyah, IOP Conference Series: Earth and Environmental Science, 187, 012082 (2018).

[16] A. Ennaoui, M. Lux-Steiner, A. Weber, D. Abou-Ras, I. Kötschau, H.-W. Schock a , R. Schurr, A. Hölzing, S. Jost, R. Hock, T. Voß, J. Schulze, A. Kirbs "Cu_2ZnSnS_4 thin film solar cells from electroplated precursors: Novel low-cost perspective" Thin Solid Films 517 (2009) 2511–2514.

[17] H. Araki, Y. Kubo, K. Jimbo, W. Shwe Maw, H. Katagiri, M. Yamazaki, K. Oishi, and A.Takeuchi "Preparation of Cu_2ZnSnS_4 thin films by sulfurization of co-electroplated Cu-Zn-Sn precursors" Phys. Status Solidi C 6, No. 5, 1266–1268 (2009).

[18] J. J. Scragg, Dominik M. Berg, Phillip J. Dale "A 3.2% efficient Kesterite device from electrodeposited stacked elemental layers" Journal of Electroanalytical Chemistry 646 (2010) 52–59.

[19] S. M. Pawar, B. S. Pawar, K. V. Gurav, D. W. Bae, S. H. Kwon, S. S. Kolekar, and J. Hyeok Kim "Fabrication of Cu_2ZnSnS_4 Thin Film Solar Cell Using Single Step Electrodeposition Method" Japanese Journal of Applied Physics 51 (2012) 10NC27.

[20] T. Chtouki, L.Soumahoro, B. Kulyk, H. Bougharraf, H. Erguig, K.Ammous, B.Sahraoui "Comparative Study on the Structural, Morphological, Linear and Nonlinear Optical Properties of CZTS Thin Films Prepared by Spin-Coating and Spray Pyrolysis" Materials Today: Proceedings 4 (2017) 5146–5153.

[21] A. Moumen, B. Hartiti, P. Thevenin, M. Siadat "Synthesis and characterization of CuO thin films grown by chemical spray pyrolysis" Opt Quant Electron (2017) 49:70.

[22] A. Kotbi, B. Hartiti, S. Fadili A. Ridah, P. Thevenin "Copper indium gallium disulphide ($CuInGaS_2$) thin films deposited by spray pyrolysis for solar cells: influence of deposition time in controlling properties of sprayed $CuInGaS_2$ absorbers" Opt Quant Electron (2016) 48:524.

[23] J. Henry, K. Mohanraj, G. Sivakumar "Electrical and optical properties of CZTS thin films prepared by SILAR method" Journal of Asian Ceramic Societies (2015).

[24] F. Jiang, H. Shen and W. Wang "Optical and Electrical Properties of Cu_2ZnSnS_4 Film Prepared by Sulfurization Method" Journal of ELECTRONIC MATERIALS, Vol. 41, No. 8, 2012.

[25] K. Wang, B. Shin, K. B. Reuter, T. Todorov, D. B. Mitzi, and S. Guha "Structural and elemental characterization of high efficiency Cu_2ZnSnS_4 solar cells Applied physics lettres 98 05 19 12 (2011).

[26] P.A. Fernandes, P.M.P. Salomé, A.F. da Cunha "Growth and Raman scattering characterization of Cu_2ZnSnS_4 thin films" Thin Solid Films 517 (2009) 2519–2523.

[27] Z. Guan, W. Luo and Z. Zou "Formation mechanism of ZnS impurities and their effect on photoelectrochemical properties on a Cu2ZnSnS4 photocathode" Cryst Eng Comm, 2014, 16. 2929.

[28] F. Biccari, R Chierchia, M. Valentini, P. Mangiapane, E. Salza, C. Malerba, C. L. Azanza Ricardo, L. Mannarino, P. Scardi, A. Mittig "Fabrication of Cu_2ZnSnS_4 solar cells by sulfurization of evaporated precursors" Energy Procedia 10 (2011) 187 – 191.

[29] P. A. Fernandes, P.M.P. Salom, A.F. da Cunha "Study of polycrystalline Cu_2ZnSnS_4 films by Raman scattering" Journal of Alloys and Compounds Volume 509, Issue 28, 2011, Pages 7600-7606.

[30] Yan-L. Zhou, Wen-H. Zhou, M. Li, Yan-F. Du, and Si-Xin Wu " Hierarchical Cu_2ZnSnS_4 Particles for a Low-Cost Solar Cell: Morphology Control and Growth Mechanism" J. Phys. Chem. C 2011, 115, 19632–19639.

[31] Prashant K. Sarswat, Michael Snure, Michael L. Free, Ashutosh Tiwari "CZTS thin films on transparent conducting electrodes by electrochemical technique" Thin Solid Films 520 (2012) 1694–1697.

[32] Y. Sun, K., H.Zheng , H.Wang, J.Liu, H.Yan, M.Zhu "Ethylene glycol based dip coating route for the synthesis of Cu_2ZnSnS_4 thin film" Materials Letters 92 (2013) 195–197.

[33] D. Tang , Q. Wang , F. Liu , L. Zhao, Z. Han, K. Sun , Y. Lai , J. Li, Y. Liu "An alternative route towards low-cost Cu_2ZnSnS_4 thin film solar cells" Surface & Coatings Technology (2013).

[34] S. Zhou, R. Tan, X. Jiang, X. Shen , W. Xu, W. Song "Growth of CZTS thin films by sulfurization of sputtered singlelayered Cu–Zn–Sn metallic precursors from an alloy target" Mater Sci: Mater Electron (2013) 24:4958–4963.

[35] Minlin Jiang, Yong Li, Rabin Dhakal, Prem Thapaliya, Michael Mastro, Joshua D. Caldwell, F. Kub, and X. Yan "Cu_2ZnSnS_4 polycrystalline thin films with large densely packed grains prepared by sol-gel method" J. of Photonics for Energy, 1(1), 019501 (2011).

[36] J.Tauc, R.Grigorivic, and A.vancu "Optical properties and Electronic structure of Amorphous Germanium" phys. Stat. sol .15,627(1996).

[37] S. A. Vanalakar, S. J. Yeo, P. S. Patil, J. Y. Kim, and J. H. Kim "Structural, Optical, Surface Morphological and Electrical Properties of Cu_2ZnSnS_4 Thin Film Synthesized by Drop Casting Technique" Z. Phys. Chem. 2014; 228(9): 917–926.

[38] S. Kahraman, S. Çetinkaya, M. Podlogar, S. Bernik, H.A. Çetinkara , H.S. Güder "Effects of the sulfurization temperature on sol gel-processed Cu_2ZnSnS_4 thin films" Ceramics International Volume 39, Issue 8, December 2013, Pages 9285-9292.

[39] S.M. Bhosale, M.P. Suryawanshi, M.A. Gaikwad, P.N. Bhosale, J.H. Kim, A.V. Moholkar "Influence of growth temperatures on the properties of photoactive CZTS thin films using spray pyrolysis technique" Materials Letters Volume 129, 2014, Pages 153-155.

[40] Z. El Khalidi, S. Fadili, B. Hartiti, A. Lfakir, P. Thevenin, M. Siadat "Behavior of NiO thin films sprayed at different annealing time" Opt Quant Electron (2016) 48:427.

[41] I.Sta, M.Jlassi, M.Hajji, M.F. Boujmil,R. Jerbi, M.Kandyla, M. Kompitsas and H.Ezzaouia "Structural and optical properties of TiO_2 thin films prepared by spin coating" Journal of Sol-Gel Science and Technology 72, 421 (2014).

[42] A. Kotbi, B. Hartiti, A. Ridah, P. Thevenin "Characteristics of $CuInS_2$ thin films synthesizes by chemical spray pyrolysis" Opt Quant Electron (2016) 48:75.

[43] I.S. Babichuk, V.O. Yukhymchuk, M.O. Semenenko, N.I. Klyui1, R. Caballero, O.M. Hreshchuk, I.S. Lemishko, I.V. Babichuk, V.O. Ganus1, and M. Leon "Optical and morphological properties of tetragonal Cu_2ZnSnS_4 thin films grown from sulphide precursors at lower temperatures" Semiconductor Physics, Quantum Electronics & Optoelectronics, 2014. V. 17, N 3. P. 284-290.

[44] Nabeel A. Bakr*, Ziad T. Khodair, Hussein I. Mahdi "Influence of Thiourea Concentration on Some Physical Properties of Chemically Sprayed Cu2ZnSnS4 Thin Films" International Journal of Materials Science and Applications Volume 5, Issue 6, 2016, Pages: 261-270.

[45] R.J. Deokate, R.S. Kate, S.C. Bulakhe, Physical and optical properties of sprayed Cu_2ZnSnS_4 (CZTS) thin flm: efect of Cu concentration. J. Mater. Sci. 30(4), 3530–3538 (2019).

First Au monolayer formation on Cu(110) surface

K. Sbiaai[1,(*)], M. Dardouri[1, 2,(**)], A. Fathi[1], A. Arbaoui[2], A. Hassani[1], F. Eddiai[1], A. El Azrak[1], A. Hasnaoui[1], Y. Boughaleb[2]

1. Laboratory LS3M, Polydisciplinary Faculty, 25000 Khouribga, University of Sultan Moulay Salimane, Morocco

2. Laboratory of Condensed Matter Physics (LCMP), Univ. Chouaib Doukkali, 24000 El Jadida, Morocco.

Corresponding authors: *k.sbiaai@usms.ma; **dardouri.m@ucd.ac.ma

Abstract: We study in this work the growth of a gold monolayer on a Cu (110) surface thanks to the kinetic Monte Carlo method. We considered a coverage of 0.1% of the monolayer. Indeed, the growth of a monolayer is done through the intermediary of several diffusion processes. All activation energies for each process were calculated using the static method coupled with the KMC method. This coupling is established by the ABBM anisotropic bond breaking model. The study of the results obtained in the temperature range (100-300K) at different deposition rates shows the formation of small 1D islands and certain 2D islands at high temperature (280 k-300 K).

Keywords: Morphology surface; growth conditions, surface diffusion; nucleation; anisotropic bond, kinetic Monte Carlo simulations.

1. Introduction

Thin film technology has large field of applications in opto-electronic devices, micro-electro-mechanical systems and biomedical products[1]–[6]. In the semiconductor industry, thin films must have minimal surface roughness and minimum internal defects, which are directly related to the electrical, optical and mechanical properties of microelectronic devices[7][8][9].
Surface morphology evolution is related to adatom diffusion processes and thin film growth growth conditions such as coverage, temperature and flux [10]–[12]. On other hand, adatoms behavior directly influenced by its local environment. This latter changes under the fact of many parameters as anisotropic geometry, barriers energies, coordination number, density, etc. In the case of (110) surface where we note the presence of diffusion channel, the energy barrier via hopping jump promotes one-dimensional adatom nucleation (1D islands). On the other hand, the diffusion through channel contributes in the shapes islands 2D instable[3], [13], [14][15]. In literature several investigations (theoretical and experimental) studies on the growth of thin layers have been made in order to follow the evolution and understand the growth at the atomic level. Surface scattering studies carried out under the scanning tunnel microscope (STM) showed the presence of multiple diffusion mechanisms contributing to the establishment of surfaces morphology[3], [16][17]. K. Morgenstrem argued that, the transition 1D at low temperature (175-220K) to 2D island is observed by STM on an anisotropic Ag(110) surface[18]. Consequently, these kinetic parameters guide the evolution of transition 1D/2D morphology[19].

The kinetic Monte-Carlo (kMC) simulations are used to predict thin layer growth morphologies at technologically relevant time and length scales. The kMC method performed with the algorithm developed by Bortz and al. [20] are used in our calculations). They can reproduce the stochastic nature of the growth process and allow the modulizations of various atomic processes, whose barriers are

ideally obtained from calculations of molecular static simulation or methods based on the theory of functional density (DFT)[21]–[23].
The aim of this methods is focused on understanding and identifying the diffusion mechanisms that influence and control the growth of monolayer. The transition observed in the morphology of a monolayer is described mainly by elementary diffusion processes such as single jumps and exchange mechanisms. During this growth, we have to take into account the geometric anisotropy (110) surface which has an effect on the energetic anisotropy in terms of activation and binding energies. This particularity allowed us to adopt the model of anisotropic bonding of broken bands ABBM[24].

This paper is organized as follows: In Section 2, we describe our model and simulation method. All results obtained during Au growth on Cu (1 1 0) surface are presented and discussed in Sections 3 and 4. The last section is devoted to concluding.

2. Model and simulation method

Epitaxial growth modeling requires knowledge of the energy barriers for the diffusion of an atom on a surface and their interactions with its neighbors. The diffusion phenomena in symmetry (110) are influenced by surface anisotropy. This effect is observed from the activation energy values of each process. In the face of this question of anisotropy, a simple energy model of broken bonds anisotropic (ABBM) is used in which the energy barrier depends on the number of bonds to be broken in its initial configuration to release the atom that must diffuse. For a given r_p process the rate is written by $r_p = m_p v_p . \exp\left(-\frac{E^0_{p,n}}{K_B T}\right)$, with m_p is the multiplicity, introduced by Maksym[25] who had the idea of grouping all the events corresponding to the same type of atomic process with the same energy of barrier. The Arrhenius law is assumed to estimate the adatom process rates, where, $E^0_{p,n}$ is the activation energy in parallel and normal channel, K_B is the Boltzmann's constant, and T is the substrate temperature. The attempt frequency v_p=1 THz has been chosen to optimize the computation time.
In the case of our surface (110) the binding energies are anisotropies and the energy barrier are written respectively for a parallel or normal diffusion to the dances row's.

$$E_i = E_i^0 + n_p E_p^b + n_n E_n^b, \quad i = n, p. \tag{1}$$

Or E_i^0 represents the energy barrier of isolated adatome on the surface, and n_i, E_i^b $(i = n, p)$, the numbers and energies of the parallel and normal bonds.

The total energy of metallic system as functions of constituent atoms are study by empirical Embedded-atom method (EAM) developed by Daw et al[26], [27]. The EAM potential is performed frequently for compute interaction atoms-atoms in metallic system. In present investigation, we adopted EAM method to calculate all activation barriers of an isolated adatom. This approach EAM potential calculations take into account pair and many-body classical interactions, which allows relaxing atomic positions through energy minimization. The total energy of a metal is given by the contribution of two term an embedding function $F_i(\rho_{hi})$ and pair potential $\varphi_{ij}(r_{ij})$ as:

$$E_{tot} = \sum_i F_i(\rho_{hi}) + \frac{1}{2}\sum_i \sum_{j(\neq i)} \varphi_{ij}(r_{ij}). \tag{2}$$

Where F_i represents the energy required to immerse an atom in the electron cloud, the term ρ_{hi} is the electron density due to the other atoms of the metal, the term $\varphi_{ij}(r_{ij})$ indicates the pair potential between atoms i and j, and r_{ij} is the distance between atom i and atom j.

3. Anisotropic diffusion barriers

In order to study the growth of a single layer. Table 1 represents the energy barriers required to perform a numerical simulation using the Monte Carlo Kinetic method for an Au/Cu system (110). The table in question represents the diffusion energies along channel, across channel, the normal and parallel bond energies. We have used the 'Drag'[28], [29] method to calculate these energies values of an isolated adatom.

Table 1: Diffusion barriers and binding energies of isolated adatom on (110) surface.

(110)	$E_p^0(ev)$	$E_n^0(ev)$	$E_p^b(ev)$	$E_n^b(ev)$
Au/Cu	0.22	1.31	0.24	0.04

The results in Table 1 represent the all energies required to displace an isolated gold type adatom on a symmetry (110) copper substrate. The energy to move an adatom of Au along the channel is of order 0.22 eV. Where the displacement is made through the channel, the energy is 1.31 eV. This difference in energies reflects the strong anisotropy of the diffusion barrier caused by the surface anisotropy. The adatom adsorption on (110) surface is ensured by four neighbors forming the hallow site. Then, to release adatom diffusion we need to break their four connections or bonds. The energy of the parallel and normal bonds is 0.24 eV and 0.04 eV respectively.

4. Results and discussion

In the present work, we use Kinetic Monte Carlo method to simulate Au diffusion on Cu (110) surface. We first, generated a rectangular network to mimic the substrate surface, using a box size of 100×100 atoms. Periodic boundary conditions are applied along the x and y-directions while a surface is created along the z-axis to allow deposition of atoms. The film growth processes can be strongly affected by growth conditions, such as substrate temperature, deposition rate, and anisotropic surface. In this part, firstly we will fix the adatom flux of Au at 0.5ML / s, with a coverage rate of 0.1%. This investigation is done on different temperatures (100-300K). The corresponding results are shown in Figure 1. It is then found that at very low temperatures (100-180 K), the distribution of adatoms and of some dimers is due to a lack of diffusion rate; this proves that the thermal activation is low. In this temperature domain, only diffusion along the channel is possible and normal bonds along the less dense rows are stable. When the temperature is increased, the clusters form the wires (1D) forming a small element of five adatomes (pentamer) along the dense rows. It is noted that at high temperatures diffusion is facilitated by perpendicular rows such as parallel to these rows, but normal bonds are no longer stable this increases the probability that adatoms can detach/attach edges[7] [24], [11].

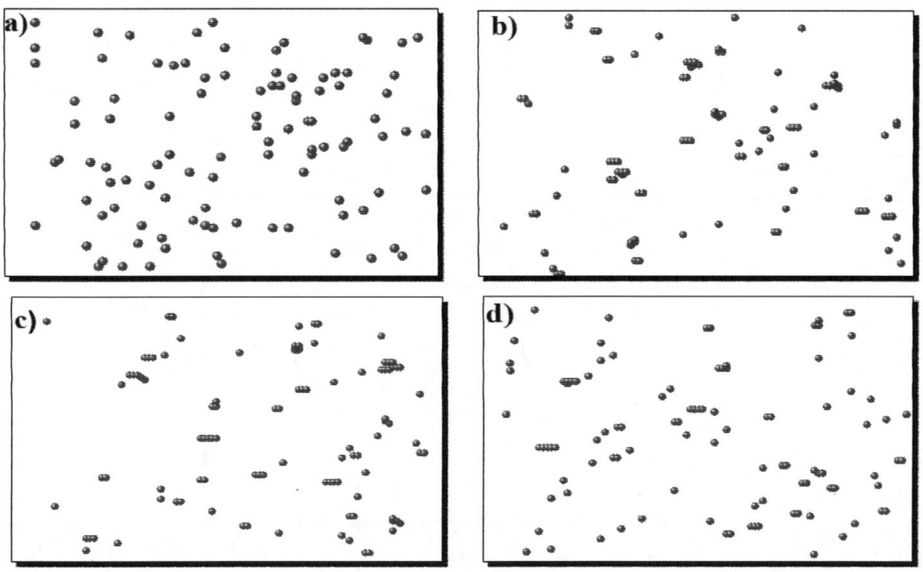

Figure 1: Snapshots from the kMC simulations for Au / Cu (110) system at flux of 0.5 ML /s for four different temperatures: a) 100K; (b) 180 K; c) 280K and d) 300K. The coverage rate is θ = 0,1%.

For the following, we set the substrate temperature to 280K then 300K at the same coverage rate which is equal to 0.1% single layer. This study is done for three deposition rates 0.1ML/s, 0.5ML/s and 1ML/s. For the two temperatures (280K and 300K) we notice that the surface topology is dominated by small islands composed of dimers, trimers and some pentamers (Figures 2 and 3).

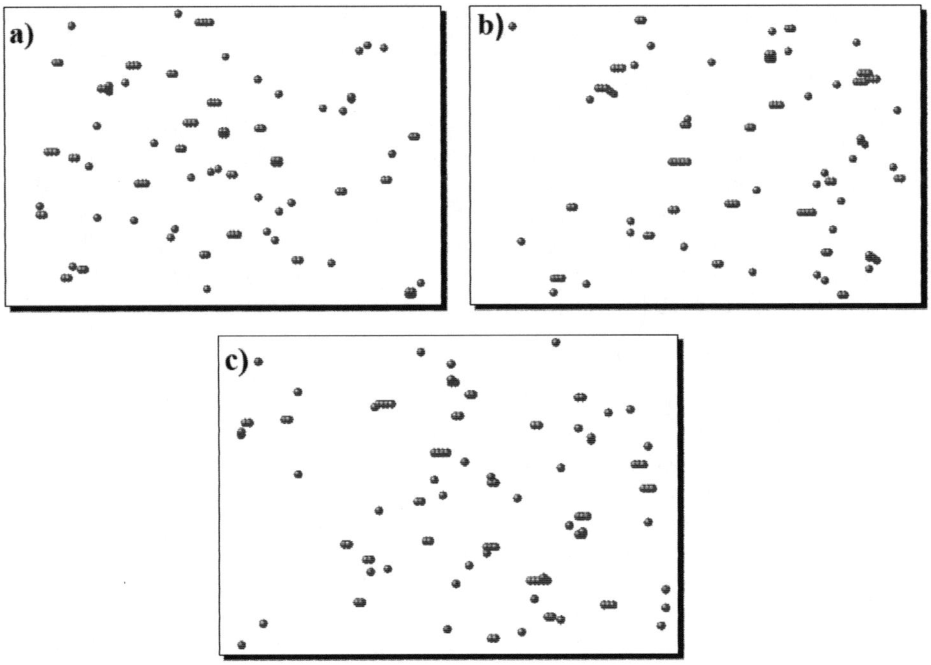

Figure 2: Snapshots from kMC simulations of Au/Cu(110) system at T=280K for different flux values: a) F = 0.1ML / s, b) 0.5ML / s, and c) 1 ML / s. The coverage rate is 0.1%.

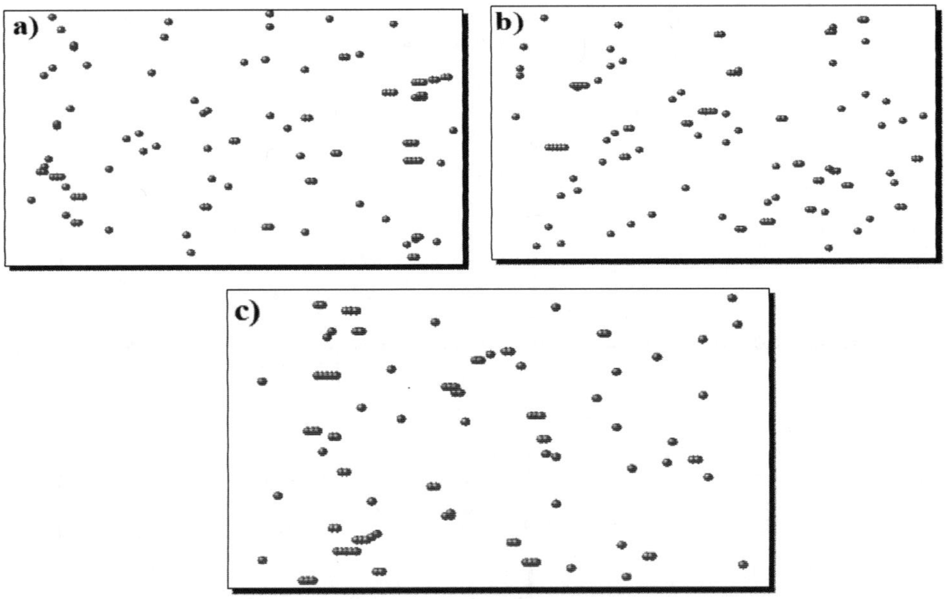

Figure 3: Snapshots from kMC simulations of Au/Cu(110) system at T=300K for different flux values: a) F = 0.1ML / s, b) 0.5ML / s, and c) 1 ML / s. The coverage rate is 0.1%.

4.1. Statistical study

In this section we summarize all the results in the form of a quantitative analysis that takes into account the influence of temperature and flux on the distribution of the elementary elements (adatom, dimer and cluster) during the hetero epitaxial growth in the Au/Cu (110) system.

a. Temperature effect

Thermal agitation has a very important effect on diffusion processes as well as on the arrangement and organization of clusters and adatoms. However; we will fix the flow and vary the temperature to better illustrate the effect of the latter on the different surface activities. We consider the Au/Cu hetero-system on surface (110). The results obtained are represented by the following histograms in figure 4. From Figure IV.10 it is noted that the distribution of adatoms is dominated for the four depositional rates 0.1, 0.5, 0.75 and 1ML/s in any temperature. As the temperature increases, the distribution of adatoms decreases with the appearance of small islands (dimers and trimers). These results allow us to conclude that temperature is not the only parameter that controls the evolution of a monolayer.

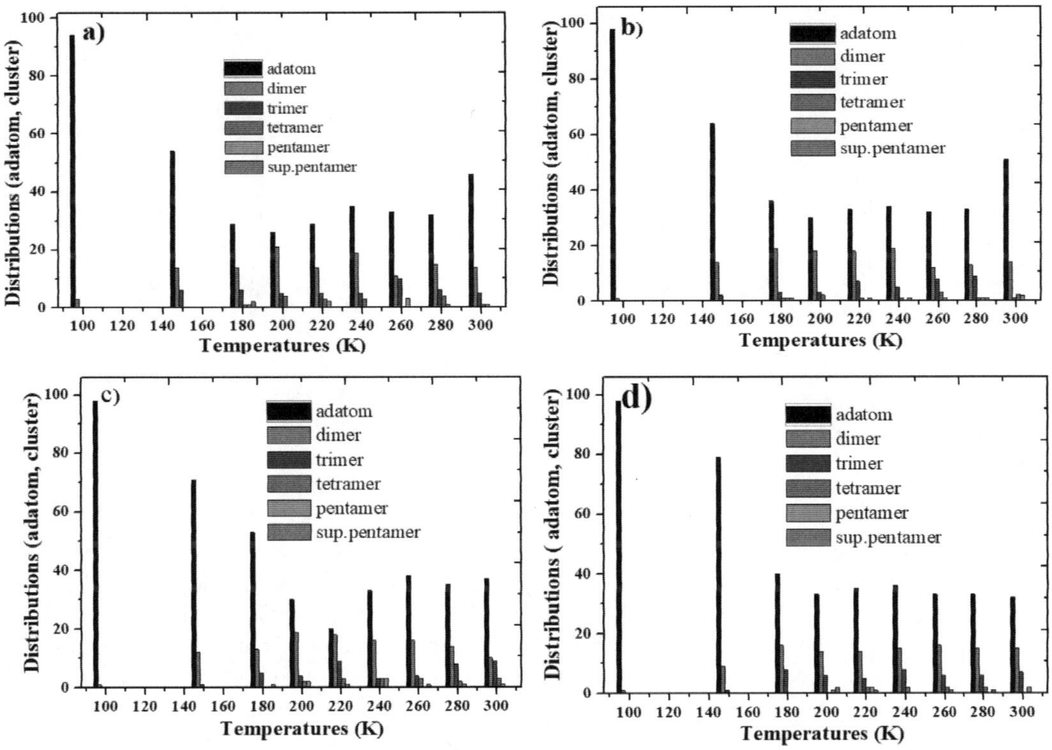

Figure 4: Distribution of the elementary elements of the Au/Cu system (110) as a function of the temperature at different fluxes: a) F = 0.1ML / s, b) F = 0.5ML / s, c) F = 0.75ML / s and d) F = 1Ml/s. The coverage rate is 0.1% monolayer.

However, the temperature has a considerable effect on the morphology of the surface by the activation of these processes nucleation, growth and coalescence.

b. Effect of deposition rate

The deposition rate of the incident atom on a substrate is another important factor in the growth of thin films. We study the effect of this factor on the distribution of elemental elements for temperatures (100, 200, 250 and 300K). According to the results presented in Figure 5, it is thus observed that the sup-pentamer clusters are formed for the flow of 0.75 ML/s in the two cases of temperatures 200K and 250K, and at 1ML/s for the temperature 200K. The distribution of the other elements is observed in all depositions rates and temperatures considered except in the case of 100K or the distribution is defined only by the adatoms. Also, when the deposition rate is decreasing and the temperature substrate is fixed, kinetic processes occur more slowly on the time scale of deposition.

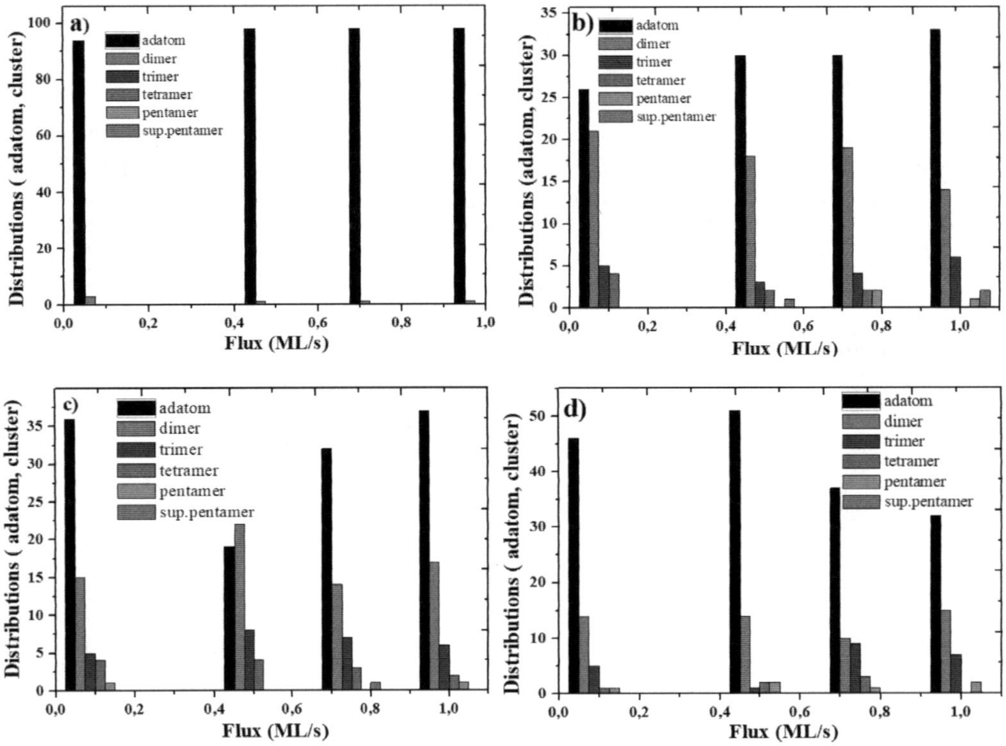

Figure 5: Distribution of the elementary elements of the Au/Cu system on (110) surface as a function of flux at different temperatures a) T = 100K, b) F = 200K, c) T = 250K and d) T = 300K. The coverage rate is 0.1% monolayer.

Figure 6 illustrate the evolution of diffusion rate $r_p/flux$ as a function of temperature(100-300K). From these results it can be concluded that the $ln(r_p/flux)$ for each flux varies linearly. As one knows, temperature activates the mobility of the surface by reducing the activation barrier. On the other hand, if the deposition rate (flux) is low, it gives the atom atom the time required to perform surface event multiplicities such as single jump and double jump, increases the probability of diffusion as well as to favor the mechanisms of growth. This result represents a good agreement with that of literature[8].

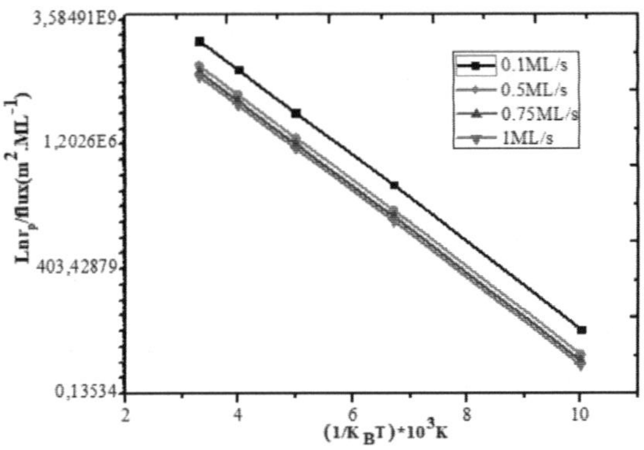

Figure 6: The Arrhenius plot of the rate $r_p/flux$ concerning diffusion of Au/Cu(110) at different flux.

5. Conclusion

In this work, we have investigated the results obtained by kinetic Monte Carlo simulations of Au heteroepitaxial monolayer growth on Cu (110) substrate. The diffusion barriers for elementary processes are determined by the molecular statics at T= 0K (based on Embedded-atom-molecular method). The anisotropic bond-breaking model (ABBM) has been used to describe the diffusion rate. The formation of small 1D and some 2D are observed at high temperature (280 k-300 K) and low deposition rates. The diffusion rate results show that this rate varies linearly as a function of the variation of temperature and flux. We conclude also that if we want the have a deep understanding of thin film growth, we have to take into account many parameters starting from the substrate properties (geometry, orientation…) until external parameters (flux, coverage …).

References

[1] A. Khalajhedayati, Z. Pan, and T. J. Rupert, "Manipulating the interfacial structure of nanomaterials to achieve a unique combination of strength and ductility," *Nat. Commun.*, vol. 7, 2016.

[2] I. Ohkubo, H. M. Christen, S. V. Kalinin, G. E. Jellison, C. M. Rouleau, and D. H. Lowndes, "High-throughput growth temperature optimization of ferroelectric SrxBa1−xNb2O6 epitaxial thin films using a temperature gradient method," *Appl. Phys. Lett.*, vol. 84, no. 8, pp. 1350–1352, 2004.

[3] Z. Zhang and M. G. Lagally, "Atomistic processes in the early stages of thin-film growth," *Science (80-.).*, vol. 276, no. 5311, pp. 377–383, 1997.

[4] C. Polop *et al.*, "The STM view of the initial stages of polycrystalline Ag film formation," *New J. Phys.*, vol. 9, no. 110, 2007.

[5] S. Cao, B. Liu, L. Fan, Z. Yue, B. Liu, and B. Cao, "Highly antibacterial activity of N-doped TiO 2 thin films coated on stainless steel brackets under visible light irradiation," *Appl. Surf. Sci.*, vol. 309, pp. 119–127, 2014.

[6] D. J. Wales and H. A. Scheraga, "Global optimization of clusters, crystals, and biomolecules," *Science (80-.).*, vol. 285, no. 5432, pp. 1368–1372, 1999.

[7] E. Cottancin *et al.*, "Size effects in the optical properties of Aun-Agn embedded clusters," *Phys. Rev. B*, vol. 62, no. 8, pp. 5179–5185, 2000.

[8] D. Deng, K. S. Novoselov, Q. Fu, N. Zheng, Z. Tian, and X. Bao, "Catalysis with two-dimensional materials and their heterostructures," *Nat. Nanotechnol.*, vol. 11, no. 3, pp. 218–230, 2016.

[9] S. Zayzoune, M. Mazroui, Y. Boughaleb, and A. Kara, "Anomalous behaviour of the diffusion coefficient in interacting adsorbates," *Surf. Sci.*, vol. 603, no. 20, pp. 3126–3130, 2009.

[10] Y. Han and J. W. Evans, "Directing anisotropic assembly of metallic nanoclusters by exploiting linear trio interactions and quantum size effects: Au chains on ag(100) thin films," *J. Phys. Chem. Lett.*, vol. 6, no. 12, pp. 2194–2199, 2015.

[11] M. Dardouri, A. Hassani, A. Hasnoui, Y. Boughaleb, A. Arbaoui, and K. Sbiaai, "Kinetic Monte Carlo Simulation of Au Monolayer Evolution on Cu(110)," *Phys. Status Solidi*, vol. 1800404, no. 110, p. 1800404, 2018.

[12] Y. G. Zhu and T. L. Wang, "Kinetic Monte Carlo simulation of the initial growth of Ag thin films," *Appl. Surf. Sci.*, vol. 324, pp. 831–836, 2015.

[13] J. Wei Wang, S. Narayanan, J. Yu Huang, Z. Zhang, T. Zhu, and S. X. Mao, "Atomic-scale dynamic process of deformation-induced stacking fault tetrahedra in gold nanocrystals," *Nat.*

Commun., vol. 4, pp. 1–8, 2013.

[14] M. Dardouri, A. Hassani, A. Hasnaoui, A. Arbaoui, Y. Boughaleb, and K. Sbiaai, "Kinetic Monte Carlo simulations of coverage effect on Ag and Au monolayers growth on Cu (1□1□0)," *J. Cryst. Growth*, vol. 522, no. October 2018, pp. 139–150, 2019.

[15] M. Mazroui, A. Asaklil, and Y. Boughaleb, "Conductivity of two-dimensional systems of interacting Brownian particles within the effective potential description," *Surf. Sci.*, vol. 409, no. 3, pp. 528–540, 1998.

[16] J. W. Evans, P. A. Thiel, and M. C. Bartelt, "Morphological evolution during epitaxial thin film growth: Formation of 2D islands and 3D mounds," *Surf. Sci. Rep.*, vol. 61, no. 1–2, pp. 1–128, 2006.

[17] T. R. Linderoth *et al.*, "Novel mechanism for diffusion of one-dimensional clusters: Pt/Pt(110)-(1 × 2)," *Phys. Rev. Lett.*, vol. 82, no. 7, pp. 1494–1497, 1999.

[18] K. Morgenstern, E. Lægsgaard, I. Stensgaard, and F. Besenbacher, "Transition from one-dimensional to two-dimensional island decay on an anisotropic surface," *Phys. Rev. Lett.*, vol. 83, no. 8, pp. 1613–1616, 1999.

[19] A. Memsouk, Y. Boughaleb, R. Nassif, and H. Ennamiri, "Dynamic scaling and self-organized criticality in diffusion fronts growth," *Eur. Phys. J. B*, vol. 17, no. 1, pp. 137–142, 2000.

[20] M. Kotrla, "Numerical simulations in the theory of crystal growth," *Comput. Phys. Commun.*, vol. 97, no. November, pp. 82–100, 1996.

[21] D. A. King and D. P. Woodruff, "Density-functional theory of epitaxial growth of metals," *Growth Prop. Ultrathin Ep. Layers*, no. Ml, pp. 490–544, 1997.

[22] M. Bon, N. Ahmad, R. Erni, and D. Passerone, "Reliability of two embedded atom models for the description of Ag@Au nanoalloys," *J. Chem. Phys.*, vol. 151, no. 6, p. 064105, 2019.

[23] F. Nita, C. Mastail, and G. Abadias, "Three-dimensional kinetic Monte Carlo simulations of cubic transition metal nitride thin film growth," *Phys. Rev. B*, vol. 93, no. 6, 2016.

[24] C. Mottet, R. Ferrando, F. Hontinfinde, and A. C. Levi, "Simulation of the submonolayer homoepitaxial clusters growth on Ag(110)," *Surf. Sci.*, vol. 417, no. 1–4, pp. 220–237, 1998.

[25] P. A. Maksym, "Fast Monte Carlo simulation of MBE growth," *Semicond. Sci. Technol.*, vol. 3, no. 6, pp. 594–596, 1988.

[26] S. M. Foiles, M. I. Baskes, and M. S. Daw, "Embedded-atom-method functions for the fcc metals Cu, Ag, Au, Ni, Pd, Pt, and their alloys," *Phys. Rev. B*, vol. 33, no. 12, pp. 7983–7991, 1986.

[27] M. S. D. and M. I. Baskes, "Embedded-atom method: Derivation and application to impurities, surfaces, and other defects in metals Murray," *Phys. Rev. B*, vol. 29, no. 12, p. 58, 1984.

[28] O. Trushin, A. Karim, A. Kara, and T. S. Rahman, "Self-learning kinetic Monte Carlo method: Application to Cu(111)," *Phys. Rev. B - Condens. Matter Mater. Phys.*, vol. 72, no. 11, pp. 1–9, 2005.

[29] M. L. Merrick and K. A. Fichthorn, "Nanostructures in thin-film epitaxy: Exploring and exploiting substrate-mediated interactions," *Mol. Simul.*, vol. 30, no. 5, pp. 273–279, 2004.

IOP Conf. Series: Materials Science and Engineering **948** (2020) 012027 doi:10.1088/1757-899X/948/1/012027

Experimental investigation on reinforced concrete beams by honeycomb sandwich panel structures: mechanical properties study

K Essaadaoui [1,*], M Ait El Fqih [2], M Idiri [1] and B Boubeker [1]

2 Laboratoire d'Igénierie et Matériaux (LIMAT), Université Hassan II Casablanca, Maroc.
[2] Laboratoire d'Ingénierie des Structures, Systèmes Intelligents & Energie Electrique (LI2SI2E), ENSAM, Université Hassan II, Casablanca, Maroc.

*khalil.essaadaoui-etu@etu.univh2c.ma

Abstract. In this work, the results of an experimental investigation on reinforced concrete by fiberglass honeycomb sandwich panel structures are presented. The experimental program included three beam specimens. Two of the beams were reinforced with different thicknesses of honeycomb structures. Flexural and uniaxial compression tests were performed. The tensile strength remains to be increases by a factor of 2.40 for compression and by 1.60 for flexion. The results obtained in terms of reinforcement are discussed. The changes are likely attributed to a modification of the adhesion forces at the honeycomb structures/concrete interface.

1. Introduction

The shear reinforcement of concrete elements with composite materials is in fact a subject of research far from being completely solved. This composite was used as an innovative strengthening and rehabilitating material and seems to be an attractive topic for practitioners and researchers [1,3]. Thus, the application of reinforcement has received much attention in the construction engineering industry. The use of reinforced structures in civil engineering increase rapidly and various type of reinforcement is used such as glass, carbon, asbestos and steel [4]. The honeycomb sandwich structures is one of the most valued structural engineering innovations developed by the composites industry [5].

The results of an experimental investigation on Reinforced Concrete (RC) T-beams retrofitted in shear with prefabricated L-shaped Carbon Fiber Reinforced Polymer (CFRP) plates were presented by Mofidi et al. [6]. They show that the performance of the specimens strengthened with partially and fully embedded L-shaped CFRP plates in the beam flange was superior to that of the beams strengthened Externally-Bonded (EB) with FRP sheets and L-shaped CFRP plates with no embedment. Rena et al. [7] studied one of the most devastating failure models in reinforced concrete structures i.e. the diagonal tension failure. Numerical model was used capable of dealing with both static and dynamic crack propagation.

Moreover, different forms simulation models presented in the literature as reviewed in the paper by Nayalet. al [8]. The model developed was used in the study of Wahalathanri et al. [9] and it was applicable for both reinforced and fiber reinforced concrete with only minor changes. Also, this method indicates similarity to the tension stiffening model that is needed for Abaqus concrete damaged plasticity model (CDP). This tension stiffening model was originally based on the homogenized stress-strain relationship developed by Gilbert et al [10] which accounts for tension stiffening, tension softening and local bond slip effects. Two descending portions of the tensile stress strain graph have accurately captured the response caused by primary and secondary cracking phenomena. The layered tension stiffening parameters used is replaced with a single set of stiffening parameters applicable to the entire tensile zone by the Nayal et al. [8].

Mofidi et al studied the behavior of T-beams, shear-reinforced with unidirectional Carbon Fiber Polymer CFP in EB the form of "U". Twelve beams of dimensions of 152 mm × 406 mm × 4520 mm were tested, whose objectives were to study experimentally and analytically the effect of the parameters that influence the shear strength of the Fiber Reinforcement Polymer FRP and to propose a set of equations of design to calculate the FRP contribution to the shear strength of reinforced concrete beams. The set of results to be achieved that the FRP contribution to shear strength was significantly higher for reinforced concrete beams without transverse reinforcement than for beams with transverse reinforcement [6].

The aim of this study is to present a new technique of the concrete reinforcement of beam by using honeycomb structures and to predict what extent a reinforced concrete structure can resist in the elastic mode. First material preparation with honeycomb sandwich panel structures and test setup is conferred. Results and discussion in term of reinforcement is presented.

2. Materials preparation and test setup

Specimens of rectangular concrete beams were prepared by mixing Portland cement with strength of 45 MPa (CPJ45), gravels, send and water with normalized proportion of ingredients. Honeycomb sandwich panel structures are introduced in the center of the rectangular concrete beam (Fig 1-a). Crushed sand with a granular class 0/5 was used. The gravel used in our concrete formulation was a granular class of 6.3/10. The mix proportions of concrete, namely the ratio of gravel, sand, cement, and the water / cement ratio was 0.55. The beams specimens were cured in the forms for 24 hours. At this time the specimens and cylinders were taken from the molds and were cured under water at 20° for 28 days. Then, they were stored to air-dry in a room at constant humidity and temperature until the time of test. Al used specimens had 400 mm in height and an overall length of 100 mm with a span length of 100 as can be seen in Fig 1-b (not reinforced one is noted S-Ref). Two of specimens were reinforced by honeycomb sandwich panel structures (noted S-HCt1 and S-HCt2) with two different thicknesses 3.22 (S-HCt1) and 5.76 mm (S-HCt2) respectively (Fig. 1-c).

Figure 1: a- Honeycomb sandwich panel structure in the concrete beam, b- Rectangular concrete beams, and c- Honeycomb sandwich panel structures.

Moreover, recently the use of honeycomb sandwich materials was very pronounced especially to strengthen structures [11-12]. It's formed by adhering two high-rigidity thin-face sheets with a low-density honeycomb core possessing less strength and stiffness. In our experimental study, the composite was made using fiberglass reinforced honeycomb. The illustrative scheme of sandwich panel concrete, composite (fiber glass / honeycomb / fiberglass) is sketched in Fig. 2. The fibre material was glass of the E type, while honeycomb and epoxy employed as matrix material. The ultimate strength parallel to the strong direction was specified by the supplier, whereas the volume fraction of fibres and the ultimate strain corresponding to the two principal material directions were not precisely known. The V_{f_s}/V_{f_w} ratio of the two volume fractions of fibers, where the subscripts s and w stand for strong and weak directions, respectively, was estimated in the ratio of 2 to 1 by the supplier. The young's modulus parallel to the strong direction was established from uniaxial tests performed on small specimens. Because fiberglass behaves in tension in a linear elastic manner up to failure, we can determine the ultimate tensile strain from Hooke's law knowing the ultimate stress and the Young's modulus.

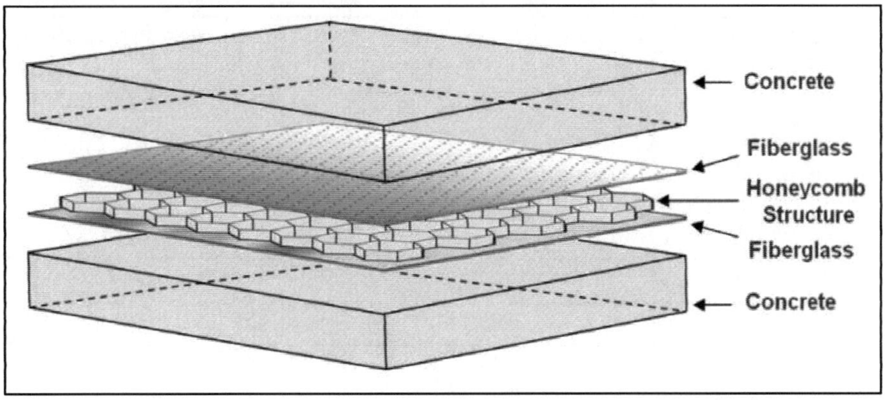

Figure 2: Scheme of sandwich panel concrete / composite (fiber glass / honeycomb / fiber glass) / concrete.

The tests of the prepared concrete were carried out using hydraulic jack of capacity 1500 kN. They were tested on the loading frame under axial compressive and flexion loading. The axial load is applied through hydraulic jack of 1500 kN capacity. The sectional properties of the specimens for uniaxial load are 400 mm x 100 mm x 100 mm. Flexion were tested under three-point bending until failure (Fig. 3-a and Fig. 3-b). The rate of loading was 0.1 mm/min, and the tests were continued up to 3 mm mid-span deflection. Two point loads with a distance of 300 mm were applied the load to the top of the specimens at the middle one-third of the span. The sectional properties of the specimens are shown in Fig. 4. The width of the composite was 3.22 mm and 5.76 mm. Moreover, The Young's modulus $\mathbf{E}$ two honeycomb structures are obtained by the following relation:

$$E = E_{FG}\left(1 - \frac{c^3}{h^3}\right) \qquad eq.1$$

Where E_{FG} is the fiberglass Young's modulus, c is the honeycomb thickness and h designate the sandwich thickness. According to equation 1, the Young's modulus of S-HCt1 is 70 GPa and for S-HCt2 71 GPa.

Figure 3: Scheme of sandwich panel concrete / composite (fiber glass / honeycomb / fiber glass) / concrete.

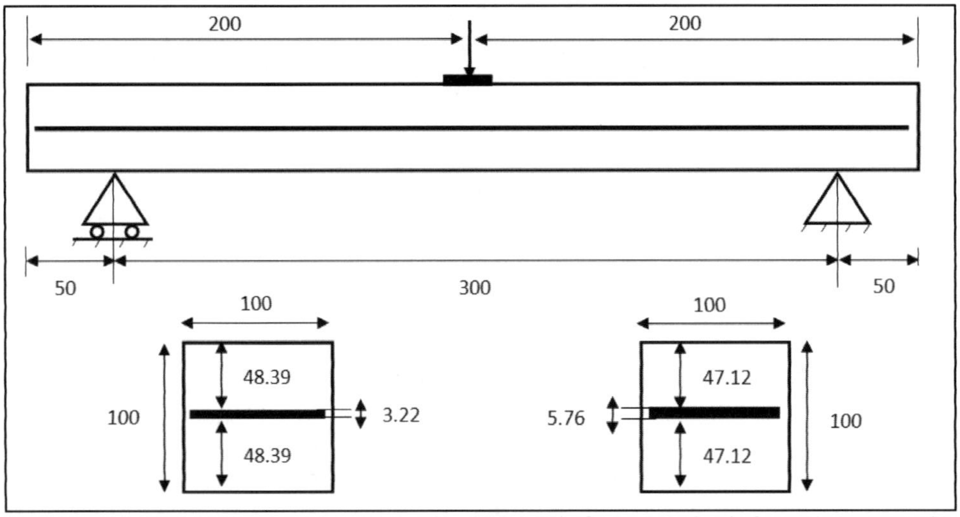

Figure 4: Test setup and specimens dimensions for the unreinforced and reinforced concrete beams

3. Results and discussion

The experimental results giving stress-strain curve on beams in axial load and flexion are presented in Fig. 5-a and Fig. 5-b respectively. Generally, the results of the uniaxial load and flexion show that the mechanical properties of the reinforced beams increase with the used composites. Indeed, and for uniaxial load, the tensile strength increase from 30 020 MPa for unreinforced beam to 32 495 MPa and 33 198 MPa for the beam reinforced by honeycomb sandwich structures with respectively 3.22 and 5.76 mm of thickness. We have registered that the reinforced beam resists sharp stresses and can reach up to a factor of 2.40 for compression. On the other hand, the same behavior was depicted for flexion and remains to be increases by a factor of 1.60. The comparison of results for compression and flexion are resumed in figure 6. A similar phenomenon was experimentally observed by Mofidi et al. [6] with beam strengthened by fiber-reinforced polymer. They affirm that the use of polypropylene fiber will enhance strength and behavior of reinforced concrete also improves resistance against impact loading and fire. The residual ultimate strength is larger than the corresponding strength of beams without polypropylene fibers by more than 60 %. No sudden failures are observed in all beams containing polypropylene fibers. Mofidi et al. [4] investigated that CFRP (carbon fiber–reinforced polymer) does provide improved impact resistance with increasing volumes of fibers. A CFRP mixture does provide reductions in permeability provided that the water-cement ratio remains below 0.5. Increased percentages of fibers further decreased the permeability provided the mixture remained workable. The study indicates a reduction in plastic shrinkage with increasing amounts of fibers. The polypropylene fibers decrease plastic shrinkage provided the water-cement ratio remains below 0.5. Wear resistance of PFRC has not been widely studied, but one study found an increase in the wear resistance with increasing fiber contents.

Table 1: Compressive strength, flexural strength and Young's modulus value

	Flexural strength (MPa)	Compressive strength (MPa)	Young's modulus (MPa)
S-Ref	25	5.20	30020
S-HCt1	46.11	7.89	32495
S-HCt2	59.68	8.37	33198

As reported by Table 1, the compressive strength of unreinforced concrete is 25 MPa while for a reinforced concrete, the resistance is respectively 46.11 MPa and 59.68 MPa for a concrete reinforced with honeycomb sandwich panel structures with thickness of 3.22 mm and thickness of 5.75 mm. Also the bending strength for an unreinforced concrete varies from 5.20 MPa to 8.34 MPa, while for a concrete reinforced by honeycomb sandwich panel structures with thickness of 3.22 mm the value was about 7.89 MPa, for a reinforced concrete by the composite with thickness of 5.76 mm the resistance was 8.37 MPa. Finally, it is very clear that the concrete reinforced with honeycomb sandwich panel structures with thickness 5.76 mm is the best one comparing it with the other two, because the greatest value of mechanical properties.

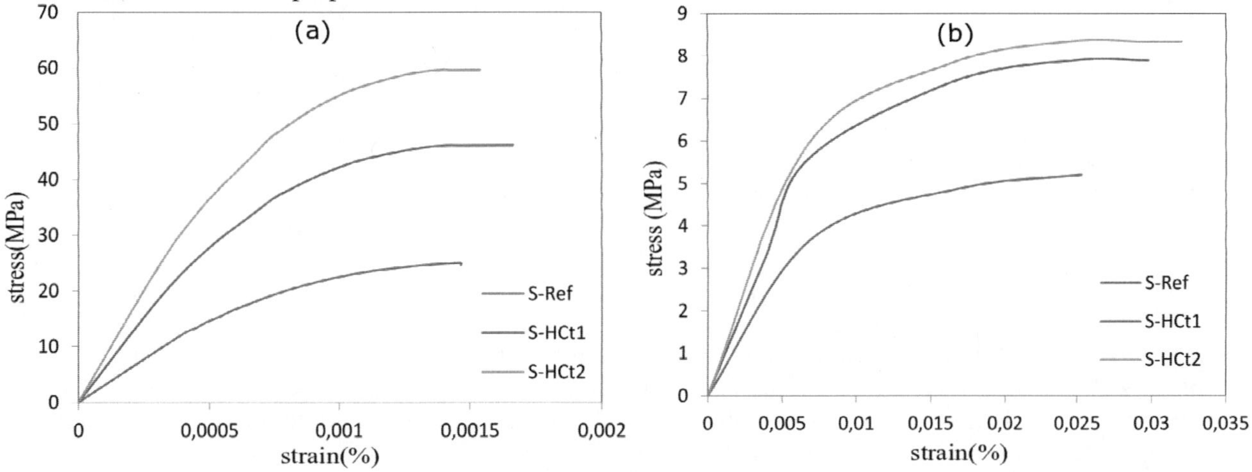

Figure 5: Experimental stress-strain curves for studied specimen; a- compression, b- flexion

Experimental results show that the fiber-reinforced composite controls concrete cracking, including limiting crack opening and transforming the fragile behavior of concrete into ductile behavior. Concrete always has micro cracks due to different causes such as shrinkage, thermal stresses etc. when a cracked concrete is subjected to tension, the tension lines bend and the stresses at the ends of the crack increase. This inflection is due to the fact that the load cannot be transmitted through the crack. For a concrete without fibers subjected to tensile stress, its low tensile strength is quickly reached or even exceeded. A crack then appears and spreads freely leading to a brittle fracture. In the case of a concrete with fibers, stressed in tension, part of the stress is taken up by the fibers bridging the cracks, this prevents the material from breaking brittle because the propagation of cracks would require greater energy. Indeed, since the maximum length of a micro crack is of the order of the largest aggregate size and the average length of a fiber is greater than the size of the aggregate, each micro crack propagating in the matrix can be bridged by one or more fibers. Thus, micro cracks can only be enlarged by the elastic or plastic elongation of the fibers or the rupture of the fibers. These tensile and shear stressed fibers absorb some of the energy required for crack propagation and prevent uncontrolled propagation of concrete cracking. More energy and effort would then be required to propagate this cracking. The greater the quantity of fiber and the greater the anchoring length, the more effective the fibers.

4. Conclusion

The experimental test is performed on simple three-point bending tests on rectangular beams and crush tests on both sides of rectangular beam after bending fracture with reinforcement by fiber reinforced honeycomb structure. Of glass in the form of a sandwich. The results obtained show that adding the composite can significantly increase the compression beam capacity by about 24%. The ultimate strength and modulus of elasticity of reinforced concrete reinforced by the composite differ from the ultimate strength and modulus of elasticity of the concrete matrix. Nevertheless, the presence of the composite in the concrete increases the deformation at break and thus confers a greater ductility to the concrete. As with bending, the composite provides a slight increase in the bending strength of reinforced concrete with a 16% increase and some ductility of post-fracture behavior of the composite. The presence of fibers plays an important role in composite bending behavior. Comparing the two reinforcements by the composite thickness 3.22 and composite thickness 5.76, the results show that increasing the thickness of the reinforcement increases the contribution vis-à-vis the shear capacity and especially in compression.

References

[1] A reference R. Seracino, NM. Jones, MSM. Ali, MW. Page, DJ. Oehlers. J. of Composites for Construction.11, 4 (2007)
[2] P. Wiwatrojanagul, BN. Ayudhya, R. Sahamitmongkol. Thammasat.Int. J. Science and Technology.17, 1 (2012)
[3] A. Mofidi,O. Chaallal,M. ASCE, Y. Shao.J. Compos.Constr. 18, 1 (2014)
[4] A. Mofidi, O. Chaallal, Int.J. of Concrete Structures and Materials. 8, 2 (2014)
[5] C. Lu, M. Qi, S. Islam, P. Chen, S. Gao, Y. Xu, X. Yang.Int. J. of Precision Engineering and Manufacturing –GreenTechnology.5, 1 (2017)
[6] A. Mofidi, S. Thivierge, O. Chaallal, M. ASCE, Y. Shao.J. Compos.Constr. 18, 2 (2014)
[7] RC. Yu, L. Saucedo, G. Ruiz. Int. J. of Fracture.169, 2 (2011)
[8] R. Nayal, HA. Rasheed. J. of Material in Civil Engineering.18, 6 (2006)
[9] B.L. Wahalathantri, D.P. Thambiratnam, T.H.T. Chan, S. Fawzia, In Proceedings of the First International Conference on Engineering, Designing and Developing the Built Environment for Sustainable Wellbeing, 260-264, (2011).
[10] R. Gilbert, R. Warner.J. of Structural Engineering.133, 6 (2007)

[11] SD. Yu, WL. Cleghorn.J. of Soud and Vibration.284, 1-2 (2005)
[12] B. Wang, M. Yang. J. of Materials Processing Technology.105, 1-2 (2000).

IOP Conf. Series: Materials Science and Engineering **948** (2020) 012028 doi:10.1088/1757-899X/948/1/012028

Correlation and diffusion breaking in the failure process of composite materials

A. Hader[a,c], H. Sbiaai[a], M. Tanasehte[a,*], R. Bakir[a] and Y.Boughaleb[a,b]

[a] LBGIM, Ecole Normale Supérieure, University Hassan II. Casablanca, Morocco
[b]LPMC , University Chouaib Doukkali. El Jadida, Morocco
[c] Centre régional des métiers d'éducation et de formation Casablanca- Settat/ établissement Settat, Morocco
*mohammed.tanasehte-etu@etu.univh2c.ma

Abstract: Fiber bundle model is one of the important statistical and theoretical physics approaches to investigate the fracture and breakdown of heterogeneous material extensively used both by the engineering and physics community. In this paper, using the correlation function and its properties, we study the diffusion of the failure process in composite materials that consist of a set of with randomly oriented fibers loaded by an external constant total load. The investigation is based on fiber bundle model where the fibers are randomly oriented; each fiber is subject only the cosine component of the external load, and when it fails, its load is shared by surviving neighboring fibers according to the local load sharing rule (LLS). The failure process ensures an advancing interfacial fracture and the areas of the damaged regions increases with time until a final crack of material. By using the correlation function of the fraction of broken fibers, we study the diffusion properties of micro crack created in the composite materials. The results show that the correlation decreases exponentially with time and decreases with both of applied load and temperature. We calculate then the effective diffusion coefficient cracks which increase exponentially with applied load and linearly with the temperature. The failure time t_f of the materials decrease with temperature as a power law. Obtained results are compared with the one of the classical model consist of a set of parallel fibers.

1. Introduction

The failure of composite materials has recently attracted much technological and industrial interest, it has been widely studied in statistical and theoretical physics, several authors propose that the failure of a composite materials subjected to an increasing external load shares many features with thermodynamic and statistic phase transitions [1-2]. In particular, a stressed composite can be considered to be in a metastable state and the point of global failure can be seen as a nucleation process in a first order transition near a spin odal [2-3]. Thus, the power law behavior observed experimentally in the acoustic emission during failure [4] has been compared with the mean-field scaling expected close to a spin odal point [5]. In analogy with spin odal nucleation [6], scaling behavior can only be seen when long-range interactions are present, as it is the case for elasticity but should not be observable when the stress transfer function is short ranged. This observation is confirmed in statistical fracture models with short-range elastic forces which usually do not show scaling [1;7;8].
The fiber bundle models (FBM) is a very important statistical models of material failure [9;10], which have been extensively studied during the past years. This model consists of a set of parallel fibers having statistically distributed strength. The sample is loaded parallel to the fibers direction, and the fibers fail if the load on them exceeds their threshold value. In stress controlled experiments, Each fiber is linearly elastic up to a threshold length which is usually taken randomly from some distribution function: Uniform, Weibull, Gaussian, Gumbel [11-12], when a fiber fails, the load carried is redistributed among the intact ones following two different transfer load type *GLS* and *LLS* [7;11;12]. Among the several

theoretical approaches, one simplification that makes the problem analytically tractable is the assumption of global load-sharing (*GLS*), which means that after each fiber breaking the stress is equally distributed on the intact fibers neglecting stress enhancement in the vicinity of failed regions [13-15]. The second is local load-sharing (*LLS*) model, the released stress is distributed equally only to the nearest surviving neighbors [16-17].

The relevance of FBM is manifold, in spite of their simplicity, these models capture the most important aspects of material damage and due to the analytic solutions, and they provide a deeper understanding of the fracture process. Furthermore, they serve as a basis for more realistic damage models having also practical importance [1;3;14]. The very successful micro mechanical models of fiber reinforced composites are improved variants of FBM taking into account stress localization (LLS) the effect of matrix material between fibers , and possible nonlinear behavior of fibers [15-21].

Previous studies of FBM addressed the macroscopic constitutive behavior, the reliability and size scaling of the global material strength, and the avalanches of fiber breaks preceding ultimate failure [21-23]. In classical fiber bundle model, the fibers have the same direction as the applied load. This choice is made in order to simplify the calculations [3;24]. Nevertheless, and in reality, it not reflects the anisotropy of the composite materials.

In order to cheek the effect of the anisotropy on failure process, we have considered, in this paper, a package of fibers randomly arranged in all directions according to the vertical one [1;24]. These materials are subject to a constant external load in the framework of a local load-sharing rule in two dimensions. Throughout the paper, we have used uniform distribution for breaking elongation threshold of each fiber. In this investigation, we propose to study the failure process by using the correlation function associated to the number of broken fiber per time unity [3], and we evaluate behavior of the diffusion failure coefficient as function of two physics parameter, load and temperature, we also study failure time behavior versus temperature. Each result will be compared with those obtained in the classical model when the external load has the same direction to fibers [3].

2. Fiber bundle model

As mentioned before, in our approach we use the fiber bundle model with other arrangement fibers. We consider composite materials represented as a discrete set of N_0 fibers randomly oriented in all directions according to the vertical one, placed on a regular square lattice of size *L*. Each fiber *i* is inclined according to the vertical axis with a fixed angle θ_i $(0 < |\theta_i| < \frac{\pi}{2})$. θ_i is randomly dispersed with the probability density *p(x)* and the uniform distribution *P(x)* [1-3]:

$$P(\theta) = \int_0^\theta p(y)dy = \theta \quad (1)$$

The fibers, which are placed vertically ($\theta_i = 0$) have the same initial length l_0 and are mounted with well-fixed clamps belonging to the two lateral surfaces of the system. However, the ones, which are weakly inclined, $(0 < |\theta_i| \leq \frac{\pi}{4})$ have different initial lengths, which are chosen in such a way that these fibers also remain attached to the two lateral surfaces by clamps: $l_0 = l_i \cos\theta_i$. Finally, the fibers, which are strongly inclined, $(\frac{\pi}{4} < |\theta_i| < \frac{\pi}{2})$ have the same initial length l_0. The fibers are mounted between two hard clamps and the matrix ensures the cohesion of the system, we suppose that each fiber keep the same direction during the application of load (θ_i is constant) [8]. The fibers can solely support longitudinal deformation under an external total constant load parallel to the vertical direction [1]. Thus, each fiber is subjected to an imposed tensile constant load f, which leads to an axial longitudinal displacement, by the effect of its axial component f_{θ_i}:

$$f_{\theta_i} = f\, cos\theta_i \quad (2)$$

This cosine component ensures an axial elongation x_i which is linked by the Hooke's law:

$$f_{\theta_i} = k\, x_i \tag{3}$$

where k denotes the stiffness, which is the same for all the fibers. However, the sinus component is mitigated by the matrix-fiber interface reaction. For this raison, the meaning action is arisen from the cosine force component [1].
On the other hand, In order to take into account of temperature effect, we introduce the corresponding local elongation ζ_i of each fiber:

$$\xi_i = \gamma l_i \sqrt{K_B T} \tag{4}$$

K_B is the Boltzmann constant, and γ is coefficient of proportionality between elongation and square root of one-dimensional thermal energy (fiber). l_i is the initial length of fiber i.
The temperature has a quasi-static effect on the failure process in the composite material, yet that will affect the lifetime of the system [25-27]. Then, the actual total elongation arising x of each fiber is written as:

$$x = x_i + \zeta \tag{5}$$

If the arising elongation x of fiber i exceeds the broken threshold elongation x_c, the fiber cracks and its load is shared by its neighboring remains fibers and so one [1;4]. In this paper the broken threshold elongation x_c is randomly chosen and given with the probability density $p(x)$ and the uniform distribution $P(x)$[1]:

$$P(x) = \int_0^x p(y)dy = x \quad \text{with} \quad p(x) = 1 \tag{6}$$

The fibers behave linearly elastic until they fail at a failure threshold elongation. The randomness of breaking thresholds is assumed to represent the disorder of heterogeneous materials.
During this process when some threshold value of elongation is exceeded the stress carried by the broken fibers is shared by the intact neighboring fibers and so on. Thus each of the other fibers undergoes another elongation δl due to the force released by the LLS share and the expression of total elongation x' of each fiber becomes: $x' = x + \delta l$
In our model the mean threshold elongation is $<x_c> \approx 0.1 < l_i >$, Our calculations are done for a load range $0 < f \leq 0.65 f_c$ and temperature range $0.025T_c \leq T \leq 0.05T_c$ where f_c represents the mean critical load at which the system crack totally and T_c represents a critical temperature value at zero applied load ($f=0$) [1].
We propose a quantitative study of failure diffusion in composite material by using the correlation function associated to number of broken fiber per time unity [3]:

$$G(\tau) = < \frac{dN_b}{dt}(t+\tau)\frac{dN_b}{dt}(t) > \tag{7}$$

$N_b(t)$ is the number of broken fibers at time t. The basic idea is that there is an analogy between the crack diffusion process and the adatoms diffusion in the scaling tunneling microscopy (STM) one [27-30], so we propose to evaluate variation of the diffusion coefficient of failure as function of two different physics parameter force and temperature. The results will be compared with those finding in the classical fiber bundle model [3].

III- Results and Discussions

We consider a composite material of square size L with $N_0 = L \times L$ intact fibers randomly oriented in different directions according to the (OZ) axis. Each fiber i is randomly oriented with a fixed angle θ_i

$(0 < |\theta_i| < \frac{\pi}{2})$, and supports an external constant load *f*. Initially, the material resists to the applied load, but at time t_r when the elongation x for a fiber become greater than their critical elongation, micro cracks appear in the system [1]. Nevertheless the length *l(i)* of the other intact fiber *i* varies differently with time according to the local load sharing (LLS) of load. When a fiber breaks, the load carried by it is redistributed equally among its neighbors, then this breaking and redistribution dynamics continues. In general, whenever a fiber breaks, the load carried by that fiber is equally shared by the surviving fiber(s) nearest to it [1;16;17].

On the other hand, Figure 1 presents an image system (3D) calculated for material of size $L = 20$ at time $t = 0{,}25t_f$. The applied load and the considered temperature are respectively f/f_c =0,3 and T/T_c =0,03. The blue spots represent the broken fibers [1].

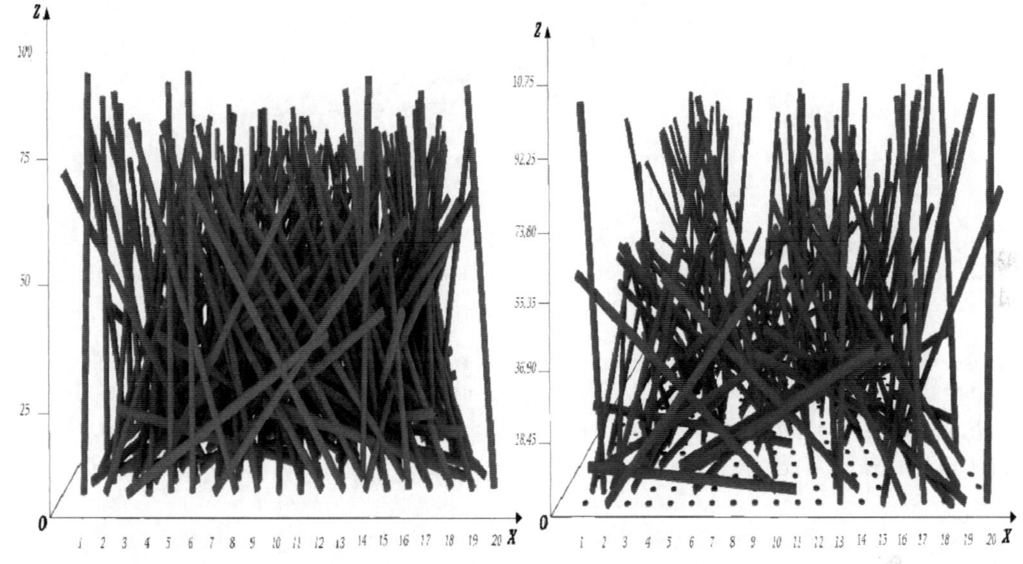

Figure1: illustration of the evolution system of size $L=20$ and for time $t=0{,}25t_f$ under f/f_c =0,3 .

As mentioned before, by correlation function associated to number of broken fiber per time unity, defined in equation (6), we study the diffusion of failure process in a composite material with randomly oriented fibers [3 ;30].

Firstly, we have calculated the time evolution of the intact fibers fraction $\rho_A = \frac{N_A}{N_0}$, where N_A is the number of surviving fibers. Results are plotted in figure 2. The fraction ρ_A is saturated the vicinity of *0,5*, for this system with randomly oriented fibers, all the considered applied stress of range values $(0 < f / f_c < 0.65)$ cannot break totally the system in spite of that the load will be shared equally by the nearest surviving fiber(s) and so on. These load shares accumulate on all the fibers on the perimeter and grow with time since their respective appearance on the perimeter of the damaged regions, in addition, when the applied load value increases, the saturated intact fiber density decreases [1], but not reaching the value *0*, because the fibers which are more inclined $(\frac{\pi}{4} < |\theta_i| < \frac{\pi}{2})$ resist to the applied load and cannot cracked. In contrary, in the FBM with parallel fibers the system crack totally and the broken fraction tend to the value *1* [1;17;31].

Nevertheless, the fibers which are less inclined $(0 < |\theta_i| < \frac{\pi}{4})$ are the most affected by the applied load than the others ones more inclined which can undergo any load. For this last case The only felled charges are transferred to the fibers by the matrix under LLS mechanism [1]. The heterogeneous material with randomly oriented fibers is more resistant than the one of the classical model where the fibers have a same direction and the corresponding observed avalanche breaking is characterized by two successive regimes [1;13;17]

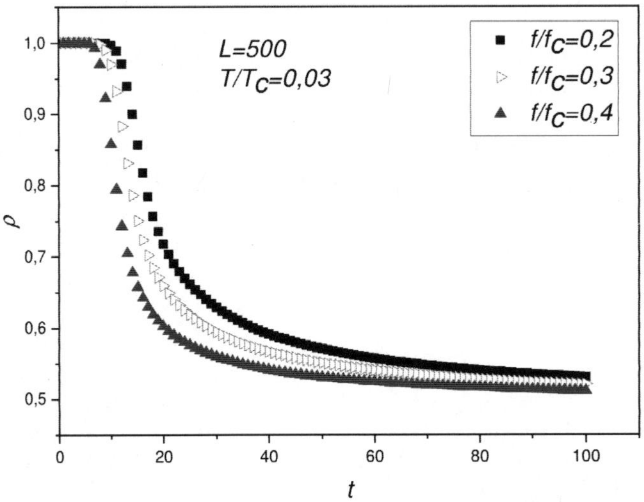

Figure 2: time variation of intact fibers for different applied load and for a temperature value T/T_c =0.03.

We analyze the correlation process observed in the failure process for the considered materials by using the equation (6). Figure 3 present time variation of the normalized crack correlation function for different values of applied load calculated for system size *L=500* at temperature $T/T_c = 0.03$. The correlation function *G(t)* decreases exponentially with time as:

$$G(t) = G_0 \exp\left(-\frac{t}{\tau}\right) \quad (8)$$

where τ is the characteristic time of the correlation function which correspond to the $0{,}37G_0$
The correlation tends to approximately of 0,5. By analogy, this behavior of *G(t)* is also observed in the surface adatoms diffusion process by using the scaling tunneling microscopy (STM) [26-28].
Also, the same behavior was found in the classical model [3] , but the correlation function of the latter case exponentially decreases going through great values and tends to 0. Comparing the two curves, we remark that in the system with randomly oriented fibers, the correlation function decreases rapidly in time compared with that found in the classical model, This strong fall of G (t) reflects a weak correlation between broken fibers, in other words, broken fibers which are randomly directed are less correlated in their breaks compared to those having the same direction in the classical system [3]. This can be explained on the one hand by the fact that inclined fibers resist more to the force, on the other hand, they are not all subject to force.
In order to check the effect of the external load on the correlation function, we have calculated for the same intervals of time, the normalized correlation function *G(t)* versus applied load for system size *L=500* at temperature $T/T_c = 0.03$ .The corresponding results are plotted in figure 4.

Correlation function decreases with applied load with a less impressive monotony even if the load is applied on all the system and even if this correlation decrepit the total undamaged region of system, the same curve shape as that found in the classical case but with small correlation values which correspond to large force values [3], which confirm that the anisotropy in the directions of fibers increases resistance of the system and paralyzes the diffusion of the crack in the material.

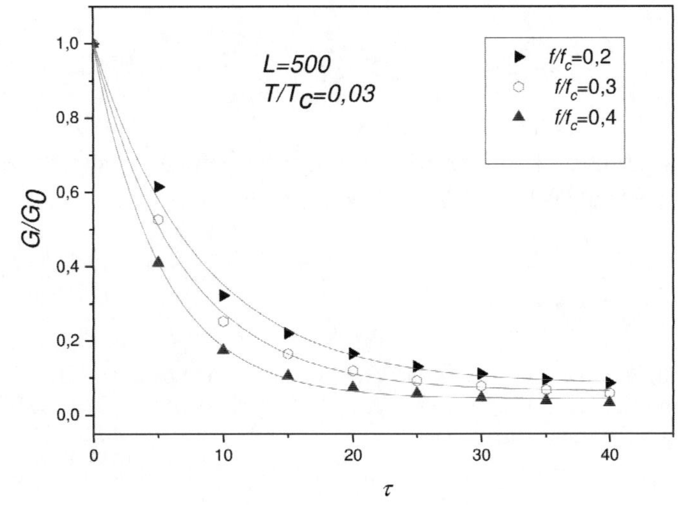

Figure3: time variation of correlation function for different applied load

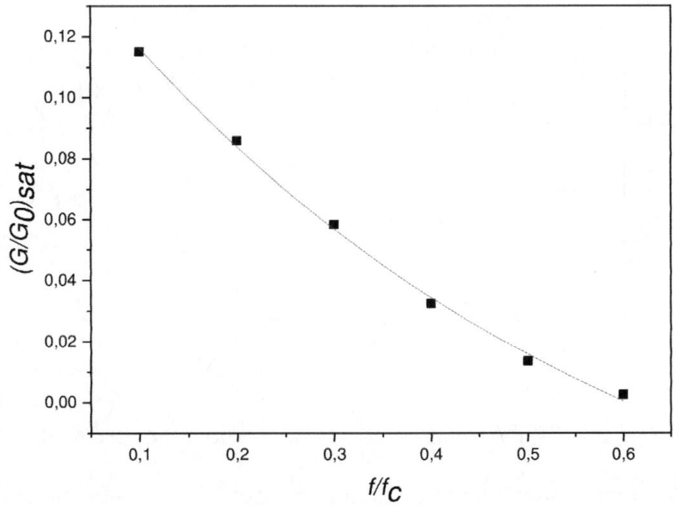

Figure 4: correlation variation versus applied load

Diffusion of micro cracks in a composite material is one of interesting process to understand the behavior of the failure process. The effective diffusion coefficient D_{eff} can be calculated by using the characteristic time τ of the crack correlation function G. The method is based on the proportionality relation between D_{eff} and τ used in the fluctuations of the adatoms surface in the scanning tunneling microscopy investigations [28-29]:

$$D_{eff} \propto \frac{1}{\tau} \tag{9}$$

We have study the behavior of The effective diffusion coefficient as function of external load for system size $L=500$ at temperature $T/T_c = 0.03$, corresponding results are plotted in figure 5.

The effective diffusion coefficient increase exponentially with applied load:

$$D_{eff} \propto \exp\left(\frac{f}{f_r}\right) \tag{10}$$

However, for the classical model [3], the effective diffusion coefficient increases as a power law with applied load, that either lead to breakdown of the whole bundle:

$$D_{eff} \propto \left(\frac{f}{f_c}\right)^{\frac{7}{4}} \tag{11}$$

In our case, the crack diffusion process is less accelerated due to anisotropy in the directions of the fibers which further amortize the diffusion of the crack within the system. These results are in agreement with the experimental results [31].

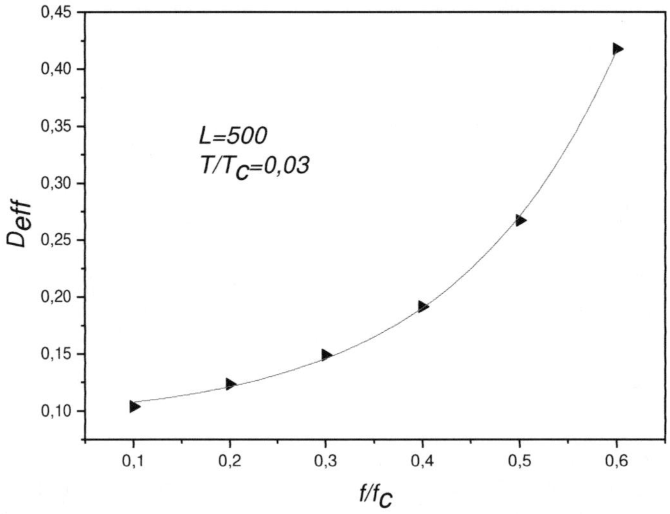

Figure5: diffusion coefficient variation versus applied load.

Temperature has a very important effect on the failure process in composite materials, The influence of this parameter on the failure in fiber bundle model has been widely studied numerically [1;32;33], it catalysis appearance micro cracks in the material. These defects or micro cracks induce strong fluctuations on material strength, which influence lifetime of the system and its transition from elastic mode to plastic one [1].

The strength of each fiber is characterized by a critical length which is randomly varies. The variation of temperature drives the cracking process. To make in evidence this point, we have investigated the time evolution of the crack correlation function for system size of $L = 500$ subject to the load value $f/f_c = 0{,}3$ and for a two different values of temperature (figure 6).

the quasi-static effect of temperature accelerates the decrease of the fibers breaks correlation, the elongation of the fibers reaches its critical elongation more quickly, leading more to a reduction of the number of intact fibers, which affects the correlation of the fibers that break. Thus, an increase in temperature activates the spread of cracking and the correlation between fibers decreases faster over time but without reaching the total destruction of the material. The similar result is observed for the classical model but with decreasing monotony slower until the total failure of the system [3].

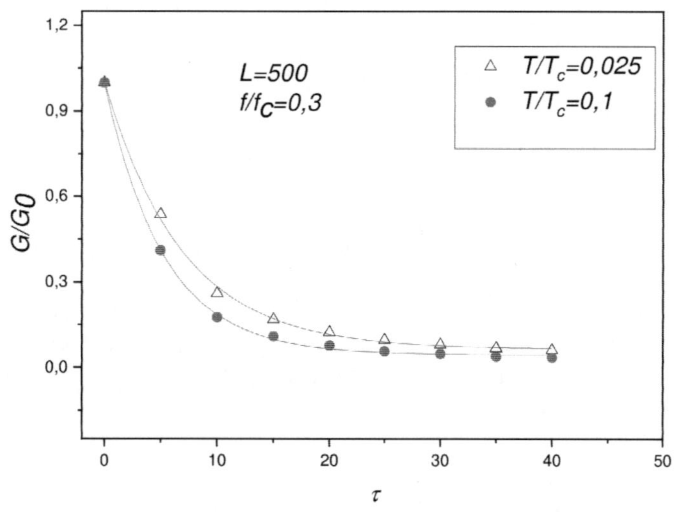

Figure 6: time variation of correlation function for different temperature value

Figure 7 presents the variation of the effective diffusion coefficient D_{eff} with temperature for system size of $L = 500$ subject to the load value $f / f_c = 0{,}3$. The obtained profile shows that the crack diffusion process increases linearly with thermal noise as:

$$D_{eff} \propto \frac{2}{3}\left(\frac{T}{T_c}\right) \quad (12)$$

proving our precedent remark. However, this variation is different comparing to the classical model, where D_{eff} exhibits a height increase with avalanche phenomena due to the avalanche breaking of the fibers, that lead to breakdown of the whole bundle [3].

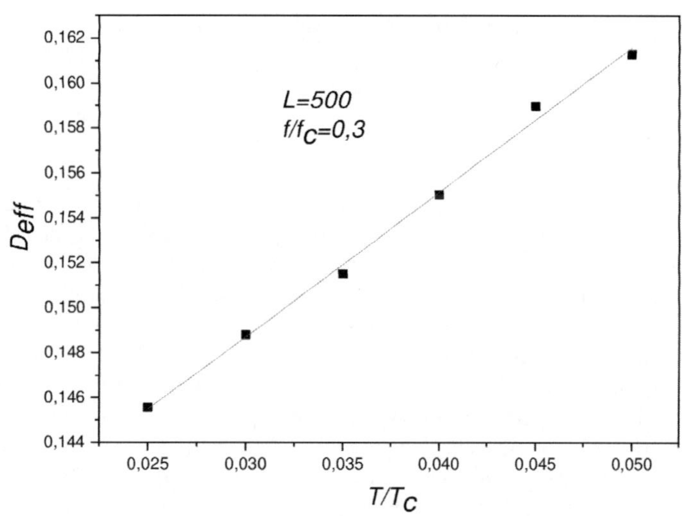

Figure 7: diffusion coefficient behavior with temperature

The lifetime one of the more investigated parameter in the failure process of composite materials, In figure 8, we present the variation of this parameter versus temperature under an external load $f = 0.3 f_c$, obtained results are fitted by a linear function indicating that the lifetime deceases with the temperature as power law:

$$\frac{t_f}{t_r} \propto \left(\frac{T}{T_c}\right)^{-\frac{1}{3}} \tag{13}$$

where t_r is the critical time when the cracks start. This scaling law is observed at each value in the used range of applied load which proves its universal character [1]. However for the classical model with local load transfer rule *LLS*, the lifetime material decreases linearly with the temperature [3].
For the classical model with global load transfer rule *GLS,* Pradhan and al found that the life time decreases exponentially with temperature [34]. It means that temperature effect is influenced by applied load transfer rule and fibers direction.

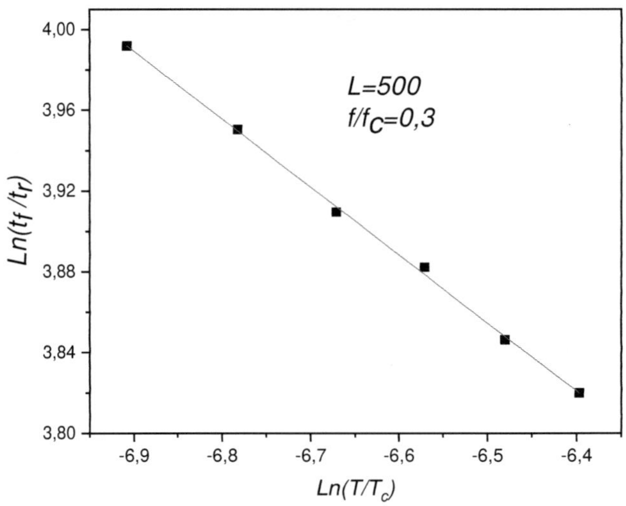

Figure8: Lifetime as function of temperature

3. Conclusion

In summary, the fiber bundle model of composite material shows some precursors which can help to study the failure phenomena loading situations. Here we use an approach for describing and study the diffusion of failure process within composite materials with randomly oriented fibers. The investigation is based on the calculation of the crack correlation function using the fiber bundle model. Each fiber is subject only the cosine component of the external load, and when it fails, its applied load is shared by surviving neighboring one according to the local load sharing rule (LLS) [8]. The results show that the correlation decreases exponentially with time and decreases with both applied load and temperature. the effective diffusion coefficient D_{eff} of the cracks increase exponentially with applied load and linearly with and temperature, however for the classical model, this coefficient increases with the applied load and exhibits a power law , on the other hand, versus temperature, D_{eff} exhibits an increase avalanche phenomena due to the avalanche breaking of the fibers.
Life time t_f decrease with temperature following a power law. However, for the classical model with local load transfer rule *LLS,* this time decreases linearly with the temperature following a linear law.

Yet, for the classical model with global load transfer rule *GLS*, Pradhan and al found that failure time decrease exponentially with temperature in GLS sharing load, which confirm that the temperature effect is influenced by applied load transfer rule and fibers direction.

References:

[1] H.sbiaai, A. Hader, R.Bakir, I.Achik, I. Tarras , Y.Boughaleb physica A , 476, 84-90 (2017)
[2] A. Buchel, J.P. Sethna, Phys. Rev. Lett. 77, 1520 (1996); Phys. Rev. E 55, 7669 (1997).
[3]. A. Hader, I. Achik, H. Sbiaai, R. Bakir, and Y. Boughaleb MOL. CRYST. LIQ. CRYST. 628.79-85 (2016)
[4]. S. Zapperi, P. Ray, H.E. Stanley, A. Vespignani, Phys.Rev. Lett. 78, 1408 (1997); Phys. Rev. E 59, 5049 (1999).
[5]. C. Unger, W. Klein, Phys. Rev. B 29, 2698 (1984); *ibidem*
[6]. M. Kloster, A. Hansen, P.C. Hemmer, Phys. Rev. E 56, 2615 (1997).
[7] Local load sharing fiber bundle in higher dimension .sautana Sinha . Jonas T kjellstadli and Alex Hansen phys rev E92.020401
[8]. H.E. Daniels, Proc. R. Soc. London A 183, 405 (1945).
[9]. W.A. Curtin, Phys. Rev. Lett. 80, 1445 (1998).
[10] SO. Ramas,P.-P. Cortet, S. Cilibreto and Vanel, Phys.Rev.Lett.110, 165506 (2013).
[11] L. Laurson and S. Zapperi. Roughness and multiscaling of planar crack fronts. *Journal of* Statistical Mechanics: Theory and Experiment, 2010(11):P11014, 2010.
[12] D. Amitrano, Eur. Phys. J.B. SpecialTopics 205, 199-215(2012)
[13] A.Hader ,I. Achik, K. Sbiaai, Y. Boughaleb, Superlattices and Microstructures S0749-6036(2014)00087-10.
[14]. S.L. Phoenix, Raj, Acta Metall. Mater. 40, 2813 (1992).
[15]. W.A. Curtin, N. Takeda, J. Comp. Matls. 32, 2042 (1998).
[16] Guarino, A. et al. Eur. Phys. J. B. 26, 141 (2002)

[17] A. Hader , Y. Boughaleb I. Achik, K. sbiaai Optical Materials 36 (2014) 3–7 Optical Materials
[18]. C. Moukarzel, P.M. Duxbury, J. Appl. Phys. 76, 1 (1994).
[19]. R.L. Smith, S.L. Phoenix, J. Appl. Mech. 48, 75 (1981);D.G. Harlow, S.L. Phoenix, J. Mech. Phys. Solids 39, 173(1991).
[20]. I.J. Beyerlein, S.L. Phoenix, J. Mech. Phys. Solids 44, 1997 (1996).
[21]. . D.G. Harlow, S.L. Phoenix, J. Composite Mater. 12, 195(1978).
[22] P.C. Hemmer, A. Hansen, J. Appl. Mech. 59, 909 (1992).
[23]. R. da Silveira, Phys. Rev. Lett. 80, 3157 (1998).
[24] D. Krajcinovic, M.A.G. Silva, Int. J. Solids Struct. 18,551 (1982).
[25]. A. Hader, I. Achik, H. Sbiaai, R. Bakir, and Y. Boughaleb Mol. Cryst. Liq. Cryst. 628.79-85 (2016)
[26]. A. Memsouk, Y. Boughaleb, R. Nassif, and H. Ennamiri, Eur. Phys. J. B, vol. 1735, pp. 137–142, Sep. 2000.
[27]. Carpinteri, A., & Pugno, N. (2005). *Nature Mater.*, *4*, 421
[28]. Qiming, L., & Tringides, M. C. (1996). *Surf. Sci.*, *365*, 495.
[29]. Saadouni, K., Hader, A., Jaouad, A., & Boughaleb, Y. (2004). *Surf. Sci.*, *572*, 366.
[30]. S. Zayzoune, M. M. Mazroui, Y. Boughaleb, and A. Kara, Surf. Sci. - Surf. SCI, vol. 603, pp. 3126– 3130, Oct. 2009.
[31] H. Nachad, thesis INSA-Lyon, (2004)
[32] Scorreti, R., S. Ciliberto and A. Guarino, 2001, Europhys.Lett. 55, 626.

[33] Ciliberto, S., A. Guarino and R. Scorretti, 2001, Physica D 158, 83.
[34] S. Pradhan and B. K. Chakrabarti. International Journal of Modern Physics B Vol. 17, No. 29 (2003) 5565-5581

.

IOP Conf. Series: Materials Science and Engineering **948** (2020) 012029 doi:10.1088/1757-899X/948/1/012029

Modeling of crack diffusion in composite materials

I. Achik[a,b,*], A. Hader[a,c], H. Sbiaai[a], S.Boufass[a], M. Tanasehte[a,] and Y. Boughaleb[a,d]**

a Ecole Normale Supérieure, Université Hassan II, Casablanca, Morocco
b LPMC, faculty of sciences ben M'sik Casablanca, Morocco
c Centre régional des métiers d'éducation et de formation Settat, Morocco
d LPMC , University Chouaib Doukkali. El Jadida, Morocco
*achikimad@yahoo.fr ** mohammed.tanasehte-etu@etu.univh2c.ma

Abstract. Our aim is to investigate the crack diffusion created at single region of composite materials by using the fiber bundle model. So, we have applied an external single crack in one fiber of the composite material, and we then continue to increase this load at a very slow rate until the considered fiber breaks and its load is redistributed to its neighboring intact ones. This breaking and redistribution dynamics repeat itself and this process ensures an advancing interfacial fracture and the area of the damaged region increases with time until a final crack of material. Our calculations are done in the context of the local load-sharing rule. The results show that the damaged region area increases with time by following the Lifshitz-Slyozof law with an exponent growth x=2. This permits us to deduce the behavior of the crack diffusion with the applied load. The corresponding results of the life time materials exhibit an exponential decreasing with the applied load and a linear decreasing with temperature.

1. Introduction

Focusing on the use of naturally different substances in the onset frucster has ever been the issue of several surveys in the engineering community for its mere technological developments. In fact, the somatogenetic handling, sometimes meant as an indispensable separation, is apparently causing a vast waste of time for extreme macroscopic blockage of the solid, while depending on the amplitude of the applied stress [1–3]. That is why scientists are actually devoting a great deal of time to tackle many theories and experiments for the clarification of the unpredictable fractures, e.g., the velocity of propagation, the roughness, and the onset of precursors.

In the contrast, the fiber bundle model describes a collection of elastic fibers under load. The fibers fail successively and for each failure, the load distribution among the surviving fibers changes. Even though very simple, this model captures the essentials of failure processes in a large number of materials and settings. This model has been studied extensively since it can be analyzed to an extent that is not possible for more complex materials [4–6]. The statistical distribution of the magnitude of avalanche phenomenon in fiber bundles is well studied [7–9], and the failure dynamics under constant load has been investigated through recursion relations which in appears a phase transitions and associated critical behavior in this model [10]. However, there are two extreme rules which consider a bunch of fibers hanging from a rigid ceiling and a platform is connected to the ends of these fibers and a load hangs from that plat form. Once the fibers begin to fail, we can distinguish two transfer load type: Global load sharing (GLS) in which the load of a broken fiber is equally shared with all intact fibers in the whole system. We assume a long-range interaction among the fibers and we treat analytically the problem by using the mean-field approximation [3]. The other type consists in a local load sharing (LLS), in which the terminal load of the failed fiber is distributed equally to all the nearest neighboring intact fibers.

Hence, one creates the initial applied load localized at an arbitrarily chosen central site in the frame work of the local load-sharing fiber bundle model. So, initially, no load is applied on any fiber except for the one at the central site. As the applied load increases beyond the failure threshold of this central fiber, it breaks and the load carried by it is redistributed among its nearest neighbors and so on.
In this paper, we study the crack diffusion in composite materials subject to a local load-sharing fiber bundle model in two dimensions under an external load applied at a single point. By the use of the local load-sharing rule, the redistributed load remains localized along the boundary of the broken patch.

2. THE MODEL

The fiber pack product identifies accumulation flexible materials below load. Under some conditions when the length of the fibers of the composite material exceeds its breaking critical value, the material cracks [11]. So, for every single disappointment, the strain circulation one of the remaining materials changes. Ergo, that creates a feedback method that could make the sum total disappointment of the system. Although quite simple, that product reflects the necessities of disappointment procedures in a significant number of components and settings. We provide here evaluation the fiber pack product with various fill redistribution elements from the function of see of data and mathematical science as opposed to components technology, with an emphasis on ideas such as for example criticality, universality and fluctuations.

We contemplate a deal of size M consisting of a big quantity of materials held at equally ends. We examine equal-load-sharing versions in which the fill formerly moved by an unsuccessful fiber is distributed similarly by closest friend unchanged materials [12–15]. The pack middle is at the mercy of an area regular additional tension similar to the fibers direction. The strain moved by that fiber is redistributed similarly among their closest remaining neighbor(s). In that way, the materials which are just confronted with the strain, state, following an avalanche, has a somewhat minimal fill than the people that are accumulating fill gives from the sooner problems and surviving. Within our function we think that the original regional load f_0 is likely to be corresponding to the unchanged materials which stimulate an original elongation δl_0 distributed by the Hooke's legislation:

$$\delta l_0 = f_0 / k \tag{1}$$

where k denotes the stiffness, which is assumed to be the same for all the fibers. The local elongation of fibers i have time-dependent fluctuations due to the presence of thermal noise load and to load transfer following breaking events given by:

$$\xi_i = \gamma l_0 \sqrt{K_B T} \tag{2}$$

Where l_0 is the initial length of fibers and K_B is the Boltzmann constant and is coefficient of proportionality. The presence of the thermal noise will affect the lifetime for a constant applied stress controlled by the temperature T of the system [2,9]. So, the actual stress arising δl_i of fiber i is written as:

$$\delta l_i = \delta l_0 + \xi_i \tag{3}$$

3. THE RESULTS

Originally, we used a lot on a single fiber devoted to the material. We then keep on to boost force at a really gradual charge before the focused fiber pauses; force moved by its redistributed similarly among their four neighbors. Because we keep on to boost force just on the materials which are presently holding a nonzero fill, that breaking and redistribution character replicate itself. This method guarantees a developing interfacial fracture and the location of the ruined location raises. That subject has been commonly learned around years, equally theoretically and experimentally [16-17] in the Plexiglas try out two dishes were taken and condition was presented by sandblasting and then were joined together creating a clear stop by having a simple plane. Therefore, when we continue to increase the external load, the material cracks totally at time t_f. So, the corresponding results of the calculation of the lifetime are plotted in fig 1. We remark clearly that the lifetime decreases exponentially with the applied load. The same result has been found in the two cases when we applied the external load on all intact fibers of the materials (LLS rule [2] and GLS rule [3]).

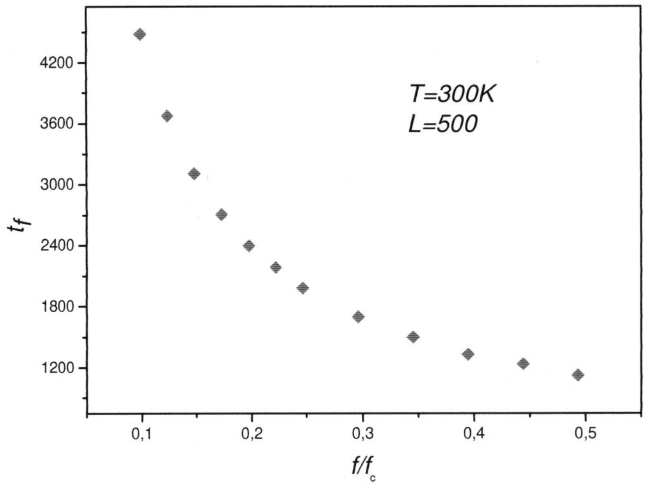

Fig 1: the lifetime versus applied load for system of seize *L=500*

To be able to investigate the crack diffusion created in the midst of the material, we have determinate the full-time evolution of the region *A* of the damaged region versus both applied load *f* and temperature *T*. The corresponding answers are plotted in fig 2. The obtained results reveal that the damaged region increases with applied load and they are more consisting with the Lifshitz-Sloyosof law:

$$A = B(T)t^x \tag{4}$$

where *x* is crack diffusion exponent and *B* is a constant which depends only on the system temperature and related to the diffusion coefficient of the created crack by:

$$D \propto B(T)^{1/x} \tag{5}$$

So the calculated value of the diffusion exponent of the created crack is $x \approx 2$.

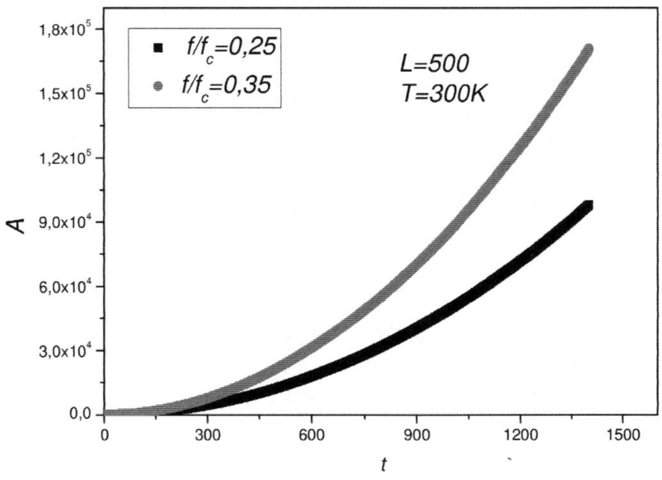

Fig 2: the time evolution of the era of the damaged region for two different values of the applied load

Utilising the formula (5), we have determined the diffusion coefficient of the created crack versus applied load. The corresponding results are plotted in fig 3. Therefore, we are finding that the diffusion coefficient increases linearly with the applied load.

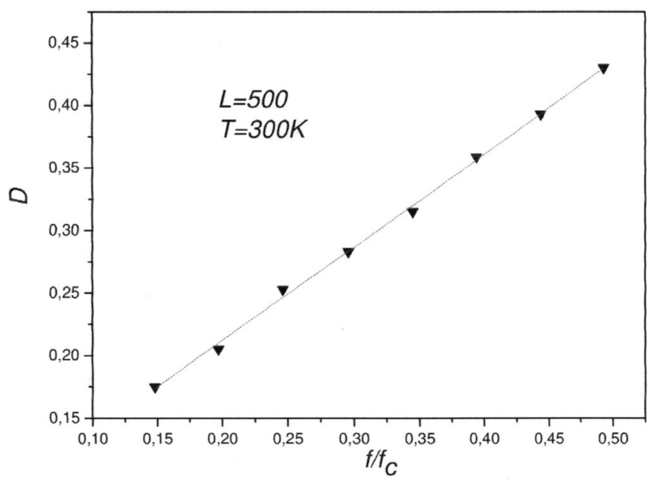

Fig 3: the diffusion coefficient versus applied load for system of seize *L=500*

4. CONCLUSION

To sum up, we have investigated the crack diffusion created at single point of the composite materials by using the fiber bundle model in the local load sharing rule. So, we have considered a fiber bundle on which we have applied an external load at a single point, and we then continue to increase this load at a very slow rate until the centered fiber breaks and its load is redistributed to its neighboring intact fibers. This breaking and redistribution dynamics repeat itself and this process ensures an advancing interfacial fracture and the area of the damaged region increases with time until a final crack of material. The equivalent effects display that place reveals a Lifshitz-Sloyosof law with an exponent growth $x=2$. The discovered law enables us to deduce the conduct of the crack diffusion coefficient. We are finding that the developed split in the midst of the material rises linearly with used fill.

References:

[1] F. Raischel, F. Kun, H.J. Herrmann, Phys. Rev. E 73, 066101(2006).
[2] A. Hader, Y. Boughaleb, I. Achik and K. Sbiaai. Opt.Mat. 36, 3-7(2013)
[3] A. Lahyani, Y. Boughaleb, M. Qjani, S. Ouaskit and R. El Guerjouma, Mediterranean Journal of electronics and communications. 6, 23-26 (2010).
[4] SO. Ramas,P.-P. Cortet, S. Cilibreto and Vanel, Phys.Rev.Lett.110, 165506(2013)
[5] I.Tarras, N.Moussa, M.Mazroui, Y.Boughaleb, A.Hajjaji Modern Physics Letters B, 2013, 27(4), 1350028
[6] D. Amitrano, Eur. Phys. J.B. Special Topics 205, 199-215(2012)
[7] S. Pradhan, A. K. Chaudra, Bihas K. Chakrabarti, Phys. Rev. E. 88, 012123 (2013).
[8] S. Pradhan, A. Hansen, and P. C. Hemmer, Phys. Rev. Lett. 95, 125501 (2005); Phys. Rev. E 74, 016122 (2006).
[9] A. Hader, I. Achik, K. Sbiaai, Y. Boughaleb, Superlattices and Microstructures S0749-6036(2014)00087-1
[10] S. Pradhan and B. K. Chakrabarti, Phys. Rev. E 65, 016113 (2001); S. Pradhan, P. Bhattacharyya, and B. K. Chakrabarti, Phys. Rev. E 66, 016116 (2002); P. Bhattacharyya, S. Pradhan, and B. K. Chakrabarti, Phys. Rev. E 67, 046122 (2003).
[11] H. A. Daniels, Proc. R. Soc. London A 183, 405 (1945).
[12] F. T. Peirce, J. Text. Ind. 17, 355 (1926).
[13] H. E. Daniels, Proc. R. Soc. London Ser. A 183, 405 (1945).
[14] R. L. Smith, Ann. Probab. 10, 137 (1982).
[15] S. L. Phoenix and R. L. Smith, Int. J. Solids Struct. 19, 479 (1983)
[16] D. Bonamy and E. Bouchaud, Phys. Rep. 498, 1 (2011).
[17] A. Delaplace, J. Schmittbuhl, and K. J. M°aløy, Phys. Rev. E 60,1337 (1999).
[18] Qiming Li, M.C.Tringides, Surf. Sci 365, 495 (1996)
[19] K. Saadouni, A. Hader, A. Jaouad, Y. Boughaleb Surf. Sci 572, 366 (2004)

Pedestrian crossing system for the mechanical energy harvesting using piezoelectric materials

Zakaria MALKI, Chouaib ENNAWAOUI, Abdelowahed HAJJAJI*, Ikram NAJIHI Mohamed EL JOUAD, Yahia BOUGHALEB

Laboratoire des Sciences de l'Ingénieur Pour l'Energie, Ecole Nationale Des Sciences Appliquées El Jadida, UCD, Route d'Azzemour, Nationale No. 1, El Haouzia, 24000 El Jadida, Maroc

*Corresponding authors: hajjaji.a@ucd.ac.ma

Abstract: In recent years, researchers have been developing solutions to reduce our energy consumption by optimizing existing systems in different ways. Indeed, it is not enough just to create or use renewable energies but we must also not waste the available energy. This and systems are developed to recover unused energy of human body movement, which will be used to power either the equipment itself or another.

This paper proposes a pressure-type generator that collects human mechanical energy by stepping, a prototype is already designed and manufactured. The average output voltage can reach to 20.9 V by step for F =100N,and finally, we did a series of tests to prove that the device can have great power output performance in the low frequency environment of the human foot movement.

1-Introduction

The global increase of the word population, and the development of various electronic devices, human life becomes more and more inseparable from electrical energy [1-4]. The world's energy demand is expected to reach 18 billion tonne oil equivalents (t.o.e.) by 2035 under current policies. (Energy harvesting from human motion: materials and techniques) which has forced the scientific research community to look for other sources of energy, especially renewable ones, namely, solar energy, wind energy, thermal energy, piezoelectric [5-8].

In this paper, we will be interessted to the energy harvested by the piezoelectric materials, because of its high electrical conversion coefficient and no external power supply being required, beside, vibration energy can be harvested from many sources arround us, such, roads, railways, buildings, and biological system movements, without any production of pollution. (Optimization design and experimental investigation of piezoelectric energy harvesting devices for pavement [3].

Recently, a new renewable energies has been under studies, kinetic energy from human body movements. This source of energy come from fact that human body offers a number of exploitable movements to generate energy : bending of legs,3 oscillation of the arms,4 breathing, … (Energy harvesting from human motion: materials and techniques), this idea of harvesting the energy of the human's body motion had been encouraged by the amount of amount of generated energy when human hell falls to the ground, which is about 67 W for a 68kg person [9-15].

Encourage by several projects in this direction [16-19]., our work comes to enrich the data by the comparison of use of many piezoelectric materials, also to try to have more performed system by the optimization in his parameters, in order to increase the harvested energy.

2- Pedestrian crossing system description

The movement of the human is considered periodic because, the left leg and the right leg alternate, and the action of "touching the ground-supporting-lifting the leg-swinging-touching the ground" is repeated regularly with a certain cycle. The resulting force of pressure on the earth is estimated between 2-3 times the weight of the human body [20, 21].

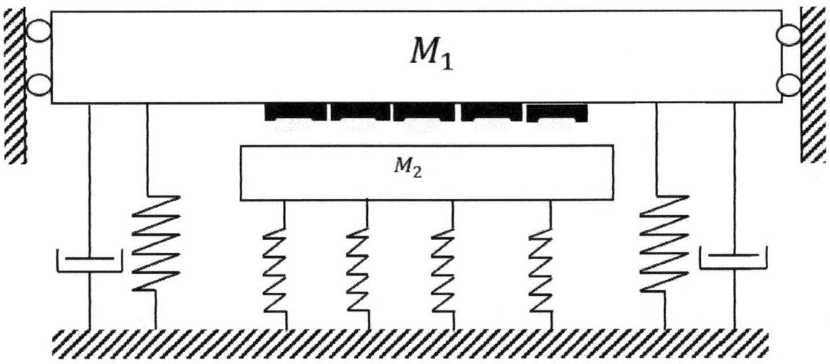

Figure 1 : Schema of the harvester system.

The harvester system (figure 1) has as objective, use of the piezoelectric effect in order to convert the mechanical energy of the human being that it is to walk, run or jump in exploitable electric energy, this conversion will be through the following elements of the system:

- Six piezoelectric patches.
- Two centered springs of length 12cm x outside diameter of 9.1cm x inside diameter 8.8cm.
- Eight substrate springs, more rigid than the other two and must not vibrate with the same dimensions but with a length of 14 cm.
- Three plates: - A plate above 35 cm wide, 55 cm long and 4 mm thick on which the materials have been glued.- One below dimension 40 cm x 65 cm x 1.5 cm on which all the springs are fixed. - A plate in the middle of dimension 22.5 cm x 32.5 cm x 2 mm attached to the centered springs.The upper plate adheres to the top of the system and is connected to 8 springs. These springs remain in their original form when the system is not deformed.

In order to design the energy harvesting system using load zirconate titanate (PZT) patches. The patches are manufactured by APC International limited, USA [12]. The PZT-5H disc was shown in figure 2.

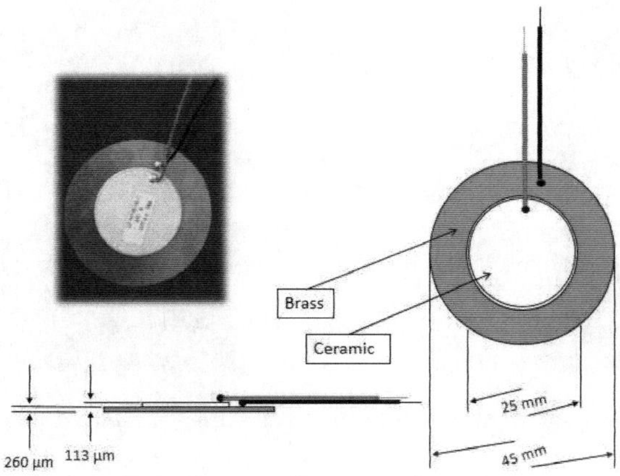

Figure 2 : PZT patches.

The material used in our work is PZT. This smart ceramic have a high piezoelectric coefficient [22-24]. In order to investigate the feasibility of PZT for energy harvesting application, the electromechanical proprieties as shown in Table 1.

Table 1. Electromechanical properties of the piezoelectric material.

Material properties	Value
Material type	PZT-5H
Piezoelectric constant d_{33} (pC/N)	630
Relative dielectric constant $\mathcal{E}$ (SI)	3300
Young's modulus (N/m²)	51.10^9
Density (10^3 Kg/m^3)	7.6
Thickness (µm)	113

Figure 3 shows the system compositions relatively inexpensive and is easily scalable in size and have a capacity to generate from milli-watts to many watts depending on system size and human flow. The prototype of mass springs system has been manufactured to verify the prediction model, design, analysis, and resulting performance .

IOP Conf. Series: Materials Science and Engineering **948** (2020) 012030 doi:10.1088/1757-899X/948/1/012030

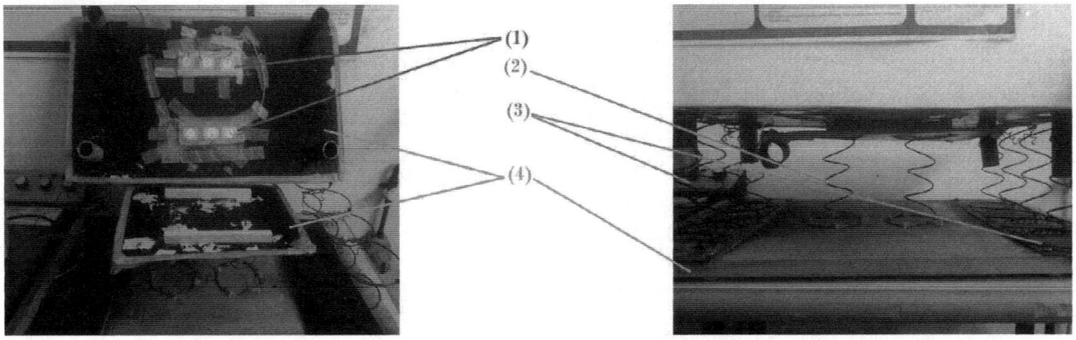

Figure 3: Complete experiment configuration for spring's system prototype.

During deformation, the moment the foot passes over the system, the one above will strike the springs compress then the plate in the middle. After the passenger crosses the system, the plate and the outer springs return to their normal state thanks to their rigidity. Since the part of the mechanical energy generated by the passing passenger is stored in the two centered springs, the springs release this stored energy in order to bring about a back and forth movement of the plate in a repetitive manner until the springs return to their original shape, and these vibrations cause beats between the two plates. Therefore, by using the springs, the vibration time is extended which helps us to harvest more energy. The spring characteristics as shown in Table 2.

Table 2. Characteristics of the springs.

Spring properties	Value
Spring stiffness K (N/m)	72
Spring high L (m)	0.3
Average diameter D (m)	0.143
Number of useful turns n (SI)	7

3- Theoretical model of power harvested by piezoelectric materials

For simplicity, in all calculations only a single piezoelectric film energy harvester is considered. Based on the potential thermodynamics and energy exchange principles, the piezoelectric phenomenon can be formulated with the following equations *(1)* [25]:

$$S = s^E . T + d_t . E \qquad (1)$$

$$D = d_t . T + \varepsilon^T . E$$

S, T, D, E, ε^T, d, and s^E are respectively strain, stress, electric displacement and electric field, the dielectric permittivity, the piezoelectric coefficient, and the compliance.

By developing these equations, thus taking into consideration the one below [26, 27]:

$$I = A . \frac{\partial D}{\partial t} \qquad (2)$$

We can deduce the theoretical expression of the current I generated by a piezoelectric material which feeds a load R is:

$$I = \text{j}\frac{d_{33}*Y*w*.A*S_3}{1+j*\left(\mathcal{E}_{33}-\frac{d^2}{s_{33}}\right)*\frac{R}{e}*w*A} \qquad (3)$$

With d_{33}, w, A and $\mathcal{E}_{33}$ are respectively piezoelectric coefficient (C/N), pulsation (rad/s), active surface (m^2) and permittivity (F/m).

By using this expression and bringin it into : $P = R.I^2$, and as the ratio $\frac{d^2}{s_{33}}$ is negligible compared to the permittivity $\mathcal{E}_{33}$, the expression of the harvested power becomes :

$$P = \frac{R}{2} * \frac{[d_{33}*Y*w*.A*S_{M3}]^2}{1+\left[\mathcal{E}_{33}*\frac{R}{e}*w*A\right]^2} \qquad (4)$$

The simulation of the equation *(4)*, will allow us to optimize at the level of several parameters, choice of springs, frequency, ... but the most important is the choice of materials to be used, this is why we present you below the curves of the simulation of this harvested power as a function of the resistance for several materials.

4-Experimental results of energy harvesting system:

The energy harvesting system contained the piezoelectric material, mechanism and electric circuits. The external circuit was integrated into the system to optimize the harvested voltage output. To obtain the optimum harvesting energy from the harvester energy system, the value of external load resistance was chosen as equal to the internal resistors of the material. In this section, the prototype was tested as shown in Figure 4. The output voltage of the energy harvester system was recorded by the oscilloscope, Gw INSTEK GDS-2074A.

Figure 4: Pedestrian crossing system for the harvester during applied force.

To perform the energy harvesting paramters, figure 5 show the evolution of the harvested electrical voltage at the output as a function of the applied force during the experience.

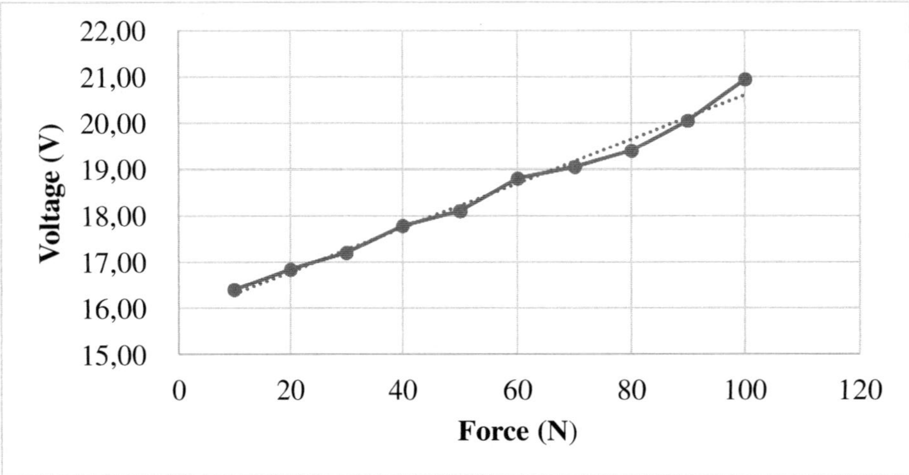

Figure 5: The harvested electrical voltage at the output as a function of the force.

Figure 5 shows the variation of the harvested voltage as a function of time for a compression force between 0 and 100 N. During excitation the maximum output voltage reaches a value of the order of 20.9 V. The experimental results show that the harvested voltage increases while increasing the force during human passing. Figure 6a and 6b show the power harvested by using PZT as a function of resistrance and frequency at 100N.

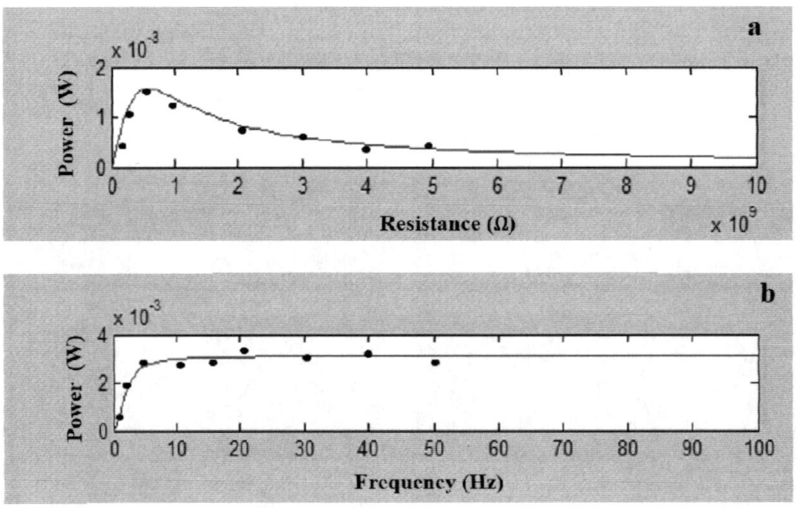

Figure 6: The power harvested by using PZT as a function of resistrance and frequency at 100N.

The theoretical model is presented for the sake of comparison with the experimental results obtained with the spring's mass prototype. The power harvested by the system increased significantly until a maximal power density that matches with an optimal load resistance, after that the power decreased. On the other hand, the harvested power increases linearly with frequency but becomes constant at higher frequencies.

The maximum power harvested is 3.08 mW achieved with F=100 N of force at 9 Hz with the optimal resistance 0.85 GΩ. Relating to the theoretical model and the experience, an increase in force and frequency, the rest of electromechanical proprieties remains constant should yield an increase in the materials voltage and therefore an increase in the power harvested.

The results obtained in this work contribute to demonstrate the ability of the Pedestrian Crossing System to improve the conversion efficiency of the energy produced during the human passage on pedestrian crossing into electrical energy for the development of micro generators.

5-Conclusion:

The aim of this paper is established a complete study of an energy harvesting system based on the use of piezoelectric materials and the Pedestrian Crossing System. The authors conclude that the combination of the two parameters (the force amplitude and the operating frequency) has a strong influence on the electrical conversion of piezoelectric ceramics for mechanical energy harvesting. The PZT material was used to harvest the mechanical energy by the piezoelectric effect. This element needed mechanical vibrations to produce energy, so we proposed a new system to harvest energy by using the Pedestrian Crossing System. According to our experimental results, we have found that the maximum of the harvested voltage could reach 20.9 V with applied force value is 100N. This article gives the experimental results that agree quite well with the theoretical predictions. This value demonstrated the excellent potential of this system for mechanical energy harvesting. It must be mentioned that this power value can be improved, for example, by using more ceramic buzzers or using another type of springs having lower spring stiffness in order to improve and generate more vibration fluctuations.

IOP Conf. Series: Materials Science and Engineering **948** (2020) 012030 doi:10.1088/1757-899X/948/1/012030

References:

[1] Deng, F., Cai, Y., Fan, X., Gui, P., & Chen, J. (2019). Pressure-type generator for harvesting mechanical energy from human gait. Energy, 171, 785-794.

[2] Invernizzi, F., Dulio, S., Patrini, M., Guizzetti, G., & Mustarelli, P. (2016). Energy harvesting from human motion: materials and techniques. Chemical Society Reviews, 45(20), 5455-5473.

[3] Wang, C., Zhao, J., Li, Q., & Li, Y. (2018). Optimization design and experimental investigation of piezoelectric energy harvesting devices for pavement. Applied energy, 229, 18-30.

[4] Hajjaji, A., Pruvost, S., Sebald, G., Lebrun, L., Guyomar, D., & Benkhouja, K. (2009). Mechanism of depolarization with temperature for< 0 0 1>(1− x) Pb (Zn1/3Nb2/3) O3–xPbTiO3 single crystals. *Acta materialia*, *57*(7), 2243-2249.

[5] Belhora, F., Cottinet, P. J., Hajjaji, A., Guyomar, D., Mazroui, M. H., Lebrun, L., & Boughaleb, Y. (2013). Mechano-electrical conversion for harvesting energy with hybridization of electrostrictive polymers and electrets. *Sensors and Actuators A: Physical*, *201*, 58-65.

[6] Ennawaoui, C., Hajjaji, A., Azim, A., & Boughaleb, Y. (2016). Theoretical modeling of power harvested by piezo-cellular polymers. *Molecular Crystals and Liquid Crystals*, *628*(1), 49-54.

[7] Wang, J., Shi, Z., Xiang, H., & Song, G. (2015). Modeling on energy harvesting from a railway system using piezoelectric transducers. *Smart Materials and Structures*, *24*(10), 105017.

[8] Meddad, M., Eddiai, A., Cherif, A., Hajjaji, A., & Boughaleb, Y. (2014). Model of piezoelectric self powered supply for wearable devices. *Superlattices and Microstructures*, *71*, 105-116.

[9] Hajjaji, A., Guyomar, D., Pruvost, S., Touhtouh, S., Yuse, K., & Boughaleb, Y. (2010). Temperature/electric field scaling in Ferroelectrics. *Physica B: Condensed Matter*, *405*(13), 2757-2761.

[10] Ennawaoui, C., Lifi, H., Hajjaji, A., Elballouti, A., Laasri, S., & Azim, A. (2019). Mathematical modeling of mass spring's system: Hybrid speed bumps application for mechanical energy harvesting. *Engineering Solid Mechanics*, *7*(1), 47-58.

[11] Belhora, F., Cottinet, P. J., Guyomar, D., Petit, L., Lebrun, L., Hajjaji, A., ... & Boughaleb, Y. (2013). Optimization of energy harvesting conversion using the hybridization of electrostrictive polymers and electrets. *Sensors and Actuators A: Physical*, *189*, 390-398.

[12] El Fatnani, F. Z., Mazroui, M. H., & Guyomar, D. (2018). Optimization of pyroelectric conversion of thermal energy through the PZT ceramic buzzer and natural convection. *The European Physical Journal Plus*, *133*(12), 519.

[13] Pirisi, A., Mussetta, M., Grimaccia, F., & Zich, R. E. (2013). Novel speed-bump design and optimization for energy harvesting from traffic. *IEEE Transactions on Intelligent Transportation Systems*, *14*(4), 1983-1991.

[14] Wang, J., Qin, L., Wei, P., & Tang, L. (2018). Modeling and analysis of multilayer piezoelectric-elastic spherical transducers. *Journal of Intelligent Material Systems and Structures*, *29*(11), 2437-2455.

[15] Zhang, Y., Bowen, C. R., Ghosh, S. K., Mandal, D., Khanbareh, H., Arafa, M., & Wan, C. (2019). Ferroelectret materials and devices for energy harvesting applications. *Nano Energy*, *57*, 118-140.

[16] Ennawaoui, C., Lifi, H., Hajjaji, A., Samuel, C., Rguiti, M., Touhtouh, S., AZIM, A. & Courtois, C. (2019). Dielectric and mechanical optimization properties of porous poly (ethylene□co□vinyl acetate) copolymer films for pseudo□piezoelectric effect. *Polymer Engineering & Science*, *59*(7), 1455-1461.

[17] Delnavaz, A., & Voix, J. (2014). Flexible piezoelectric energy harvesting from jaw movements. *Smart Materials and Structures*, *23*(10), 105020.

[18] Hajjaji, A., Benayad, A., Sebald, G., Pruvost, S., Guiffard, B., Qiu, J., & Guyomar, D. (2008). Synthesis and characterization of 0.65 Pb (Mg1/3Nb2/3) O3–0.35 PbTiO3 fibers with Pt core. *Materials Research Bulletin*, *43*(3), 493-501.

[19] Hajjaji, A., Pruvost, S., Sebald, G., Lebrun, L., Guyomar, D., & Benkhouja, K. (2008). Composition dependence of 90° domain switching in Pb (Mg1/3Nb2/3) 1– xTixO3 system. *Solid state sciences*, *10*(8), 1020-1027.

[20] Eddiai, A., Meddad, M., Sbiaai, K., Boughaleb, Y., Hajjaji, A., & Guyomar, D. (2013). A new technique for maximizing the energy harvested using electrostrictive polymer composite. *Optical Materials*, *36*(1), 13-17.

[21] Duc, C., Stoclet, G., Soulestin, J., & Samuel, C. (2020). Poly (ethylene oxide)/PEDOT: PSS Blends: An Efficient Route to Highly Conductive Thermoplastic Materials For Melt-State Extrusion Processing?. *ACS Applied Polymer Materials.*

[22] Lifi, H., Ennawaoui, C., Hajjaji, A., Touhtouh, S., Laasri, S., Yessari, M., & Benjelloun, M. (2019). Sensors and energy harvesters based on (1–x) PMN-xPT piezoelectric ceramics. *The European Physical Journal Applied Physics*, *88*(1), 10901.

[23] Rjafallah, A., Hajjaji, A., Guyomar, D., Kandoussi, K., Belhora, F., & Boughaleb, Y. (2019). Modeling of polyurethane/lead zirconate titanate composites for vibration energy harvesting. *Journal of Composite Materials*, *53*(5), 613-623.

[24] Lifi, H., Ennawaoui, C., Hajjaji, A., Touhtouh, S., Benjelloun, M., & Azim, A. (2017). Elaboration, characterization and thermal shock sensor application of pyroelectric ceramics PMN–xPT. *Sensor Letters*, *15*(9), 751-757.

[25] El Fatnani, F. Z., Guyomar, D., Mazroui, M. H., Belhora, F., & Boughaleb, Y. (2016). Optimization and improvement of thermal energy harvesting by using pyroelectric materials. *Optical Materials*, *56*, 22-26.

[26] Meddad, M., Eddiai, A., Guyomar, D., Belkhiat, S., Cherif, A., Yuse, K., & Hajjaji, A. (2012). An adaptive prototype design to maximize power harvesting using electrostrictive polymers. *Journal of Applied Physics*, *112*(5), 054109.

[27] Yang, L., Chi, S., Dong, S., Yuan, F., Wang, Z., Lei, J., & Wang, J. (2020). Preparation and characterization of a novel piezoelectric nanogenerator based on soluble and meltable copolyimide for harvesting mechanical energy. Nano Energy, 67, 104220.

IOP Conf. Series: Materials Science and Engineering **948** (2020) 012031 doi:10.1088/1757-899X/948/1/012031

Prediction of Mechanical Performance of Natural Fibers Polypropylene Composites: a Comparison Study

M. Rouway[1,5,6] , M. Nachtane [2], M. Tarfaoui [3,4], N. Chakhchaoui [5,6], LH. Omari [1], F. Fraija [1], O. Cherkaoui [5]

[1] Hassan II University, Faculty of Sciences Aïn Chock, LERDYS and LPMMAT Laboratories, BP 5366 Maarif, 20100 Casablanca, Morocco
[2] University of Bordeaux, I2M Laboratory, UMR CNRS 5295, 33405, Talence Cedex, France
[3] ENSTA Bretagne, IRDL, UMR, CNRS 6027, F-29200, Brest, France
[4] University of Dayton, Dayton, OH 45469-0168, United States
[5] REMTEX Laboratory, Higher School of Textile and Clothing Industries (ESITH), Casablanca, Morocco
[6] BGIM Laboratory, Higher Normal School (ENS), Hassan II University, Casablanca, Morocco.
[6] Author to whom any correspondence should be addressed
marwane.rouway-etu@etu.univh2c.ma

Abstract. The future trends of wind turbine blade materials are mostly centered around utilizing lighter materials with improved life cycle and low cost. In modern wind turbine blade, using green composite or biocomposite is a sustainable solution, especially for small wind turbine blade with natural fiber and biodegradable polymer. In this work, four natural fibers (Alfa, Flax, Sisal and Hemp) are used with Polypropylene (PP) thermoplastic polymer. The objective is to evaluate the elastic muduli of composite by two methods; Analytical Mori Tanaka theory and numerical with Digimat MF. The results show a good validation between the two methods. The comparison of the mechanical behaviour of the transversely isotropic of natural fibers indicate that the hemp have a good performance with high Young and shear modulus regarding to their low density, which can offer an excellent candidate for the manufacturing of small wind turbine blades in rural and isolated areas.
Keywords: Biocomposite, Small wind turbine, Natural fiber, Mori-Tanaka

1. Introduction

The biocomposites are fast growing attention in industries because of their excellent advantages such as cheap, recyclable, biodegradable and renewable which can be used in manufacturing of wind or tidal turbine blades [1–3]. Examples for the natural fibers are sisal, flax, hemp, jute and cellulose fiber. The advantages of natural fibers are: low costs, available and environmental friendly. However, their disadvantages are the quality variations, high moisture uptake and low thermal stability of the raw fibers [4]. An interesting option for developing countries is small turbines, producible on-site, and made from bioresources [5]. Most blades are simply cut or shredded for incineration before being permanently buried in landfills, which poses environmental problems since the blades are made from glass and carbon fibres that are not biodegradable [6].

Nevertheless, research interests on natural fiber reinforced composites happened much later in time than high-strength synthetic fibers. Relevant analytical solutions for the prediction of the mechanical properties of natural fiber composites are relatively limited in the reported studies. Several papers therefore present more specialized analytical expressions based on the Mori-Tanaka method for the elastic stiffness of nanocomposites. These expressions are then established for specific composites, containing inclusions with isotropic or anisotropic material properties, with a specific geometric shape, and with either aligned or random orientations. Qiu and Weng [7] have presented a general model for spherical inclusions with anisotropic material properties, also for random oriented needles (or circular fibres), thin discs and spherical inclusions with anisotropic material properties. Tandon and Weng [8,9] have presented a model for the elastic moduli of a composite with unidirectionally aligned isotropic and 3D random oriented inclusions applicable for spheroidal, fibre-like and thin discs with an aspect ratio α.

Several homogenization techniques are used in profit of composite in renewable energy production [10–12], and in order to evaluate the mechanical properties of the fiber-reinforced composite. The required quantities are the isotropic mechanical properties of the matrix (E_m, ν_m, ρ_m) and reinforcing fibers (E_f, ν_f, ρ_f) which can be isotropic or transversely isotropic. Chamis approach [13] defines the elastic properties of a unidirectional lamina made of anisotropic fibers in an isotropic matrix. Hahn approach [14] allows to obtain the transversely isotropic mechanical properties of a fibrous composite made by fibers with circular cross-section randomly distributed in a normal plane to the direction of the oriented fibers. Hashin–Rosen approach [15] represents a simple approach to compute the mechanical properties of a unidirectional fiber-reinforced composite with an isotropic properties of the fiber and matrix. Halpin–Tsai approach [16,17] evaluates the mechanical properties of the three-phase multiscale composite by using the Hill's elastic moduli and a semi-empirical approach.

2. Mathematical model

The homogenization concept used in this work is the Mori Tanaka (MT) approach, in order to evaluate the effective transversely isotropic properties of two-phase polymer and fiber composite. There are four natural fibers used in this numerical simulation by Digimat MF (version 2019). Also, the obtained numerical results by Digimat are completely agreed with the analytical model based on Eshelby tensor theory as described below.

Representative volume element (RVE) was constructed maintaining a fiber volume fraction rang (10% - 50%) embedded in polymeric matrix. In the current model, the natural fibers and the polypropylene were treated as isotropic material. The properties of the natural fibers and polypropylene used as inputs are summarized in Tables 1 and 2.

Table 1. Properties of natural fibers [18–21]

Natural fibers	**Alfa**	**Flax**	**Sisal**	**Hemp**
Young modulus, E_f (GPa)	19.4	27.6	22	70
Poisson ratio, ν_f	0.34	0.45	0.32	0.4
Density, ρ_f (g/cm^3)	1.52	1.38	1.2	1.35

Table 2. Properties of polypropylene as matrix resin [22,23]

Properties	**Value**
Elastic modulus, E_m (GPa)	1.6
Poisson's ratio, ν_m	0.4
Density, ρ_m (g/cm^3)	0.92

The Eshelby tensor theory is based on considering a homogenous linear elastic solid with volume v and surface area S, with an inclusion volume v_0 and a surface area S_0, as shown in Figure 1. The volume v outside the inclusion is called the matrix. The Eshelby tensor S_{ijkl} expresses the constrained strain inside the inclusion $e^c{}_{ij}$ to its eigenstrains $e^*{}_{kl}$:

$$e^c{}_{ij} = S_{ijkl}\, e^*{}_{kl} \tag{1}$$

Since this tensor relates two strain tensors, the Eshelby tensor satisfies the minor symmetry condition, i.e. $S_{ijkl} = S_{jikl} = S_{ijlk}$, and does not satisfy the major symmetry condition, i.e. $S_{ijkl} \neq S_{klij}$ [24].

Moreover, for fibre-like spheroidal inclusions the Eshelby tensor can be expressed as follows [8],

$$S_{1111} = \frac{1}{2(1-v_0)}\left\{1 - 2v_0 + \frac{3\alpha^2 - 1}{\alpha^2 - 1} - \left[1 - 2v_0 + \frac{3\alpha^2}{\alpha^2 - 1}\right]g\right\} \tag{2}$$

$$S_{2222} = S_{3333} = \frac{3}{8(1-v_0)}\frac{\alpha^2}{\alpha^2 - 1} + \frac{1}{4(1-v_0)}\left[1 - 2v_0 + \frac{9}{4(\alpha^2 - 1)}\right]g$$

$$S_{2233} = S_{3322} = \frac{1}{4(1-v_0)}\left\{\frac{\alpha^2}{2(\alpha^2 - 1)} - \left[1 - 2v_0 + \frac{3}{4(\alpha^2 - 1)}\right]g\right\}$$

$$S_{2211} = S_{3311} = \frac{1}{2(1-v_0)}\frac{\alpha^2}{\alpha^2 - 1} + \frac{1}{4(1-v_0)}\left\{\frac{3\alpha^2}{\alpha^2 - 1} - (1 - 2v_0)\right\}g$$

$$S_{1122} = S_{1133} = \frac{1}{2(1-v_0)}\left[1 - 2v_0 + \frac{1}{\alpha^2 - 1}\right] + \frac{1}{2(1-v_0)}\left[1 - 2v_0 + \frac{3}{2(\alpha^2 - 1)}\right]g$$

$$S_{2323} = S_{3232} = \frac{1}{4(1-v_0)}\left\{\frac{\alpha^2}{2(\alpha^2 - 1)} + \left[1 - 2v_0 - \frac{3}{4(\alpha^2 - 1)}\right]g\right\}$$

$$S_{1212} = S_{1313} = \frac{1}{4(1-v_0)}\left\{1 - 2v_0 - \frac{\alpha^2 + 1}{\alpha^2 - 1} - \frac{1}{2}\left[1 - 2v_0 - \frac{3(\alpha^2 + 1)}{4(\alpha^2 - 1)}\right]g\right\}$$

Where

$$g = \frac{\alpha}{(\alpha^2 - 1)^{3/2}}\left\{\alpha(\alpha^2 - 1)^{1/2} - \cosh^{-1}\alpha\right\} \tag{3}$$

And $\alpha = l/d$ is the aspect ratio of the fibre length l and the fibre diameter d. Note that the aspect ratio is applied for indicating the size of the inclusion in this case. The aspect ratio is explicitly included in several models for fibre-like inclusions in a matrix.

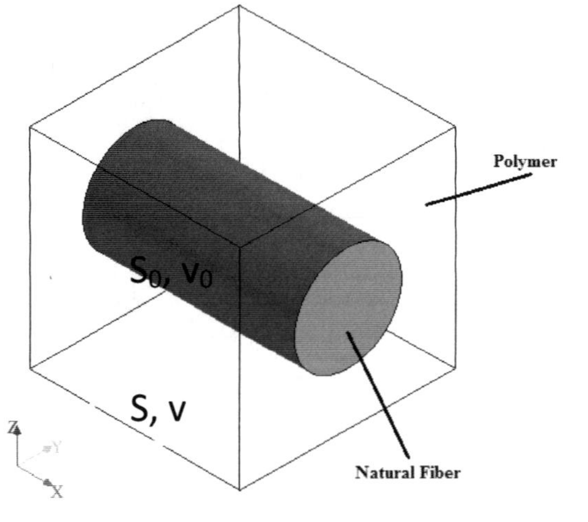

Figure 1. Representative volume element RVE of natural fiber reinforced polymer

2.1 Tensor notation

The original paper by Mori and Tanaka, describing their method, is from 1973 [25]. However, in the derivation of the Mori-Tanaka method presented in this work, we follow the derivation by Fisher and Brinson [26]. In the Mori-Tanaka method, it is assumed that the composite is comprised of N phases. Phase 0 is the matrix, and the remaining N-1 phases are inclusion phases. The matrix phase has stiffness C_0 and a volume fraction V_0, whereas the *rth* inclusion phase has stiffness C_r and a volume fraction V_r. The quantities C_0 and C_r are generally fourth-order elasticity tensors, with certain symmetry properties. The elasticity tensors satisfy the minor symmetry condition, i.e $C_{ijkl} = C_{jikl} = C_{ijlk}$. In addition, the elasticity tensor will satisfy the major symmetry condition, i.e. $C_{ijkl} = C_{klij}$.

Figure 2 shows a multi-phase composite with inclusions, as well as a comparison material. The average stress for the comparison material is given by Hooke's law,

$$\sigma_0 = C_0\, \varepsilon_a \tag{4}$$

Whereas for the composite with inclusions, the average stress is given as

$$\bar{\sigma} = C\, \varepsilon_a \tag{5}$$

Due to the inclusion, the average strain of the matrix of the composite will be perturbed, reading

$$\bar{\varepsilon}_0 = \varepsilon_a + \bar{\varepsilon}_0{}^{pt} \tag{6}$$

Where the over-score represent the volume average of the quantity, and $\bar{\varepsilon}_0{}^{pt}$ is the perturbation strain.

The average strain of the *rth* inclusion is perturbed by the amount $\varepsilon_r{}^{pt}$,

$$\bar{\varepsilon}_r = \bar{\varepsilon}_0 + \varepsilon_r{}^{pt} = \varepsilon_a + \bar{\varepsilon}_0{}^{pt} + \varepsilon_r{}^{pt} \tag{7}$$

Given that the stress in the *rth* inclusion can be given as $\bar{\sigma}_r = C_r\, \bar{\varepsilon}_r$, and using the equivalent method, the stress can be expressed in terms of the matrix stiffness,

$$\bar{\sigma}_r = C_r\, \bar{\varepsilon}_r = C_0\, (\bar{\varepsilon}_r - \varepsilon_r{}^{*}) \tag{8}$$

The perturbed strain and the eigenstrain for a single ellipsoidal inclusion, can be related using the Eshelby tensor, reading

$$\varepsilon_r{}^{pt} = S_r\, \varepsilon_r{}^{*} \tag{9}$$

Using the above expressions, one finds that

$$\bar{\varepsilon}_r = \bar{\varepsilon}_0 + \varepsilon_r{}^{pt} = \bar{\varepsilon}_0 + S_r\, \varepsilon_r{}^{*} \tag{10}$$

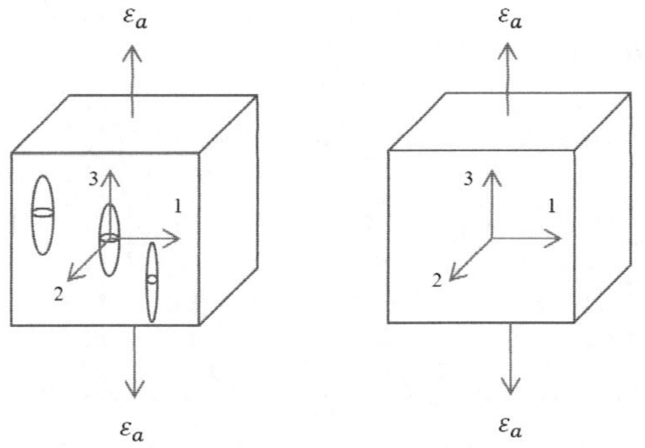

Figure 2. Schematic of Mori-Tanaka method. a) Multiphase composite material. b) Comparaison material [26]

Now, solving for ε_r*in (8)

$$\varepsilon_r{}^* = C_0{}^{-1}\,(C_0 - C_r)\bar{\varepsilon_r} \tag{11}$$

And inserting into (10),

$$\bar{\varepsilon_r} = \bar{\varepsilon_0} + S_r\,C_0{}^{-1}(C_0 - C_r)\bar{\varepsilon_r} \;\rightarrow\; \bar{\varepsilon_r} = A_r{}^{dil}\bar{\varepsilon_0} \tag{12}$$

Where

$$A_r{}^{dil} = \left[I + S_r\,C_0{}^{-1}(C_r - C_0)\right]^{-1} \tag{13}$$

Hence, the quantity $A_r{}^{dil}$ for the *rth* inclusion contains the Eshelby tensor, which depends on the shape of the inclusion.

Furthermore, it is required that the volume-weighted average phase strain must equal the far-field applied strain. From this, a strain-concentration factor can be established, that accounts for the inclusion interaction by relating the average matrix strain in the composite to the uniform applied strain. The factor reads

$$A_0 = \left[V_0 I + \sum_{r=1}^{N-1} V_r A_r{}^{dil}\right]^{-1} \tag{14}$$

In the *rth* inclusion, the strain concentration factor in the non-dilute composites can be written as

$$A_r = A_r{}^{dil} A_0 \tag{15}$$

An effective stiffness for the composite for a unidirectionally aligned composite, can then be defined as

$$C_{C,aligned} = V_0\,C_0 A_0 + \sum_{r=1}^{N-1} V_r C_r A_r = \left(V_0 C_0 + \sum_{r=1}^{N-1} V_r C_r A_r{}^{dil}\right)\left(V_0 I + \sum_{r=1}^{N-1} V_r A_r{}^{dil}\right)^{-1} \tag{16}$$

2.2 *Matrix notation*

Since the previous expressions include fourth-order tensors, it will be advantageous to reduce the size of the involved quantities for implementation and calculations. Considering the involved quantities, the stress and strain second-order tensors are symmetric. Moreover, the fourth-order tensors have at least minor symmetry properties (e.g. the Eshelby tensor), or both minor and major symmetry properties (e.g. the elasticity tensors for the matrix and the inclusions). Applying the symmetry properties of the strain tensors and the Eshelby tensor, we find that

$$\begin{pmatrix} \varepsilon^c_{11} \\ \varepsilon^c_{22} \\ \varepsilon^c_{33} \\ \varepsilon^c_{12} \\ \varepsilon^c_{23} \\ \varepsilon^c_{31} \end{pmatrix} = \begin{pmatrix} S_{1111} & S_{1122} & S_{1133} & 2S_{1112} & 2S_{1123} & 2S_{1131} \\ S_{2211} & S_{2222} & S_{2233} & 2S_{2212} & 2S_{2223} & 2S_{2231} \\ S_{3311} & S_{3322} & S_{3333} & 2S_{3312} & 2S_{3323} & 2S_{3331} \\ S_{1211} & S_{1222} & S_{1233} & 2S_{1212} & 2S_{1223} & 2S_{1231} \\ S_{2311} & S_{2322} & S_{2333} & 2S_{2312} & 2S_{2323} & 2S_{2331} \\ S_{3111} & S_{3122} & S_{3133} & 2S_{3112} & 2S_{3123} & 2S_{3131} \end{pmatrix} \begin{pmatrix} \varepsilon^*_{11} \\ \varepsilon^*_{22} \\ \varepsilon^*_{33} \\ \varepsilon^*_{12} \\ \varepsilon^*_{23} \\ \varepsilon^*_{31} \end{pmatrix} \tag{17}$$

Using engineering shear strains in the above relation, we obtain the expression

$$\begin{pmatrix} \varepsilon_{11}^c \\ \varepsilon_{22}^c \\ \varepsilon_{33}^c \\ \gamma_{12}^c \\ \gamma_{23}^c \\ \gamma_{31}^c \end{pmatrix} = \begin{pmatrix} S_{1111} & S_{1122} & S_{1133} & S_{1112} & S_{1123} & S_{1131} \\ S_{2211} & S_{2222} & S_{2233} & S_{2212} & S_{2223} & S_{2231} \\ S_{3311} & S_{3322} & S_{3333} & S_{3312} & S_{3323} & S_{3331} \\ 2S_{1211} & 2S_{1222} & 2S_{1233} & 2S_{1212} & 2S_{1223} & 2S_{1231} \\ 2S_{2311} & 2S_{2322} & 2S_{2333} & 2S_{2312} & 2S_{2323} & 2S_{2331} \\ 2S_{3111} & 2S_{3122} & 2S_{3133} & 2S_{3112} & 2S_{3123} & 2S_{3131} \end{pmatrix} \begin{pmatrix} \varepsilon_{11}^* \\ \varepsilon_{22}^* \\ \varepsilon_{33}^* \\ \gamma_{12}^* \\ \gamma_{23}^* \\ \gamma_{31}^* \end{pmatrix} \tag{18}$$

The coefficients of the Eshelby tensor in the two latter expressions are dependent on the geometry of the inclusion. In a similar way, the general Hooke's law for linear elastic solids can be expressed [2],

$$\begin{pmatrix} \sigma_{11} \\ \sigma_{22} \\ \sigma_{33} \\ \sigma_{12} \\ \sigma_{23} \\ \sigma_{31} \end{pmatrix} = \begin{pmatrix} E_{1111} & E_{1122} & E_{1133} & 0 & 0 & 0 \\ E_{1122} & E_{2222} & E_{2233} & 0 & 0 & 0 \\ E_{1133} & E_{2233} & E_{3333} & 0 & 0 & 0 \\ 0 & 0 & 0 & E_{1212} & 0 & 0 \\ 0 & 0 & 0 & 0 & E_{2323} & 0 \\ 0 & 0 & 0 & 0 & 0 & E_{3131} \end{pmatrix} \begin{pmatrix} \varepsilon_{11} \\ \varepsilon_{22} \\ \varepsilon_{33} \\ \gamma_{12} \\ \gamma_{23} \\ \gamma_{31} \end{pmatrix} \tag{19}$$

The five transversely isotropic constants of unidirectional natural/polypropylene ply can be predicted analytically through the Mori-Tanaka theory as:

$$E_{11} = \frac{1}{S_{1111}} \quad , \quad E_{22} = E_{33} = \frac{1}{S_{2222}} \tag{20}$$

$$\nu_{12} = \nu_{13} = -E_{11}S_{2211} \quad , \quad \nu_{23} = -E_{22}S_{3322} \tag{21}$$

$$G_{12} = G_{13} = \frac{1}{S_{3131}} \quad , \quad G_{23} = \frac{1}{S_{1212}} \tag{22}$$

Where E_{11} is the axial Young modulus, E_{22} is the in-plane Young modulus, ν_{12} is the transverse Poisson ratio, ν_{23} is the in-plane Poisson ratio, G_{12} and G_{13} are the transverse shear modulus, G_{23} is the in-plane shear modulus.

Then these expressions are implemented in Matlab program.

3. Results and discussions

From the isotropic properties of natural fibers (Alfa, Flax, Sisal and Hemp) and polypropylene resin, the analytical model of Mori Tanaka calculate the transversely isotropic with six elastic constants; E_{11} the axial Young modulus, $E_{22} = E_{33}$ the in-plane Young modulus, ν_{23} the in-plane Poisson ratio, $\nu_{12} = \nu_{13}$ the transverse Poisson ratio, G_{23} the in-plane shear modulus, $G_{12} = G_{13}$ the transverse shear modulus, and finally the density ρ. The natural fiber is considered as aligned fibre-like inclusions, the analytical results are compared with the mean field homogenization MFH of Digimat 2019, the comparison shows very well agreement as seen in Table 3.

Table 3. Comparison between the elastic moduli obtained by Matlab code and Digimat MF

	Alfa fiber					
	V_f	**0.1**	**0.2**	**0.3**	**0.4**	**0.5**
E_{11} (GPa)	Digimat MF	3.3807	5.1612	6.9416	8.7218	10.502
	Matlab code	3.3804	5.1607	6.9409	8.7210	10.501
E_{22}=E_{33} (GPa)	Digimat MF	1.9986	2.3791	2.8172	3.3546	4.0433
	Matlab code	1.9986	2.3791	2.8172	3.3545	4.0433

ν_{23}	Digimat MF	0.50271	0.52834	0.53370	0.53033	0.52172
	Matlab code	0.50270	0.52833	0.53369	0.53032	0.52171
$\nu_{12}=\nu_{13}$	Digimat MF	0.3931	0.3864	0.37996	0.37370	0.36764
	Matlab code	0.3931	0.38642	0.37996	0.37371	0.36765
G_{23} (GPa)	Digimat MF	0.665	0.77833	0.91842	1.096	1.3285
	Matlab code	0.665	0.77833	0.91842	1.096	1.3285
$G_{12}=G_{13}$ (GPa)	Digimat MF	0.6781	0.80673	0.96488	1.164	1.4225
	Matlab code	0.6781	0.80673	0.96488	1.164	1.4225
ρ (g/cm^3)	Digimat MF	0.98	1.04	1.1	1.16	1.22
	Matlab code	0.98	1.04	1.1	1.16	1.22

Figure 2 shows the composite elastic constants for the four material systems considered. As observed, the longitudinal and transversal Young modulus (E_{11}, E_{22}) of composite increase linearly with the fibre volume fraction (10% - 50%) with an order of hemp, flax, sisal and then Alfa. The curves of flax, sisal and Alfa are closely, but for hemp fiber which have high value of E_{11}. The same remarks have been found for longitudinal and transversal shear modulus (G_{12}, G_{23}). In figure 2-e) the transverse Poisson ratio ν_{12} increase for flax fiber, and decrease for Alfa and Sisal fibers, but the variation remains constant for Hemp fiber. This is due to the Poisson ratio 0.45, 0.4, 0.34 and 0.32 for Flax, Hemp, Alfa and Sisal fibers respectively. In figure 2-f) the transversal Poisson ratio increase with volume fraction from 10% to 30%, and then decrease slowly, and have a maximum value at 30%. In figure 2-g) shows the linear increase of density of natural fibers with volume fraction. As demonstrated, the lightest natural fiber is the sisal, and the heaviest natural fiber is the Alfa. Flax and hemp fibers have close values between sisal and Alfa fibers.

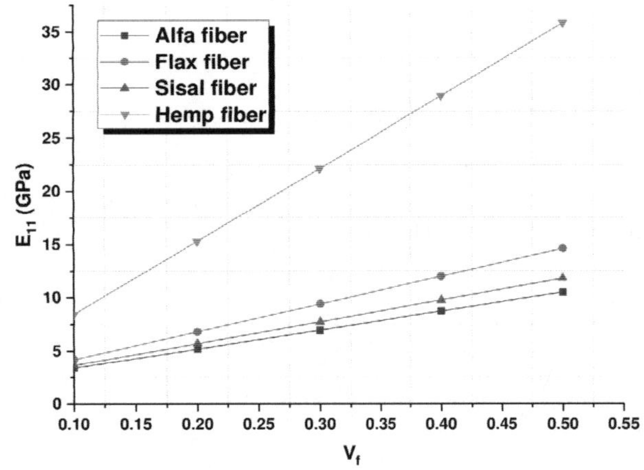

a)

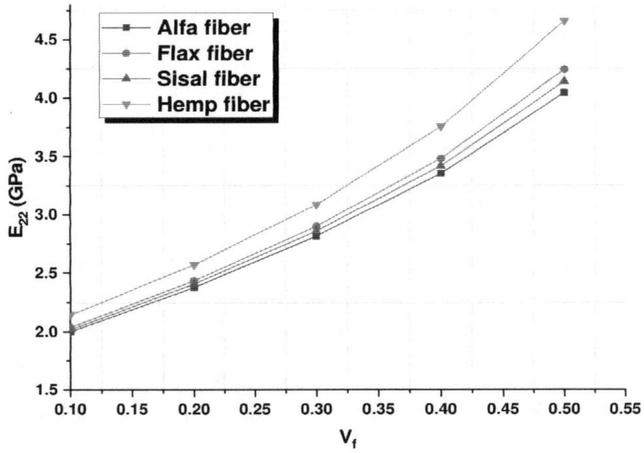

b)

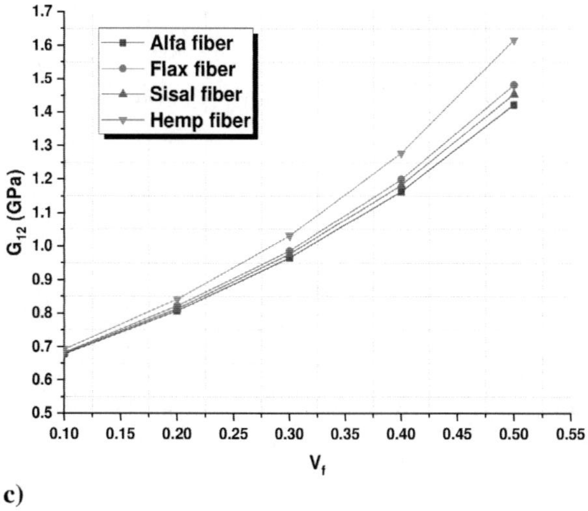

c)

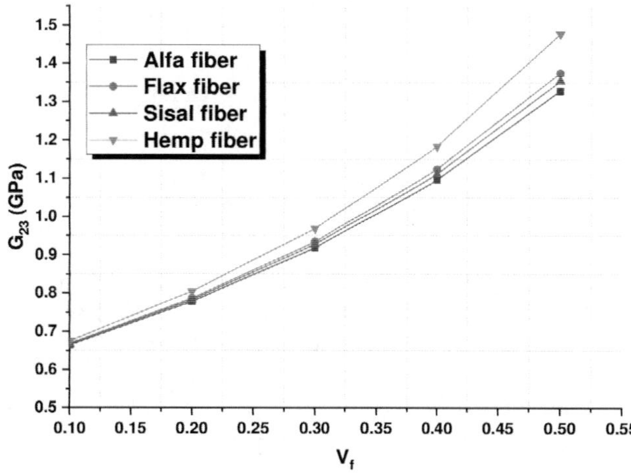

d)

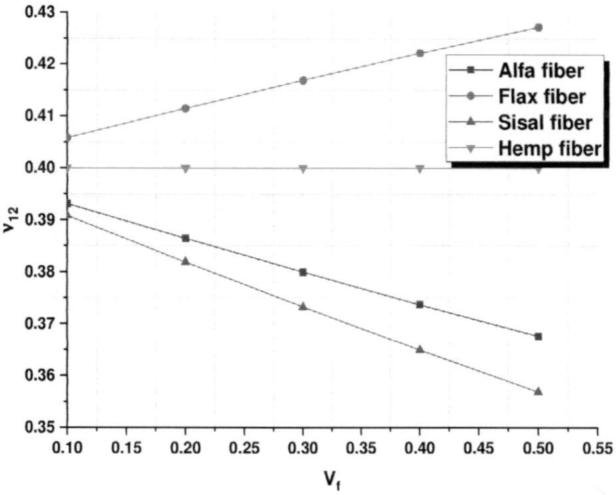

e)

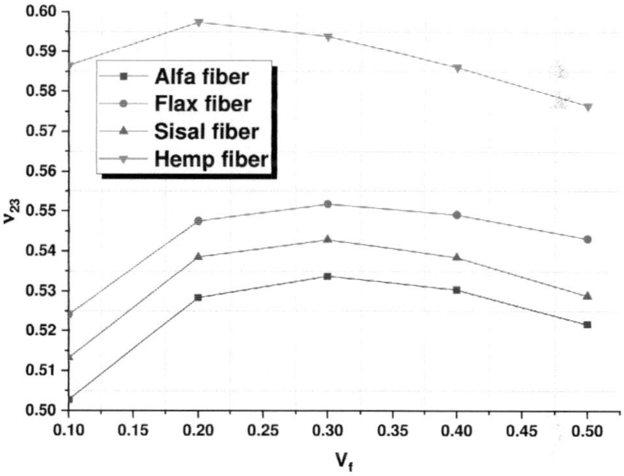

f)

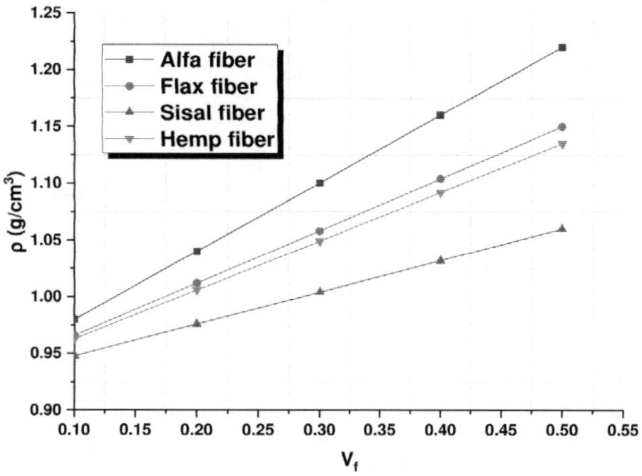

g)

Figure 2. Transversely isotropic constants of natural fibers (Alfa, flax, sisal and hemp) in function of fiber volume fraction V_f. a) Axial Young modulus $E_{11}(GPa)$, b) In-plane Young modulus $E_{22}(GPa)$,

IOP Conf. Series: Materials Science and Engineering **948** (2020) 012031 doi:10.1088/1757-899X/948/1/012031

c) Transverse shear modulus $G_{12}(GPa)$, d) In-plane shear modulus $G_{23}(GPa)$, e) Transverse Poisson ratio v_{12}, f) In-plane Poisson ratio v_{23}, g) density $\rho\ (g/cm^3)$.

4. Conclusion

In conclusion, continuous efforts are needed for the development of numerical and analytical models suitable for the prediction of various behaviors of natural fiber composites before their practical applications. From Mori Tanaka and Digimat, we found that the natural fiber of hemp has good mechanical properties according to their light weight in comparison to other fibers, which can be used for fabrication of small wind turbine blades in rural and isolated zone; but the mechanical behavior is not the only important parameter in manufacturing process. The availability and ease of fiber processing plays a fundamental role, which Alfa and Sisal natural fibers offers, especially in Morocco.

References

[1] Nachtane M, Tarfaoui M, Goda I, Rouway M. A review on the technologies, design considerations and numerical models of tidal current turbines. Renew Energy 2020;157:1274–88. https://doi.org/10.1016/j.renene.2020.04.155.

[2] Nachtane M, Tarfaoui M, Ait Mohammed M, Saifaoui D, El Moumen A. Effects of environmental exposure on the mechanical properties of composite tidal current turbine. Renew Energy 2020;156:1132–45. https://doi.org/10.1016/j.renene.2020.04.126.

[3] Boudounit H, Tarfaoui M, Saifaoui D, Nachtane M. Structural analysis of offshore wind turbine blades using finite element method: Wind Eng 2019. https://doi.org/10.1177/0309524X19849830.

[4] Kalagi G, Patil R, Nayak N. Natural fiber reinforced polymer composite materials for wind turbine blade applications. Int J Sci Dev Res 2016;1:28–37.

[5] Piggott H. A Wind Turbine Recipe Book: the Axial Flux Windmill Plans. Scoraig wind; 2009.

[6] Rinnovabiliti. ARGO Biodegradable wind turbines, wind energy research evolves // Energies Division // Environment Communications | Energies News | Communications Reports // Environment Energies Communications Technical Society of Services and Advanced Consulting n.d. https://www.argoit.com/en/sezione_id,3/newssez_id,115/biodegradable-wind-turbines-wind-energy-research-evolves/communications.html (accessed January 30, 2020).

[7] Qiu YP, Weng GJ. On the application of Mori-Tanaka's theory involving transversely isotropic spheroidal inclusions. Int J Eng Sci 1990;28:1121–1137.

[8] Tandon GP, Weng GJ. The effect of aspect ratio of inclusions on the elastic properties of unidirectionally aligned composites. Polym Compos 1984;5:327–333.

[9] Tandon GP, Weng GJ. Average stress in the matrix and effective moduli of randomly oriented composites. Compos Sci Technol 1986;27:111–132.

[10] El Moumen A, Tarfaoui M, Lafdi K. Computational Homogenization of Mechanical Properties for Laminate Composites Reinforced with Thin Film Made of Carbon Nanotubes. Appl Compos Mater 2018;25:569–88. https://doi.org/10.1007/s10443-017-9636-2.

[11] Tarfaoui M, Lafdi K, El Moumen A. Mechanical properties of carbon nanotubes based polymer composites. Compos Part B Eng 2016;103:113–21. https://doi.org/10.1016/j.compositesb.2016.08.016.

[12] Rouway M, Boulahia Z, Chakhchaoui N, Cherkaoui O, Omari LH, Fraija F. Graphene and Carbone nanotubes reinforced ceramic nanocomposite TiO2-MgO: Experimental and numerical study. Mater Today Proc 2020. https://doi.org/10.1016/j.matpr.2020.04.177.

[13] Chamis CC. Thermoelastic properties of unidirectional filamentary composites by a semiempirical micromechanics theory. Union Carbide Corp Cleveland Ohio Carbon Products Div; 1974.

[14] Hahn HT. Simplified formulas for elastic moduli of unidirectional continuous fiber composites. J Compos Technol Res 1980;2:5–7.

[15] Bacciocchi M, Luciano R, Majorana C, Tarantino AM. Free vibrations of sandwich plates with damaged soft-core and non-uniform mechanical properties: modeling and finite element analysis. Materials 2019;12:2444.

[16] Halpin JC. Effects of environmental factors on composite materials. Air Force Materials Lab Wright-Patterson AFB OH; 1969.

[17] Tsai SW. Structural Behavior of Composite Materials. Philco Corp Newport Beach Ca Space and Re-Entry Systems; 1964.

[18] Pickering KL, Efendy MGA, Le TM. A review of recent developments in natural fibre composites and their mechanical performance. Compos Part Appl Sci Manuf 2016;83:98–112. https://doi.org/10.1016/j.compositesa.2015.08.038.

[19] Suardana NPG, Piao Y, Lim JK. Mechanical properties of hemp fibers and hemp/pp composites: effects of chemical surface treatment. Mater Phys Mech 2011;11:1–8.

[20] Joseph PV, Joseph K, Thomas S. Effect of processing variables on the mechanical properties of sisal-fiber-reinforced polypropylene composites. Compos Sci Technol 1999;59:1625–40. https://doi.org/10.1016/S0266-3538(99)00024-X.

[21] Zhong Y, Kureemun U, Lee HP. Prediction of the mechanical behavior of flax polypropylene composites based on multi-scale finite element analysis. J Mater Sci 2017;52:4957–4967.
[22] Kellersztein I, Dotan A. Chemical surface modification of wheat straw fibers for polypropylene reinforcement. Polym Compos 2016;37:2133–41. https://doi.org/10.1002/pc.23392.
[23] Notta-Cuvier D, Lauro F, Bennani B, Nciri M. Impact of natural variability of flax fibres properties on mechanical behaviour of short-flax-fibre-reinforced polypropylene. J Mater Sci 2016;51:2911–25. https://doi.org/10.1007/s10853-015-9599-3.
[24] Agarwal BD, Broutman LJ, Chandrashekhara K. Analysis and performance of fiber composites. John Wiley & Sons; 2006.
[25] Mori and Tanaka. Average stress in matrix and average elastic energy of materials with misfitting inclusions - ScienceDirect n.d. https://www.sciencedirect.com/science/article/abs/pii/0001616073900643 (accessed January 14, 2020).
[26] Fisher FT, Brinson LC. Nanomechanics of nanoreinforced polymers. Handb. Theor. Comput. Nanosci., American Scientific Publishers; 2006, p. 253–360.

The effect of fibers-matrix interaction on the composite materials elongation

M Tanasehte[1,*], A Hader [1,4], H Sbiaai [1], I Achik[1,2] and Y Boughaleb [1,3]

[1] LBGIM, Ecole Normale Supérieure, University Hassan II. Casablanca, Morocco
[2] LPMC , Faculty of sciences Ben M'sik . University Hasan II. Casablanca, Morocco
[3] LPMC , University Chouaib Doccali. El Jadida, Morocco
[4] Centre régional des métiers d'éducation et de formation Casablanca- Settat/ établissement Settat, Morocco

* Corresponding author: mohammed.tanasehte-etu@etu.univh2c.ma

Abstract. Failure process in composite materials is the basis of various actual researches. It study is the almost treated with fiber bundle model (FBM), in the framework of local load sharing rule (LLS), were the breakdown mechanisms is controlled by different parameters. Almost tentative investigates only fiber failure process and neglects the effect of the matrix, as a second crucial component of composite. Effectively, quantification of fibers-matrix interactions is not generally clear. Furthermore, the diversity of composite intrinsic proprieties complicates this quantification. The originality of our investigations is to quantify all interactions between fibers and matrix in a single parameter. Likewise, this latter gives us a different results than the ones obtained with FBM where the interaction amplitude is neglected. Moreover, interaction decelerates the failure process; it subdivides the avalanche phenomena on two consecutive regimes separated by delaying duration. These results are more similar to the ones obtained regular fibers substations. Therefore any fiber have its own elongation, the mean fiber elongation produces an elastic energy. The temporal variation of these latter presents two extremum separated by time duration; at this separated period, the material self-rearranging. These results are more similar to the ones obtained by the regular fibers substitution process.

Keywords: Fiber bundle model, elastic energy, matrix-fibers interaction, composite materials.

1. Introduction

Recently, composite materials present various benefits in different application areas. These composites constitute of two non-miscible homogeneous components and present crucial physical and chemical characteristics [1–4].Moreover, failure phenomena of composites and structures, as a perplexed process in industry and science community, produces an enormous destructions and shedding (natural catastrophes such as mine fall-in, earth tremor and mudslide) in various daily human activities [5–7]. Diverse efforts have been depleted to present an explicit physical description of breakdown process. Nevertheless the huge damages produce engineers and scientific comity to investigate for the basic mechanisms of the fracture process [8–10]. So that the loss may be minimized by furnishing a forerunner: the elongation of composite is a substantial propriety in order to make enormous structures like constructions, buildings, bridges, etc.). It has been indicates that the disorder performs a crucial character in determination of the material elongation and also in the avalanche phenomena [11–13]. The avalanche phenomena in composites generally can't be described by a basic linear equation for the crucial non-regularity and the disorder in composites. As an outcome, the theoretical method of statistical physics is mostly involved to investigate characteristics of the failure process. However, the purpose for this point is most probably the emergence of the computer as an investigation tool, rendering issues, which were over the domain of systematic theoretical researches, now accessible. Furthermore, we observed that the fluctuation rather than the average feature performs an important role in the explanation of the breakdown mechanism [14–18]. Correspondingly to the way how the applied load is subdivided between the intact fibers; different models are proved in the literature.

We can cite, numerical and analytical approaches which have been studied [19–21], for example Monte Carlo simulation [22–24], fuse networks [25–27] and molecular dynamics [24–29]. Various statistical researches are relying to fiber bundle model (FBM), they investigating breakdown of disorder materials. In the most cases, this approach shows correctly proprieties of failure states in composites. That is to say that this model (FBM) has also proved to be a valuable method to explore the dynamical fracture in composite materials [30–32]. The approach of the FBM is perspicuous and approximately basic to reach numerically persuasive statistical properties of breakdown material or to recover analytically adequate results [30].

Under a constant applied load, most composites have a time-dependent response and fails in finite time. Over its great industrial utilization, the conception of such rupture process needs several essential issues for statistical physics too. The complexity of studying breakdown mechanism appears from the conditional circumstance on the materials type, it may have a various microscopic origins, from the applied load, through thermally activate, existence of frictional interfaces and components viscoelasticity, to the interactions between matrix and fibers [31]. Preceding investigations [1-15] study only the impact of applied load, thermal noise and system size. Hence, the inventiveness of our research is studying other parameters by quantifying the fibers-matrix interactions in a single term: interaction amplitude η .

To investigate the failure processes in composite materials, we may treat fiber bundle models, with random thresholds value for fibers formed our system; the widely exploited models [6–13], are commanded by accurate ruling for how load generated by a damaged ingredient is transferred to unbroken fibers. Almost investigations, by using this approach, may be applied to an inadequate domain for perplex materials.

Generally, the FBM is assumed to be composed by a set of fibers whose break strengths are assumed to comply with a certain statistical law, such as uniform distribution [32]. The first ascertainment confirms that the fiber will break when it overcomes the threshold value. Under stress-commanded loading condition, the system is loaded parallel to the fiber orientation. The stress delivered by each damaged fiber is transferred among the surviving fibers. As a result, a series of load redistribution may produce two consecutive avalanches, which may occasionally finish after an infallible number of successive fractures, preserving the plurality of the system, or may be catastrophic, leading a macroscopic damage of the totality of the material [30]. Depend to elongation of transverse connection in the failure process [28-30]. It delivered belong the unbroken fibers may be arranged into various categories, as though we cited the equal load sharing (ELS) or global load sharing (GLS), the local load sharing (LLS), and so on.

In this paper, we investigate numerically, the effect of the matrix-fibers interaction on the failure process of the composite materials in the framework fiber bundle model by using the LLS approach. So, this paper is organized as follow: the adopted model in this research is developed in section 2. The obtained results are reported and discussed in section 3. Finally, we summarized our results in the conclusion section.

2. Fiber bundle model with random oriented fibers:

Fiber bundle model explicates the description of composite material's fracture. It also approved as a steady approach that elucidates the damage of a diverse category of disordered materials of numerous interacting constituents to different natures of external effect i.e. charge, temperature, system size, diverse interactions... etc. [32]. The fiber bundle model considers composite materials as a bundle, which matrix reinforces fibers. Then the studied system is obtained by introducing the two principal constituents in various geometrical configurations. The performant proprieties of composite materials impose new industry; specifically technology, aerospace and modern automotive, to operate them. They offer two particular benefits: they support high elongation while it relative low mass and they preserve these qualities also below extreme conditions (high value of temperature, pressure). They have the flexibility to produce modifications in conception and customization conforming to the requested implementations. The capacity to command the characteristics of our materials permits us to investigate the damage of our composite by inspecting fracture of fibers and discussing the impact of interaction between fibers and matrix. All fibers are delimited by matrix of the composite. When a fiber fails, the fiber-matrix interface disintegrates in proximity of the failure. As result of fragmental contact at the interface, the imposed force rise up against as a load restoration length, hence the fractured fiber involves to the global stress capacity of the material.

Amid the theoretical approaches to the problematic, the fiber bundle model (FBM) shows an essential role, and then it captures the focal elements of the rupture of disordered composites while it is still simple enough to simplify analytical calculations. Conferring to the charge transfer mode, fiber bundle model is separated into two groups: Global load-sharing (GLS) or democratic FBM and local load-sharing (LLS). In democratic mode intact fibers bear the charge equally and in local load-sharing the charge of the broken fiber is given to the intact neighbors. Actual developments exhibit old comportment of GLS bundle envisaging variation in applied charge. As well, giving noise-generated failure probability of fibers, old comportment is stretched in a homogeneous material under GLS with constant load [22]. In opposition to the fuse model [9], the fiber bundle model offers a satisfactory balance between being numerically well-regulated and analytically dissolvable. In its basic configuration, the global load sharing fiber bundle model, essentially each factor can be found. In other advanced models, the localized load sharing, some characteristics can be found analytically in this model, but others cannot [7-22]

We envisage a randomly material which is elaborated by discrete set of N0 fibers. Each fiber has identical initial elongation l0 and is randomly oriented in all directions confiding to the vertical one, and is introduced on a regular lattice of size L. The structure is connected at the first end and in the second end we imposed an exterior force [18]. The FBM is a relevant model that contributes to more describing breakdown in composites. Fibers are subject to an exterior force parallel to the ends of our structure. The totality of charge sustains by the fibers. While the fiber-matrix interface and matrix determine mainly the load transfer between fibers [28-29]. Each fiber i is inclined correspond to a constant angle α_i, (with $0 \leq \alpha_i \leq \pi/2$). It's randomly dispersed with the probability density p(y) and the uniform distribution [9-13].

$$P(\alpha) = \int_0^{\alpha} p(y)dy = \alpha \qquad (1)$$

The initial fibers elongation relies on fibers orientation. The fibers which are deposited vertically $(\alpha_i = 0)$ or strongly inclined $(\pi/4 \leq \alpha_i \leq \pi/2)$ have identical initial elongation l0 and are clamped with well-fixed clamps belonging to the both ends of the material. The fibers which are weakly inclined $(0 \leq \alpha_i \leq \pi/4)$ have another initial elongation. Which is chosen with similar method that these fibers also stay bounded to the two lateral areas by clamps.

The fibers are climbed between two hard clamps. The matrix assures the material cohesion and permits each fiber preserving the same direction over the application of force [16]. Each fiber i has its proper breaking threshold fi. Fibers crack individually when stretching threshold value is reached. If the elastic load proceeds readily, it overtakes fi then it cracks. Composites in the bundle form, composed of an important number of elastic fibers with low mass, are well known to be lucid templates of critical systems that show important fracture proprieties [17-22]. The fibers may merely sustain longitudinal displacement below an exterior charge. In our investigation we consider that the applied load fi on each fiber produces an axial longitudinal deformation. This component presents an axial displacement ξ which is expressed by the Hooke's law:

$$\xi_i = f_i / k \qquad (2)$$

where k denotes the stiffness, we consider that its value is the same for all fibers. In order to take into account of thermal noise in the failure process, it is supposed that the localized charge on fibers has time dependent on fluctuations, and then we introduce the corresponding local elongation ξ_T of each fiber due to the presence of thermal noise [6-11]

$$\xi_T = \chi l_0 \sqrt{K_B T} \qquad (3)$$

where K_B is the Boltzmann constant and χ is the coefficient of proportionality between elongation and thermal energy.

The presence of the thermal noise has a quasi-static effect on the failure process in the composite material, yet that will affect the lifetime of the system [9-15]. Then, the actual total length arising x of each fiber is written as:

$$l = \xi_i + \xi_T \qquad (4)$$

If the elongation of fiber i overtakes the failure threshold elongation ξ_c, the fiber fails and it charge is transferred to it neighboring intact fibers and so one [25]. In this paper, the broken threshold elongation ξ_c is randomly chosen and given with the density of probability p(x) and the uniform distribution P(x) which is the most studied threshold distributions [28-32]:

$$P(x) = \int_0^x p(y)dy = x \text{ ; with } p(x) = 1 \qquad (5)$$

The fibers behave linearly elastic until they crack at a failure threshold elongation [29].The randomness of failure thresholds elongation is presumed to show heterogeneous materials disorder. The number of intact fibers N_i is:

$$P(x) = 1 - N_i / N_0 \qquad (6)$$

Effectively the intact fiber density is:

$$\rho = \frac{N_0 - N}{N_0} \qquad (7)$$

The imposed force is considered as a constant quantity. Among the failure, the load of an individual broken fiber discharges and distributes to other intact fibers. This load may produce avalanche phenomenon where the consecutive feedback process may generate the entire material damage [28].

When forces are applied to mechanical materials, constituents of this system store an elastic potential energy if they are deformed. In our study the variation of elongation l produces an elastic energy. As cited previously k is the stiffness of our material. So we explicit elastic energy by:

$$E_e = \frac{1}{2} k l^2 \tag{8}$$

3. Results and discussions

The avalanche phenomenon observed in the failure process in composite materials can characterize clearly by different parameters. In the following part of this investigation, and for a first moment, we will give and discuss the results corresponding to the variation intact fiber density with load, matrix-fibers interaction and system size. In a second moment we will present the corresponding results characterizing the avalanche phenomenon and. In the last we present the elastic energy variation and system lifetime variation versus applied load and interaction amplitude.

Almost materials present a time-dependent response under a constant external load and its fail in a finite time. Creep breakdowns process is almost used in various technological applications, investigating creep phenomenon need different crucial topics for statistical physics [7-18].Creep fracture perplexity comes from the moment that the occurrence that it can have various microscopic origins: thermally noise, fibers-matrix interactions, friction in interfaces, components viscoelasticity and system size.

In first moment, we imposed a constant exterior force f on our material composed on a fiber bundle in a lattice of size $L \times L$. Firstly weak fibers broke and their loads are transferred to the intact neighboring fiber in the framework the LLS as redistribution load mode. We have calculated the intact fiber density ρ versus time for system with 500×500 fibers and for different values of the applied load. The matrix-fibers interaction amplitude is fixed $\eta = 0.5$. We represent the number of broken fibers N_b and the initial number of intact fibers N_0. The obtained results for different applied load and for temperature value $T = 0.03 T_c$ are plotted in figure 1.

We note that the considered material cracks as the applied load increases. However, failure process is triggered by breaking only the weak fibers and the total load is delivered on the nearly surviving fibers affecting a secondary breakdown. In the additional, the breaking process of the composite materials is different to the ones obtained by the classical model [28–30] where the effect of the matrix on the breaking fibers is neglected.

So, the introduction of the noise interaction matrix-fiber subdivided the avalanche breaking on two successively stages separated by delaying duration Δt. The temporal evolution of the intact fiber density, as presented in figure 1, is characterized by two regimes: a decreasing one with two different velocities, and a constant one. The two observed regimes are separated by a cross over time t_c. So, in the first time when the applied load is less than the threshold value ($f < f_r$), no fiber failed. While the load attaints the failure threshold value of weak fibers, they crack and their loads are redistributed to the intact neighboring fibers according to the LLS rule. That is why the density of intact fiber decreases. After a certain time, a less intact fibers number amortizes the failure process; finally intact fiber density becomes constant by achieving its constant value. For the higher regime load, the avalanche breaking is similar to the one observed in the classical model when the interaction matrix-fiber is neglected. So the interaction matrix-fiber is not an important effect in the higher load regime. We should affirm that with low force value the density reaches zero: all fibers broke, isn't the case for high forces which the density that's not reaching zero, material damaged before all fibers crack. We conclude clearly that the presence of fibers-matrix interaction make our composite more resistant.

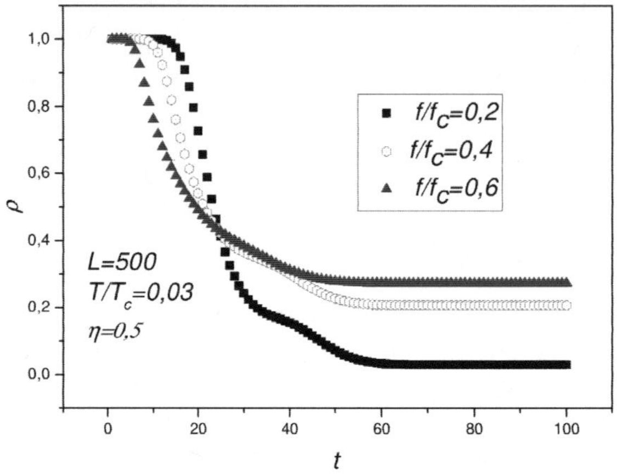

Figure 1: the time evolution of the intact fiber density for different values of applied load

In result, the extremity of the fiber has not any contact with the clamps but they undergo the stress by local sharing load of its nearly surviving fibers. Thus, the intact fibers density decreases to reach zero value for less force domain, but with the same system without including the interaction amplitude, the density value attaints 0.55 as indicated in [35]. Furthermore, the avalanche time for our studied composite is higher than one calculated for the same composite without introducing fibers-matrix interaction as established in [1]. Although, in the higher force domain, our results is analogues to the one without matrix-fiber interaction as obtained in [28-30].

The first important ingredient (fibers) constituting our system is merged as a bundle. These fibers are regularly linked by matrix which constitutes the second crucial ingredient strengthening the materials. The fibers-matrix interface is capable to sustain different charge types. It summarizes categories and figures of imposed charge: compression, elongation or intricate deformation [28].

Temperature and charge are not solely the important parameters which command the breakdown process of composites, yet various other elements may contribute the fatigue fracture of these systems: chemical effect, weather effects … etc. Actually, there has been an interim to integrate all noise modes via a unique parameter which may affect the breakdown probability of the each constituent, such a breakdown probability at any imposed force [29]. So, the breakdown process in composites should be considering the interaction between these two essential ingredients. Due to the large composite characteristics and the stochastic phenomena of these systems, we have included a randomly interaction where η is the matrix-fibers interaction amplitude.

However, interaction amplitude may contain all details referring the load sharing between fibers and matrix. So, to expose the effect of interaction in homogeneity on thermally activated breakdown, we have calculated the density of intact fiber observed during the avalanche phenomenon with three different interaction amplitudes. We applied a constant load f=0,2fc on a system constituted of 500×500 fibers below a value of temperature equal to T=0,03Tc. So, we plot the intact fiber density versus time. The numerical corresponding results are plotted in figure 2. In the first observation, the results present identical schemes as found for different forces value. Hence, the intact fibers density decreases in time and decreases with the increasing interaction amplitude. Though, if interaction amplitude increases, the intact fibers density attaints the zero in long time duration and the system breaks down slowly.

The intact fiber density evolution in time, as showed in figure 2, is characterized by two stages: a decreasing one with two different velocities, and a constant one. A cross over time tc distinguishes the two obtained stages. So, when the interaction attaints the failure threshold value of weak fibers, they fail and their charge is transferred to the intact neighboring fibers by using to the LLS mode. Effectively that the reason why intact fibers density decreases. After a certain time, a less intact fibers number mitigates the breakdown phenomena; the composite doesn't break. Finally, the intact fibers density becomes zero when all fibers crack. So the fibers-matrix interaction decelerates the failure process in the studied materials. Matrix-fibers interaction dissipates the imposed charge and the fluctuations appropriate to thermal activated. It delays the breakdown mechanism and finally, it reinforces our composite.

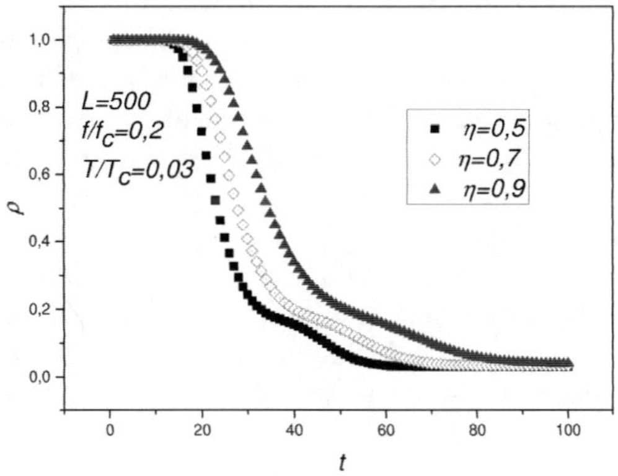

Figure 2: the time evolution of the intact fiber density for different values of noise interaction

The avalanche distribution size is related on the adequate imposed force on the material, we determine the maximal charge F/N as a maximal applied load which system can support until it fails totally. Another mode but homologue process is to apply a constant load F to the composite, thus forthwith all fibers with a threshold elongation shorter than F/N fail. The correspond fiber elongation is given by the maximal F/N value whether doesn't produce an entire failure of our composite [28-35]

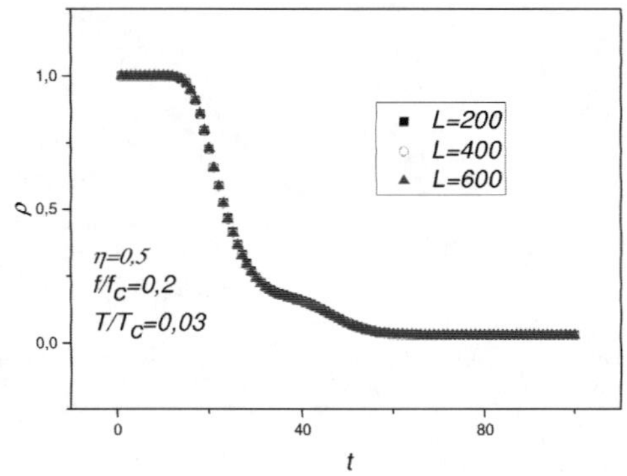

Figure 3: the time evolution of the intact fiber density for different values system sizes

To study the scaling law in the failure process on composites, we have calculated the intact fibers density evolution in time for different system sizes, under f=0,2fc with interaction amplitude η=0,5 in temperature equal to 0,03Tc. The corresponding results are presented in figure 3. Moreover, the obtained profiles collapse on a unique curve which attests that the introduction of the matrix-fibers interaction doesn't affect the scaling law detected in [38–41]

In the elementary literature of the model a homogeneous fibers assemblage is treated, with all the fibers have the same failure elongation, which is then under a constant sub-critical exterior charge. The material growth is operated by thermal noise: the local charge on fibers has thermally provoked fluctuations which can generate breakdown. The principal benefit of this elementary model is that various important macroscopic systems characteristic may be studied analytically. Then, it has been proved in [42] which the lifetime of the material depends on temperature by an Arrhenius law, even when the fibers have disordered elongation. More than that, Guarino in [31] proposed an elementary fiber bundle model, which has demonstrated very effective to get a theoretical results assuring comprehension of the applied load effect produced fractures accumulation on the creep failure process.

Our investigation is based on the consideration of an unbroken composite with N0 intact fibers distributed in random directions relative to the vertical axis. Each fiber is under an exterior charge. For the early time regime, the material resists to the force and after a threshold time value tf where the mean elongation fibers becomes greater than the threshold elongation, the breakdown process appear as micro-cracks which enhance rapidly in time producing an avalanche phenomena, we observe similar results as proved in new investigations [1-4].

As discussed previously, each fiber has an initial elongation l0. Hence when we applied a constant charge value, fibers elongate. This effect produce an elastic energy depends on fiber elongation. To more describing the breakdown process we calculate the elastic energy, obtained from elongation, as a crucial parameter. We plotted on figure 4 the temporal elastic energy variation for different applied load values. This energy pursues two stages and presents two extremums: a minimum at the time tmin and a maximum at the time tmax, the two times are separated by duration of time Δt. After attainting the threshold breakdown value, in the first stage fibers fail, the number of broken fibers increases, and as effect elastic energy increases to attain the maxima. We should note that LLS mode accelerate the breakdown mechanism, it correspond to the transfer of supplementary broken fibers charge. Although, in the second stage the mean fibers elongation decreases to a minimal value. As a consequence, the elastic energy decreases to reach a minimum value. We should note that the less broken fibers number, in this stage, neglect the LLS effect. Moreover, the intact fibers self-rearranged according to the self-criticality theory.

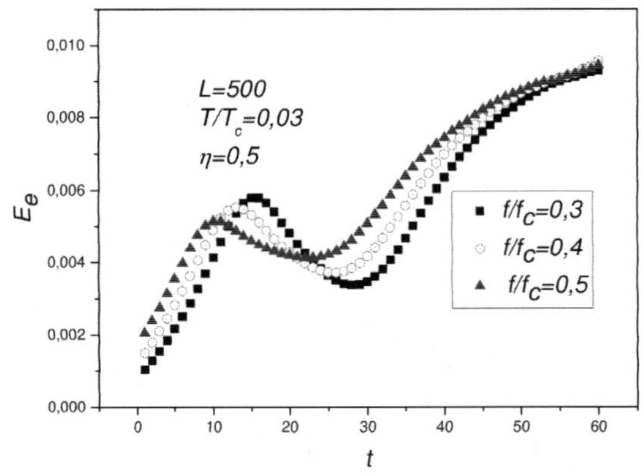

Figure 4: the time evolution of the elastic energy E_e for different applied load values.

The maximal value of elastic energy decreases with increasing applied load, however the corresponding time to maximum tmax increases with decreasing applied load. Moreover, the minimal of elastic energy increases with decreasing applied load, hence the corresponding time to minimum tmin decreases with increasing applied load. These results are more similar to the ones obtained by a new investigation in regular fibers substitution process [1]. To confirm this remark, we calculate on figure 5 the separated duration of time variation with applied load values. We observe that this separated duration decreases rapidly with increasing applied load values and changes the decreasing monotony at te (is the inflexion time at which micro-crack process changes the velocity).

Therefore, the rearrangement of fibers arises at an earlier stage of failure when the surface was obviously unbroken [16]. Matrix contributes, in both boards, to the high elongation and preserves the articular fiber from rupture. Although, random fibers can reply more appreciatively to higher charge before failure, a random disposition of each fiber within composite is more adequate in standard conditions

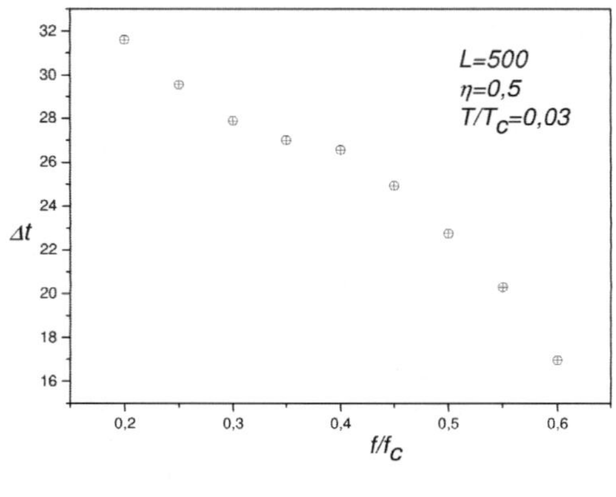

Figure 5: The variation of separated duration of time Δt versus applied load

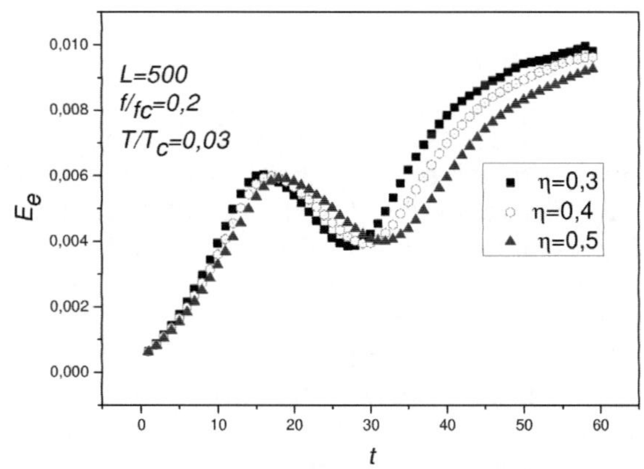

Figure 6: the time evolution of the elastic energy E_e for different interaction amplitude.

These facts will regularly be the origins for the structure of the ensemble fibers-matrix interaction on composites. So, we presented on figure 6 the temporal evolution of elastic energy for different noise values. Inversed behavior is observed than the obtained one with different applied load value: two extremums appear; the extremums has the same value for different noise values. The mean fibers elongation increases with increasing matrix-fibers interaction. However, the time corresponding to

IOP Conf. Series: Materials Science and Engineering **948** (2020) 012032 doi:10.1088/1757-899X/948/1/012032

extremum increases with increasing noise values. All observation confirm that interaction ensure the material resistance.

To finalize our study, we plot on figure 7 the variation of separated time duration Δt versus interaction amplitude. We observed that this duration increase exponentially with increasing interaction. It delayed the elongation process, and the separated time corresponding to fibers rearrangement stage. We can say that the interaction amplitude decelerates the avalanche phenomenon and enhances the resistance of our composite. As result, after the first fracture, the charge of broken fibers has been transferred on the neighboring unbroken ones, and then the interaction sustains the load for a long time by slowing down the breakdown mechanism and preserving fibers.

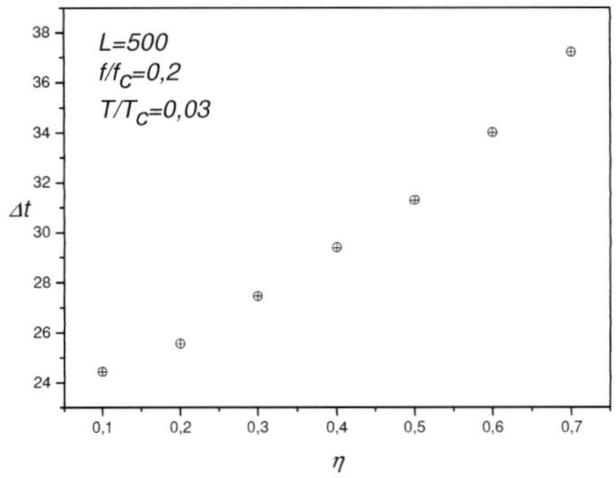

Figure 7: The variation of separated duration of time Δt versus interaction amplitude

Conclusion

In this paper, we adapt fiber bundle model, treated in LLS mode, to investigate failure process on composite materials. Evenly, previous researches study only the fibers breakdown without taking in account the impact of the intimate assemblage between fibers and matrix. As an originality of our investigation, we produce a single parameter quantifying all interactions between fibers and matrix.

To attain our aims, we pursue the temporal evolution of intact fiber density versus applied forces, fibers-matrix interaction amplitudes and system sizes. Therefore, we have proved that failure process dividing the avalanche failure on two consecutively periods separated by delaying duration. The impact of the interaction amplitude becomes fewer significant when enhancing applied load, thus the avalanche process is condensed to a one single regime. These results are different to the ones obtained without introducing interaction amplitude, but they are more similar to the ones acquired by the regular fibers substitution process. Moreover, fibers have it owing elongation; the variation of the mean elongation generates elastic energy. Effectively, we present the temporal evolution of this energy versus different applied loads and interaction amplitudes. Elastic energy increases to attain a maximum, then it decreases to reach a minimum and finally increases again.

As a general conclusion, our investigation presents the benefit of introducing matrix-fibers interaction amplitudes. Then these interaction sustains the charge for a long time, decelerates the breakdown mechanism, it blocks the breakdown of fiber and therefore it preserves the system. Hence, it also enhances the resistance of material and its lifetime. This impact expands the mechanical performance of composites. Moreover, during the time separated the two avalanches failure of fibers due to interaction amplitude, the material is self-rearranged and resists for longtime.

References:

[1] H. Sbiaai, A. Hader, S. Boufass, I. Achik, R. Bakir, and Y. Boughaleb, "The effect of the substitution on the failure process in heterogeneous materials: Fiber bundle model study," *Eur. Phys. J. Plus*, vol. 134, no. 4, pp. 2–9, 2019.

[2] W. Li, W. Chen, L. Tang, Z. Jiang, and P. Huang, "A general strength model for fiber bundle composites under transverse tension or interlaminar shear," *Compos. Part A Appl. Sci. Manuf.*, vol. 121, no. January, pp. 45–55, 2019.

[3] T. Gao, "The diffusion geometry of fibre bundles: Horizontal diffusion maps," *Appl. Comput. Harmon. Anal.*, vol. 1, no. Vdm, pp. 1–69, 2019.

[4] A. Tamura, J. ichi Hongu, and T. Matsumoto, "Theoretical elastic tensile behavior of muscle fiber bundles in traumatic loading events," *Clin. Biomech.*, vol. 69, no. August 2018, pp. 184–190, 2019.

[5] H. Kawamura, T. Hatano, N. Kato, S. Biswas, and B. K. Chakrabarti, "Statistical physics of fracture, friction, and earthquakes," *Rev. Mod. Phys.*, vol. 84, no. 2, pp. 839–884, May 2012.

[6] F. Kun, S. Zapperi, and H. J. Herrmann, "Damage in fiber bundle models," *Eur. Phys. J. B*, vol. 17, no. 2, pp. 269–279, Sep. 2000.

[7] F. Raischel, F. Kun, and H. J. Herrmann, "Fiber Bundle Models for Composite Materials," -, no. 1, pp. 1–10, 2006.

[8] M. A. Clemente, J. B. Gómez, and A. F. Pacheco, "Time to failure of dynamic local load-sharing fiber bundle models in 1 to 3 dimensions," *Phys. D*, no. xxxx, pp. 1–8, 2018.

[9] H. Wu, Z. Yu, and Y. Wang, "based on acoustic emission Experimental study of the process failure diagnosis in additive manufacturing based on acoustic emission," *Measurement*, 2018.

[10] D. P. Hao, G. Tang, Z. P. Xun, H. Xia, and K. Han, "The avalanche process of the fiber bundle model with defect in local loading sharing," *Phys. A Stat. Mech. its Appl.*, vol. 505, pp. 1095–1102, 2018.

[11] D. P. Hao, G. Tang, H. Xia, Z. P. Xun, and K. Han, "The avalanche process of the fiber bundle model with defect," *Phys. A Stat. Mech. its Appl.*, vol. 472, pp. 77–85, 2017.

[12] T. Derda, "Statistical analysis of mechanical damage in nanopillar arrays with mixed-mode load transfer," *J. Appl. Math. Comput. Mech.*, vol. 16, no. 3, pp. 5–16, 2017.

[13] A. De Luca and F. Caputo, "A review on analytical failure criteria for composite materials," *AIMS Mater. Sci.*, vol. 4, no. 5, pp. 1165–1185, 2017.

[14] A. Hader, I. Achik, H. Sbiaai, R. Bakir, and Y. Boughaleb, "Crack diffusion in failure process of composite materials," *Mol. Cryst. Liq. Cryst.*, vol. 628, no. 1, pp. 79–85, 2016.

[15] D. Amitrano and L. Girard, "Fiber bundle model under fluid pressure," *Phys. Rev. E*, vol. 93, no. 3, 2016.

[16] R. Gottardi *et al.*, "Supramolecular Organization of Collagen Fibrils in Healthy and Osteoarthritic Supramolecular Organization of Collagen Fibrils in Healthy and Osteoarthritic Human Knee and Hip Joint Cartilage," no. October, 2016.

[17] R. Yahaya, S. M. Sapuan, M. Jawaid, Z. Leman, and E. S. Zainudin, "Effect of fibre orientations on the mechanical properties of kenaf–aramid hybrid composites for spall-liner application," *Def. Technol.*, vol. 12, no. 1, pp. 52–58, 2016.

[18] S. E. Stapleton, L. Appel, J. Simon, and S. Reese, "Composites□: Part A Representative volume element for parallel fiber bundles□: Model and size convergence," *Compos. Part A*, vol. 87, pp. 170–185, 2016.

[19] C. Roy, S. Kundu, and S. S. Manna, "Fiber bundle model with highly disordered breaking thresholds," *Phys. Rev. E - Stat. Nonlinear, Soft Matter Phys.*, vol. 91, no. 3, pp. 1–7, 2015.

[20] A. Lahyani, Y. Boughaleb, M. Qjani, and R. Nassif, "Creep rupture in fiber bundle model," *2008 2nd Ict. Mediterr. Winter*, no. December 2013, pp. 1–1, 2008.

[21] B. K. Chakrabarti, "A fiber bundle model of traffic jams," vol. 372, pp. 162–166, 2006.

[22] F. Kun, F. Raischel, R. C. Hidalgo, and H. J. Herrmann, "Extensions of fibre bundle models,"

Lect. Notes Phys., vol. 705, pp. 57–92, 2006.
[23] R. C. Hidalgo, Y. Moreno, F. Kun, and H. J. Herrmann, "Fracture model with variable range of interaction," *Phys. Rev. E - Stat. Physics, Plasmas, Fluids, Relat. Interdiscip. Top.*, vol. 65, no. 4, p. 8, 2002.
[24] S. J. Zhou, D. M. Beazley, P. S. Lomdahl, and B. L. Holian, "Large-scale molecular dynamics simulations of three-dimensional ductile failure," *Phys. Rev. Lett.*, vol. 78, no. 3, pp. 479–482, 1997.
[25] P. K. V. V Nukala and S. Simunovic, "A continuous damage random thresholds model for simulating the fracture behavior of nacre," *Biomaterials*, vol. 26, no. 30, pp. 6087–6098, 2005.
[26] L. Lamaignère, F. Carmona, and D. Sornette, "Experimental Realization of Critical Thermal Fuse Rupture," *Phys. Rev. Lett.*, vol. 77, no. 13, pp. 2738–2741, 1996.
[27] D. Sornette and C. Vanneste, "Dynamics and memory effects in rupture of thermal fuse networks," *Phys. Rev. Lett.*, vol. 68, no. 5, pp. 612–615, 1992.
[28] S. Pradhan, A. Hansen, and B. K. Chakrabarti, "Failure processes in elastic fiber bundles," *Rev. Mod. Phys.*, vol. 82, no. 1, pp. 499–555, 2010.
[29] A. Hader, A. Memsouk, and Y. Boughaleb, "Universality in diffusion front growth dynamics," *Eur. Phys. J. B*, vol. 28, no. 3, pp. 315–319, 2002.
[30] A. Hader, I. Achik, A. Lahyani, K. Sbiaai, and Y. Boughaleb, "Characterization of the failure process in composite materials by the Fiber Bundle Model," *Superlattices Microstruct.*, 2014.
[31] A. Guarino, A. Garcimartín, and S. Ciliberto, "An experimental test of the critical behaviour of fracture precursors," *Eur. Phys. J. B*, vol. 6, no. 1, pp. 13–24, 1998.
[32] A. Garcimartín, A. Guarino, L. Bellon, and S. Ciliberto, "Statistical properties of fracture precursors," *Phys. Rev. Lett.*, vol. 79, no. 17, pp. 3202–3205, 1997.
[33] R. Rypl, R. Chudoba, A. Scholzen, and M. Vořechovský, "Brittle matrix composites with heterogeneous reinforcement: Multi-scale model of a crack bridge with rigid matrix," *Compos. Sci. Technol.*, vol. 89, pp. 98–109, 2013.
[34] F. Kun, F. Raischel, R. Hidalgo, and H. J. Herrmann, "Extensions of Fibre Bundle Models," in *Lecture Notes in Physics*, vol. 705, 2006, pp. 57–92.
[35] H. Sbiaai, A. Hader, R. Bakir, I. Achik, I. Tarras, and Y. Boughaleb, "Failure process in heterogeneous materials with randomly oriented fibers," *Phys. A Stat. Mech. its Appl.*, 2017.
[36] K. Saadouni, A. Hader, and Y. Boughaleb, "Study of Adatoms Diffusion Through Current Density Fluctuation Functions Study of adatoms diffusion through current density fluctuation functions," no. August 2013, 2004.
[37] K. Saadouni, A. Hader, Y. Boughaleb, and A. Ayadi, "Temporal fluctuations of current surface density in a triangular lattice," *Moroccan J. Condens. Matter*, vol. 4, no. 1, pp. 5–8, 2001.
[38] I. Achik, Y. Boughaleb, A. Hader, K. Sbiaai, and A. Hajjaji, "Scaling behavior and morphological properties of the interfaces obtained by the multilayer deposition process," *Thin Solid Films*, vol. 545, pp. 353–356, 2013.
[39] B. R. Lawn, "Fracture and deformation in brittle solids: A perspective on the issue of scale," *J. Mater. Res.*, vol. 19, no. 01, pp. 22–29, 2004.
[40] A. Petri, G. Paparo, A. Vespignani, A. Alippi, and M. Costantini, "Experimental evidence for critical dynamics in microfracturing processes," *Phys. Rev. Lett.*, vol. 73, no. 25, pp. 3423–3426, 1994.
[41] W. A. Curtin, "Size scaling of strength in heterogeneous materials," *Phys. Rev. Lett.*, vol. 80, no. 7, pp. 1445–1448, 1998.
[42] H. Nechad, A. Helmstetter, R. El Guerjouma, and D. Sornette, "Andrade and critical time-to-failure laws in fiber-matrix composites: Experiments and model," *J. Mech. Phys. Solids*, vol. 53, no. 5, pp. 1099–1127, 2005.

AUTHOR INDEX

Institute of Physics
Dirac House, Temple Back
Bristol BS1 6BE UK

ISSN:1757-8981
ISBN 978-1-7138-3197-6

CURRAN ASSOCIATES INC.
proceedings.com